AF606146

Cell Fusion in Health and Disease

ADVANCES IN EXPERIMENTAL MEDICINE AND BIOLOGY

For further volumes:
http://www.springer.com/series/5584

Thomas Dittmar · Kurt S. Zänker
Editors

Cell Fusion in Health and Disease

I: Cell Fusion in Health

Editors
Dr. Thomas Dittmar
University of Witten/Herdecke
Institute of Immunology
Stockumer Str. 10
58448 Witten
Germany
thomas.dittmar@uni-wh.de

Dr. Kurt S. Zänker
University of Witten/Herdecke
Institute of Immunology
Stockumer Str. 10
58448 Witten
Germany
kurt.zaenker@uni-wh.de

ISSN 0065-2598
ISBN 978-94-007-0762-7 e-ISBN 978-94-007-0763-4
DOI 10.1007/978-94-007-0763-4
Springer Dordrecht Heidelberg London New York

Library of Congress Control Number: 2011924883

Printed on acid-free paper

Springer is part of Springer Science+Business Media (www.springer.com)

Preface

Cell fusion is a specialized cellular event which occurs in multicellular organisms in health and disease. Known as a phenomenon in modern science for over 100 years, cell fusion takes the mandatory center stage in eutherians for the conception, development and physiology of organogenesis or, in pathophysiology, during the process of oncogenesis. The ability of two or more cells to unit and to form a new syncytial cell takes place in metazoans throughout evolution to form muscles, bones and placentae, and, even to form a tumor. This process requires migration, recognition and adhesion between the cells together with the fusion of their plasma membrane and rearrangement of their cytoplasmatic and nuclear contents. Membrane fusion arise during many cellular processes, including membrane traffic, intracellular vesicle fusion, fertilization, and infection by enveloped viruses. Fusion allows to exchange biological materials between different membrane compartments. In order to maintain the functional individuality of each of the intracellular compartments and of the cell itself, membranes do not fuse easily under normal circumstances. The process is subjected to selective control and requires the expression of special (glyco-)proteins and carbohydrates and the formation of a phospholipid interbilayer via an hourglass-shaped structure called a "stalk".

Sperm-egg fusion (fertilization) is the most prominent example of "natural" occurring membrane fusion without the deliberate addition of exogenous fusing agents such as viruses or chemicals in order to create by orchestrated and stepwise processes a zygote. Billions of sperms are deposited at ejaculation in the female reproductive tract, but only one sperm finds and fertilizes the egg. On their way, the spermatozoa ignore the thousands of cells they make contact with during their locomotion to find a single cell, namely the oocyte. Gamete fusion is an extremely important process that must emerge without error to launch life (B.M. Gadella, Utrecht, The Netherlands; J.P. Evans, Baltimore, USA).

For basic research, the nematode *Caenorhabditis elegans* has become an excellent system to study mechanisms and developmental functions in many cell fusion events at the molecular and cellular levels (L. Friedlander-Shani, B. Podbilewicz, Haifa, Israel).

There is still a considerable lack of knowledge, which molecules (fusogens/SNARE proteins) mediate vesicle fusion (B.P. Jena, Detroit, USA), fuse myoblasts to form myotubes in muscles (A. Simionescu, G.K. Pavlath, Atlanta, USA), macrophages to form osteoclasts in bone (A.K. McNally, J.M. Anderson, Cleveland, USA) and cytotrophoblasts to form syncytiotrophoplasts in placentae (B. Huppertz, M. Gauster, Graz, Austria). The chapters written by these well respected authors will throw some lights on the mystery to reveal genuine fusogens.

Until recently, cells were thought to be intregral and discrete components of tissues, and their state was determined by cell differentiation. However, under some conditions, stem cells or their progeny can fuse with cells of other types, mixing cytoplasmic and even genetic material of different (heterotypic) origins (X. Zhou, J.L. Platt, Ann Arbor, USA). The fusion of heterotypic cells could be of central importance for development, for repair of tissues (M. Alvarez-Dolado, M. Martínez-Losa, Valencia, Spain), for the production of fusion vaccines derived from dendritic and tumor cells (W. Lee,

Durjam, USA) and even for cellular reprogramming (D. Sanges, F. Lluis, M.P. Cosma, Barcelona, Spain). The chapters written by these outstanding experts will highlight the process of cell fusion in diverse biological systems. *Volume I* deals with molecular and cellular aspects of cell–cell fusion as a biological meaning to establish pluripotency or, in other words, when it takes more to make one.

For human health cell–cell fusion is a crucial and highly regulated event in the genesis and homeostasis of both form and function of many tissues. However, cell–cell fusion may also play a critical role in the development of cancer and progression of the disease. Very recently, Gao P. and Zheng J. (Virol J. (2010) 7:238) put forward an attractive working hypothesis that high-risk HPV-16 E5-inducable cell fusion might be a critical initiating event in the early stage of HPV-associated cervical cancer. In general, establishment of a role of cell fusion in cervical carcinogenesis by the HPV-16 E5 fusogenic protein to form tetraploid cells would open an intellectual window to understand additional pathogenic modes of actions for emerging virus-associated cancers.

At the cutting edge, *Volume II* brings into prominence heterogenic fusion processes in oncogenesis. The editors are very thankful to J.G. Sinkovics (Tampa, USA) that the second volume can start with a chapter, which reflects more then 50 years of clinical and experimental cancer research within a polycontextural and intelligent framework of immunology, cancer vaccines – alone or combined with chemotherapy –, oncolysis and the place of viruses in the "tree of life", mostly addressing sarcomas as a clinical entity. He nicely demonstrates that cell fusion and horizontal exchanges of genes are fundamental attributes and inherent characteristics of the living matter.

Structural studies of viral fusion glycoproteins allows to categorize viral membrane fusogens into three distinct classes. M. Backovic (Paris, France) and Theodore S. Jardetzky (Stanford, USA) describe the newly identified group of class III viral fusion proteins, whose members include fusion proteins form rhabdoviruses, herpesviruses, and baculoviruses. Before embarking on cell fusion in malignancies, we inserted a chapter written by A. Malassiné, G. Pidoux, P. Gerbaud, J.L. Frendo and D. Evian-Brion (Paris, France) on the importance of trophoblast fusion in trisomy 21, demonstrating that cell–cell fusion is increasingly of interest in non cancerous diseases, too.

Myeloma bone disease leads to progressive destruction of the skeleton and is the most severe cause of morbidity in multiple myeloma. Osteolytic lesions are not characterized by a massive presence of osteoclasts, whereas malignant plasma cells may occur as large multinucleated cells. The possibility that myeloma cells fuse and generate polykaryons in vivo is suggested by the in vitro formation of multinuclear cells that express tartrate-resistant acid phosphatase and produce pits and erosive lacunae on experimental osteological substrates (F. Silvestris, S. Ciavarella, S. Strippoli, F. Dammacco, Bari, Italy).

Findings from experimental and clinical cancer research suggest a potentially multifaceted involvement of cell fusion in different stages of tumor progression, including aneuploidy, origin of cancer stem cells (X. Lu, Y. Kang, Princeton, USA), multidrug resistance (C. Nagler, K.S. Zänker, T. Dittmar, Witten, Germany) and the acquisition of metastatic abilities (R. Lazova, A. Chakraborty, J.M. Pawelek, New Haven, USA). These distinguished authors clearly demonstrate that the century-old hypothesis that cell fusion may contribute to the initiation and progression of cancer has revitalized.

Cells of the monocyte/macrophage lineage are important for tumor cell migration, invasion and metastases formation. Fusion between macrophages and cancer cells in vitro and in animal models causes hybrids with increased metastatic potential. Expression of the macrophage antigen CD163 in rectal and breast cancer is associated with early recurrence and reduced survival time (I. Shabo, J. Svanvik, Linköping, Sweden).

Membrane vesicles are membrane-covered cell fragments generated by normal and transformed cells. Autophagosomes are the most prominent double-membrane bound vesicles. Fusion of autophagosomes with lysosomes results in the formation of autolysosomes, where the proteins and organelles are degraded. This degradation pathway is induced under nutrient deprivation, metabolic stress or microenvironment conditions to ensure energy balance, clearance of damaged proteins and adaptation to stress. Disruption of autophagy is involved in diverse human diseases including cancer.

Tumor-derived vesicles may serve as prognostic markers, they were detected in blood plasma and in other body fluids. All of them reflect the special potential of tumor cells for survival and for the expansion of the tumor. The vesicles may facilitate the escape of tumor cells from immune surveillance, they are involved in the establishment of a beneficial environment for newly formed and migrating tumor cells, influencing angiogenesis and the reorganization of the extracellular matrix (E. Pap, Budapest, Hungary).

The editors like to extend their gratitude to all authors, who have presented a review of their respective fields, but have been invited to do so from their unique point of view. All have tried to summarize informations and to provide critical reviews connoting cell–cell fusion as a fundamental biological process, upon which future therapies might be built. If these two volumes serve as a scientific reference from which to plan future research strategies – enlightening cell–cell fusion in health and diseases –, many of which have not yet been anticipated by the editors and the authors, then the publication of these two volumes has fulfilled the intended purpose.

For the current two volumes the Editors want to express a special word of thanks to Springer Publishers (Dordrecht, The Netherlands) and in particular to Tanja van Gaans and Meran Owen who have worked closely with us to achieve a rapid and comprehensive publishing standard at the state-of-the-art of cell–cell fusion in health and disease.

Witten, Germany
Autumn 2010

Thomas Dittmar
Kurt S. Zänker

Contents

Contributors

Manuel Álvarez-Dolado Laboratory of Cell-Based Therapies for Neuropathologies, Andalusian Center for Molecular Biology and Regenerative Medicine (CABIMER), CSIC, 41092 Seville, Spain, manuel.alvarez@cabimer.es

James M. Anderson Department of Pathology, Wolstein Research Building, Case Western Reserve University, Cleveland, OH 44106, USA, jma6@case.edu

Maria Pia Cosma Institució Catalana de Recerca i Estudis Avançats (ICREA), Barcelona, Spain; Center for Genomic Regulation (CRG), 08003 Barcelona, Spain, pia.cosma@crg.eu

Thomas Dittmar Institute of Immunology, Witten/Herdecke University, 58448 Witten, Germany, thomas.dittmar@uni-wh.de

Janice P. Evans Division of Reproductive Biology, Department of Biochemistry and Molecular Biology, Bloomberg School of Public Health, John Hopkins University, Baltimore, MD, USA, jpevans@jhsph.edu

Lilach Friedlander-Shani Department of Biology, Technion – Israel Institute of Technology, Haifa 32000, Israel, lilachf@technion.ac.il

Bart M. Gadella Departments of Biochemistry and Cell Biology and of Farm Animal Health, Research Program: Biology of Reproductive Cells, Faculty of Veterinary Medicine, Utrecht University, Utrecht, The Netherlands, b.m.gadella@uu.nl

Martin Gauster Institute of Cell Biology, Histology and Embryology, Medical University Graz, 8010 Graz, Austria, martin.gauster@medunigraz.at

Berthold Huppertz Institute of Cell Biology, Histology and Embryology, Center for Molecular Medicine, Medical University of Graz, 8010 Graz, Austria, berthold.huppertz@medunigraz.at

Bhanu P. Jena Department of Physiology, Wayne State University School of Medicine, Detroit, MI 48201, USA, bjena@med.wayne.edu

Walter T. Lee Division of Otolaryngology – Head and Neck Surgery, Duke University Medical Center, Durham, NC 27710, USA, walter.lee@duke.edu

Frederic Lluis Center for Genomic Regulation (CRG), 08003 Barcelona, Spain, Frederic.lluis@crg.eu

Magdalena Martínez-Losa Laboratory of Cell-Based Therapies for Neuropathologies, Andalusian Center for Molecular Biology and Regenerative Medicine (CABIMER), CSIC, 41092 Seville, Spain, magdalena.martinez@cabimer.es

Amy K. McNally Department of Pathology, Wolstein Research Building, Case Western Reserve University, Cleveland, OH 44106, USA, akm2@case.edu

Jeffrey L. Platt Departments of Surgery and Microbiology and Immunology, University of Michigan, Ann Arbor, MI 48109-2200, USA, plattjl@umich.edu

Benjamin Podbilewicz Department of Biology, Technion – Israel Institute of Technology, Haifa 32000, Israel, podbilew@tx.technion.ac.il

Daniela Sanges Center for Genomic Regulation (CRG), 08003 Barcelona, Spain, Daniela.sanges@crg.eu

Adriana Simionescu Department of Pharmacology, Emory University, Atlanta, GA 30322, USA, asimion@emory.edu

Grace K. Pavlath Department of Pharmacology, Emory University, Atlanta, GA 30322, USA, gpavlat@emory.edu

Kurt S. Zänker Institute of Immunology, Witten/Herdecke University, Witten 58448, Germany, kurt.zaenker@uni-wh.de

Xiaofeng Zhou Transplantation Biology, Department of Surgery, University of Michigan, Ann Arbor, MI 48109-2200, USA, xiazhou@med.umich.edu

Chapter 1
Introduction

Thomas Dittmar and Kurt S. Zänker

Abstract Although cell fusion is an omnipresent process in life, to date considerably less is still known about the mechanisms and the molecules being involved in this biological phenomenon in higher organisms. In Cell Fusion in Health and Disease Volume 1 international leading experts will present up-to-date overviews about the current knowledge about cell fusion-mediating molecules in *C. elegans* and mammalian cells. Further topics of the book will focus on cell fusion in physiological processes including fertilization, placentation, skeletal muscle development, and tissue repair and will sum up the use of artificial cell fusion for cellular reprogramming and cancer vaccine development. Thus, Cell Fusion in Health and Disease Volume 1 represents a state-of-the-art work for researchers, physicians or professionals being interested in the biological phenomenon of cell fusion in physiological processes and beyond.

When we talk about cell fusion the possibly most descriptive example for this process in higher organisms is the fusion between the oocyte and the sperm, which gives rise to the fertilized egg cell and the generation of a new life. However, cell fusion does not only play a role in the beginning of life, but is also a prerequisite in a plethora of processes being involved in growth, development and tissue repair. In mammals, trophoblastic cells fuse with each other, thereby giving rise to multinucleated syncytiotrophoblasts, which facilitate and ensure the nutrient exchange between the mother and the fetus. Likewise, myoblasts fuse to form multinucleated skeletal muscle fibers, whereas cells of the monocytic origin fuse to osteoclasts being participated in bone resorption (e.g., bone repair after fracture). Moreover, we know from various studies that bone marrow-derived stem cells as well as cells of the myelomonocytic lineage restore tissue function, e.g., liver, lung, by cell fusion, which raised (and still raise) expectations for autologous stem cell-based tissue regeneration strategies. In addition to these physiologically cell fusion events, artificial cell fusion protocols have been developed to reprogram stem cells, to generate hybridomas and to generate tumor vaccines. Hybridomas, derived from myeloma cell/plasma cell fusions, are the source of monoclonal antibodies. What was once developed for scientific purposes, e.g., Western Blot, immunohistochemistry, is now used in a plethora of approaches ranging from simple diagnostic tests (pregnancy test, drug tests) to routine diagnostic applications (determination of inflammatory markers in serum, blood typing, virus detection in patient samples) to clinical applications (immunosuppression for organ transplantation, use of humanized monoclonal antibodies in cancer therapy). To date, tumor cell-dendritic cell hybrids are the most promising tools for tumor vaccination strategies. Due to fusion of professional antigen presenting dendritic cells with tumor cells hybrid cells evolve being capable to initiate a anti-tumor specific immune response because of tumor antigen presentation.

T. Dittmar (✉)
Institute of Immunology, Witten/Herdecke University, 58448 Witten, Germany
e-mail: thomas.dittmar@uni-wh.de

T. Dittmar, K.S. Zänker (eds.), *Cell Fusion in Health and Disease*, Advances in Experimental Medicine and Biology 713, DOI 10.1007/978-94-007-0763-4_1, © Springer Science+Business Media B.V. 2011

In addition to these cell fusion events being crucial to maintain and ensure the body's homeostasis, the biological process of cell fusion can also be linked to various (malignant) diseases. Without the ability to fuse with the plasma membrane viruses would not be able to deliver their genome into host cells. Thus the identification and characterization of viral membrane fusion proteins and plasma membrane fusion partners is one promising approach to develop inhibitors, which specifically block the fusion of a virus with its target cells. Such approaches are currently tested, e.g., for blocking Hepatitis C virus as well as HIV infection.

The normal view of viral infections assumes that one virus (or more) infects (fuses with) only one target cell, whereas the virus-mediated cell fusion is neglected in this context (although viruses, e.g., Sendai virus, were the first tools for study the process of cell fusion and to characterize hybrid cells). Recent studies indicate that the virus-mediated cell fusion seems to be a common phenomenon in viral infections and that such processes might cause cancer due to induction of chromosomal instability. If so, this would mean that cell fusion (possibly driven by viruses) might also contribute to cancer stem cells, which have been defined as the seed for tumor growth. Whether such a process would also explain the phenomenon that tumor cells are highly fusogenic is, however, unknown. Nonetheless, the fusion of cancer cells with other (normal) cells can give rise to hybrids exhibiting new properties, such as an increased proliferation rate, an enhanced metastatic capacity, as well as an increased drug resistance towards chemotherapeutic/cytotoxic compounds. Macrophage antigens have been identified on tumor cells and both in vivo and in vitro studies revealed that tumor cell/macrophage hybrid cells possesses an enhanced metastatic capacity. Stem cell/tumor cell fusions have also been observed both in vivo and in vitro. Because tumor cell/normal cell hybrids might exhibit an increased drug resistance concomitantly with an enhanced malignity it was suggested that cancer relapses might originate from fusion events. If so, this indicates that not only tumor initiation, but also metastasis formation and cancer relapses can be linked to this biological process.

In addition to cell–cell fusion events further fusion related processes have been associated to cancer. These include microvesicles and autophagy. Microvesicles are plasma membrane fragments being shed from almost all cell types including tumor cells following activation or apoptosis. Elevated amounts of microvesicles are found in the blood of cancer patients and, because microvesicles harbor a multitude of biologically active (oncogenic) proteins and RNA species, it is currently assumed that microvesicles might be a mode of intercellular tumor cell communication. The phenomenon of autophagy (or autophagocytosis) describes the degradation of a cell's own components through the lysosomal machinery. The role of autophagy in cancer is unclear. On the one hand, autophagy can act as a tumor suppressor by degrading damaged organelles. On the other hand, autophagy can promote survival of cancer cells under conditions of poor nutrient supply as well as protecting tumor cells against therapy-induced apoptosis.

This short introduction indicates that cell fusion is not limited to a few physiologically processes, but is a common biological phenomenon, whereby cell fusion plays a pivotal role both in health and disease. Because of the complexity of cell fusion the book will be divided into two volumes. The first volume will summarize cell fusion in health, whereby the second volume will give an overview about cell fusion process being related to (malignant) disease.

We further realise this book as a platform for a summary of the latest findings on cell fusion-mediating molecules in mammals. In contrast to *C. elegans* or *D. melanogaster*, where cell fusion and the molecules to be involved in are well-characterised, only a handful of fusogenic proteins (e.g., syncytin, SNAREs, CD200, CD44, CD47 and PTPNS1) have been identified in mammals. However, as mentioned above for virus membrane fusion proteins, the knowledge about these cell fusion-mediating proteins is crucial for developing specific cell fusion inhibitors. Studies on viruses indicate that such approaches are feasible, thereby impairing viral infections. If we conclude that malignant cells could evolve from cell fusion events than the inhibition of this biological process might be one approach to prevent cancer formation and/or impair cancer progression, which in turn perquisites the knowledge about the process itself and the molecules to be involved-in.

We are thankful that so many internationally recognised experts accepted our invitation to contribute to this exciting book project. We sincerely thank them all for their interest in this important topic and that they, despite other duties and responsibilities, found the possibility to present excellent and comprehensive overviews of the most important recent findings in their field of scientific engagement within this topic. We would also like to thank Tanja van Gaans and Meran Owen from Springer Publishers (Dordrecht, The Netherlands) for their kind assistance and excellent collaboration on this project, as well as for giving the opportunity to realize this book project.

We hope that this book may encourage new scientific approaches within the field of cell fusion in health and disease as well as closer interdisciplinary collaborations on this fascinating and important issue in the future.

Chapter 2
Heterochronic Control of AFF-1-Mediated Cell-to-Cell Fusion in *C. elegans*

Lilach Friedlander-Shani and Benjamin Podbilewicz

Abstract In normal development cell fusion is essential for organ formation and sexual reproduction. The nematode *Caenorhabditis elegans* has become an excellent system to study the mechanisms and developmental functions of cell-to-cell fusion. In this review we focus on the heterochronic regulation of cell fusion. Heterochronic genes control the timing of specific developmental events in *C. elegans*. The first microRNAs discovered were found as mutations that affect heterochronic development and cell–cell fusions. In addition numerous heterochronic transcription factors also control specific cell fusion events in *C. elegans*. We describe what is known about the heterochronic regulation of cell fusion of the epidermal seam cells. The fusogen AFF-1 was previously shown to mediate the fusion of the lateral epidermal seam cells. Here we provide evidence supporting the model in which LIN-29, the heterochronic Zinc-finger transcription factor that controls the terminal fusion of the seam cells, stimulates AFF-1 expression in the seam cells before they fuse. Therefore, the heterochronic gene LIN-29 controls AFF-1-mediated cell–cell fusion as part of the terminal differentiation program of the epidermal seam cells.

2.1 Introduction

Throughout development of the nematode *Caenorhabditis elegans* about one third of the somatic cells go through cell to cell fusion (for recent reviews see [1–3]).

Cell fusion events occur during both embryonic and postembryonic development in various organs including the hypodermis, vulva, pharynx and uterus [4–6]. It was shown that the cell fusions in *C. elegans* are mediated by two fusogens EFF-1 and AFF-1 [7, 8]. Mutations that cause ectopic fusion lead to embryonic lethality. In agreement, developmental cell fusion was found to be a tightly regulated process in *C. elegans*. In addition to spatial regulation, the developmental timing of cell fusion events is also critical. Several transcription factors have been found to regulate the precise developmental stage in which fusions occur [1].

2.2 Heterochronic Genes Regulate the Timing of Developmental Events

Heterochrony is a change in the timing of a specific developmental event relative to other developmental events which are not affected [9, 10]. Heterochrony can cause evolutionary variation since a change in timing of a certain developmental event can result in speciation [9] (reviewed in [11, 12]).

B. Podbilewicz (✉)
Department of Biology, Technion – Israel Institute of Technology, Haifa 32000, Israel
e-mail: podbilew@tx.technion.ac.il

T. Dittmar, K.S. Zänker (eds.), *Cell Fusion in Health and Disease*, Advances in Experimental Medicine and Biology 713, DOI 10.1007/978-94-007-0763-4_2, © Springer Science+Business Media B.V. 2011

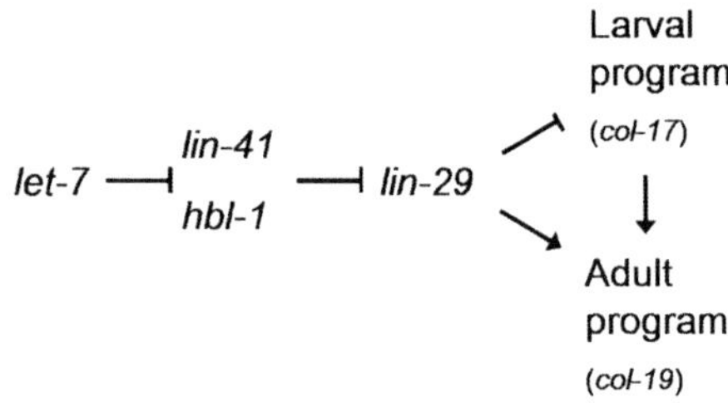

Fig. 2.1 The heterochronic pathway. Simplified model of the heterochronic gene pathway controlling the terminal seam cells differentiation

For example, two nematodes species, *Pristionchus pacificus*, and *C. elegans*, display heterochronic differences between them in several cell lineages such as vulval and gonadal lineages, although the genetic basis for those differences is not known [13].

Heterochronic mutants that control several developmental events were identified in *C. elegans*. Mutations in these heterochronic genes change the time of developmental events while other events in the organism occur in the normal timing of the wild type. Heterochronic mutations can result in two types of phenotypes. In the precocious phenotype the developmental event occurs earlier than in wild type and in the retarded phenotype the event happens in a later stage with respect to the wild type and in addition the event can be reiterated [10] (for recent reviews see [12, 14, 15]).

The heterochronic genes that were identified by genetic approaches control various developmental events among them vulva formation, dauer larva formation, aging and terminal differentiation of the hypodermal seam cells [10, 16–18]. Using epitasis analysis these heterochronic genes were organized into a model of heterochronic pathway (Fig. 2.1).

2.3 The Heterochronic Gene *lin-29* Determines the Final Fate of the Seam Cells

The transcription factor LIN-29 is the most downstream known heterochronic regulator of the seam cells terminal differentiation [18] (Fig. 2.1). By terminal differentiation the seam cells switch from larval seam cells program into adult seam cells program [18]. In *lin-29* mutants the seam cells exhibit retarded phenotype; the seam cells fail to terminally differentiate and the larval program is reiterated [10, 18]. The *lin-29* gene encodes a transcription factor that contains five $(Cys)_2$-$(His)_2$ zinc finger domains [19].

To date the heterochronic pathway is comprised of numerous genes, but here we will describe a brief summary of the pathway. *lin-4* and *let-7,* the first two members discovered in the microRNA (miRNA) family, were identified as members of the heterochronic gene pathway [20, 21]. During the mid-L1 stage *lin-4* is expressed and downregulates the LIN-14 nuclear protein, which specifies the L1 fate [10, 20, 22]. When *lin-4* is mutated, the lineage pattern of the L1 stage is reiterated [23]. In *lin-14* loss of function mutants the L2 pattern occurs precociously in the L1 stage [10]. *lin-4* also represses LIN-28 permitting transition to L3 stage fate [24]. LIN-46 and *let-7* paralogs *mir-84*, *mir-48* and *mir-241* downregulate HBL-1 to control the L2 to L3 transition [25]. In addition DAF-12 while bound to its ligand, also downregulates HBL-1 to control this transition by directly activating *let-7* miRNA family members [26]. Later during development LIN-41 and HBL-1 repress LIN-29 expression thus, specifying late larval fate. Next, *let-7* downregulates LIN-41 and HBL-1 allowing LIN-29 expression that direct the seam cells terminal differentiation at L4 to adult transition [27–29].

In addition to the seam cells terminal differentiation, *lin-29* is also required in other tissues. *lin-29* is necessary in the egg laying system for specification of the utse, regulation of genes expression in the vulval cells at the L4 stage and differentiation of the vulval cells [30]. Furthermore *lin-29* is required in a subset of the lateral seam cells for proper vulva morphogenesis and egg laying [31]. Thus, by acting in several components of the egg laying system, *lin-29* may coordinate the vulval-uterine-seam cell connection.

Additionally *lin-29* is required for the linker cell death in *C. elegans* male that occur during or just after the L4/adult transition. Since *let-7* also controls the linker cell death, it is likely that the linker cell death is regulated by the heterochronic pathway [32].

2.4 LIN-29 Controls the Terminal Differentiation of the Epidermal Seam Cells

The lateral hypodermal seam cells form two rows of cells one on each side along the body of the worm. The seam cells are hypodermal cells that synthesize and secrete the cuticle [33]. During each of the 3 larval stages (L1–L3), around the time of the molts, the seam cells divide in a stem cell manner producing one daughter cell that retains a seam cell fate and a second cell that either will fuse to the hypodermal hyp7 syncytium or will have other fate [34].

During the larva to adult molt, which is the transition from L4 to adult, the seam cells undergo terminal differentiation. The seam cells stop cell divisions, fuse with each other forming a longitudinal syncytium on each side of the worm (Figs. 2.2 and 2.3a), synthesize adult cuticle which includes secretion of the adult "alae" and stop the molting cycle [18]. The alae are a set of raised cuticular stripes that are positioned along the body of the worm above the seam cells. In addition to this morphological difference, the larval and the adult cuticle are also distinguished in their collagen gene expression [35]. *lin-29* is required for all the events of the terminal differentiation of the seam cells (Fig. 2.1).

Since *lin-29* is a transcription factor it can act by regulating either directly or indirectly the transcription of genes that are required for the terminal differentiation, therefore regulating genes involved in cell cycle exit, cell fusion, switching to the adult cuticle and in the molting cycle. It was found that *lin-29* regulates the transcription of specific collagen genes (*col-17* and *col-19*) at the L4-adult molt [36]. *lin-29* represses *col-17* and activates *col-7* and *col-19* transcription at this stage [36]. It was previously shown that LIN-29 protein binds in vitro to *col-19* and *col-17* promoter sequences [19]. In the case of *col-19* this binding of LIN-29 is to the regulatory sequence which is necessary for in vivo adult-specific activation of the collagen gene *col-19*. These results suggest a direct role of *lin-29* in regulating collagen genes which are required for seam cells terminal differentiation. Additional possible targets of LIN-29 are *nhr-23* and *nhr-25* that encode conserved nuclear hormone receptors which are essential for larval molting. *nhr-23* and *nhr-25* were shown to be downstream effectors of *let-7* and *mir-84*. A possible model is that LIN-29 represses *nhr-23* and *nhr-25* after the forth molt and by that cause exit from the molting cycle [37]. LIN-29 is the best candidate for regulating the seam cells fusion in the L4 to adult switch.

Now two important questions can be asked: how do the seam cells fuse during the terminal differentiation? And – what is the regulation mechanism of the seam cells fusion?

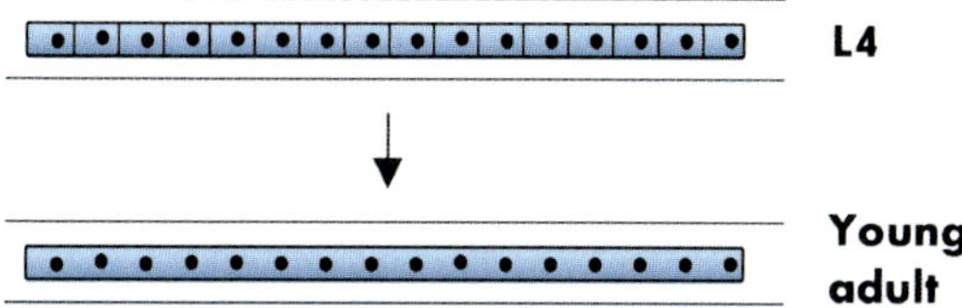

Fig. 2.2 The seam cells terminal fusion. In the L4 stage the lateral hypodermal seam cells form two rows (*left* and *right*) each containing 16 cells. During the transition from L4 to adult, the seam cells fuse with each other forming a longitudinal syncytium on each side of the worm. The seam cells fusion is part of their terminal differentiation process

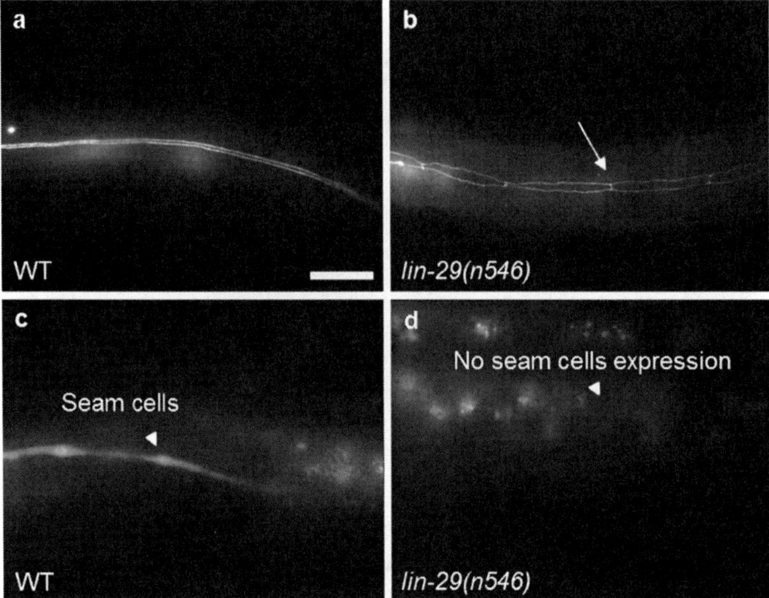

Fig. 2.3 LIN-29 controls AFF-1- mediated fusion of the seam cells. (**a**,**b**) Immunofluorescence of worms stained with MH27 antibody which recognizes an antigen in the adherens junctions of epithelial cells [4, 46, 47]. (**c**,**d**) Transgenic worms expressing *aff-1p::GFP* construct. (**a**) In wild type worms at late L4 stage, the seam cells fused forming a syncytium that is marked by two parallel lines of adherens junctions. (**b**) Young adult *lin-29(n546)* worm in which the seam cells failed to fuse. *Arrow* indicates unfused adherens junction. (**c**) *aff-1promoter::GFP* construct was expressed (*arrowhead*) in wild type worms in the seam cells at late L4 stage. (**d**) Late L4 *lin-29(n546)* mutant in which there was no *aff-1promoter::GFP* expression in the seam cells while *aff-1promoter::GFP* expression was retained in other tissues like the utse (not shown). In (**a**), (**c**) anterior is to *right*, in (**b**), (**d**) to the *left*. *Scale bar* represents 25 μm

2.5 AFF-1 Protein Mediates the Terminal Fusion of the Hypodermal Seam Cells

In order to study cell fusion in *C. elegans* forward genetic screens were performed searching for fusion failure phenotypes. Two genes were identified as encoding proteins that induce cell fusion events in *C. elegans*. One of these proteins, AFF-1 was found to be the protein necessary for the fusion of the seam cells during their terminal differentiation [8].

In addition to the seam cell fusion *aff-1* (**a**nchor cell **f**usion **f**ailure-1) is also required for the fusion of the anchor cell (AC) to the utse syncytium [8]. The AC coordinates the connection between the uterus and the vulva. The AC induces vulval precursor cells to receive vulval fates thereby inducing vulva formation [38, 39]. Next, the AC induces surrounding uterine cells to become π cells that produce the uterine cells (utse, uv1) which connect to the vulva [40, 41]. The final stage in the formation of the connection between the uterus and vulva involves the removal of the AC by cell fusion in order to enable passing of eggs through the egg-laying organ. First, eight π cells progeny fuse to form the utse syncytium, and then the AC fuses to this utse syncytium. The AC-utse syncytium is an H-shaped cell that in its middle region has a thin cytoplasmic process which is located between the vulva and the uterus and serves as the nematode's hymen. While the first egg exits the uterus this hymen is broken generating a connection between the uterine and vulval lumens [41].

Additional fusions events to which *aff-1* is required are the fusion of the vulval rings vulA and vulD that occur in the L4 stage [8]. AFF-1 ectopic expression is sufficient to induce cell fusion in cells that do not normally fuse in *C. elegans*. Moreover, AFF-1 was shown to fuse heterologous tissue culture cells. These observations indicate that AFF-1 serves as a bona fide fusogen [8].

aff-1 is required for only a part of the fusion events in *C. elegans*. The first fusogen identified in *C. elegans* was EFF-1 (**e**pithelial **f**usion **f**ailure 1). EFF-1 is essential for most of epidermal, vulval and pharyngeal cell fusion events [7, 42]. As AFF-1, EFF-1 fuses cells that normally do not fuse in vivo and also causes fusion of heterologous tissue culture cells, therefore, EFF-1 is an actual fusogen [42, 43]. Further research has shown that EFF-1 needs to be expressed in both fusing cells in *C. elegans* and in heterologous cells for cell fusion to occur. Thus, EFF-1 functions in a homotypic fusion mechanism [43].

Recently it was shown that in addition to its role in epithelial cells, EFF-1 also has a role in controlling dendrites structure in *C. elegans* by inducing dendrites retraction and autofusion [44].

eff-1 and *aff-1* genes encode type I transmembrane proteins that share only moderate sequence homology but exhibit significant similarity in their presumptive structure. EFF-1 and AFF-1 proteins show conservation in the number of cysteines and partial conservation of prolines residue number in the extracellular region. In addition the proteins contain a possible TGF-β-type-I-Receptor domain [8]. *aff-1* and *eff-1* represent the first two members of developmental eukaryotic fusogens. Together, *eff-1* and *aff-1* account for most cell fusions in *C. elegans* but not for all. For example, both sperm-egg fusion and π cell daughters fusion forming the utse are carried out in each of *eff-1* mutant and *aff-1* mutant and also in *eff-1 aff-1* double mutant [8]. These observations suggest that there are additional fusogens in *C. elegans*.

What is the regulation mechanism of the seam cells fusion? As mention above, in *lin-29* loss of function mutant worms the seam cells fail to undergo their terminal fusion (Fig. 2.3b). Thus, in order to answer the above question we recently examined the possibility that *lin-29* is regulating *aff-1* in the seam cells. We found that while in wild type worms *aff-1promoter::GFP* is expressed in the seam cells starting from the L4 stage, in *lin-29(n546)* loss of function mutant worms there is partial or no *aff-1p::GFP* expression in the seam cells during this stage (Fig. 2.3c, d) [45]. These results suggest that *lin-29* positively regulates *aff-1* expression in the seam cells during the L4 to young adult transition by transcriptional regulation. Thus, *aff-1* may be an effector of the heterochronic pathway (Friedlander-Shani and Podbilewicz, unpublished results).

References

1. Alper S, Podbilewicz B (2008) Cell fusion in Caenorhabditis elegans. Methods Mol Biol 475:53–74
2. Sapir A, Avinoam O, Podbilewicz B et al (2008) Viral and developmental cell fusion mechanisms: conservation and divergence. Dev Cell 14:11–21
3. Oren-Suissa M, Podbilewicz B (2010) Evolution of programmed cell fusion: common mechanisms and distinct functions. Dev Dyn 239:1515–1528
4. Podbilewicz B, White JG (1994) Cell fusions in the developing epithelial of C. elegans. Dev Biol 161:408–424
5. Shemer G, Podbilewicz B (2000) Fusomorphogenesis: cell fusion in organ formation. Dev Dyn 218:30–51
6. Podbilewicz B (2006) Cell fusion. WormBook 6:1–32
7. Mohler WA, Shemer G, del Campo JJ et al (2002) The type I membrane protein EFF-1 is essential for developmental cell fusion. Dev Cell 2:355–362
8. Sapir A, Choi J, Leikina E et al (2007) AFF-1, a FOS-1-regulated fusogen, mediates fusion of the anchor cell in C. elegans. Dev Cell 12:683–698
9. Gould SJ (1977) Ontogeny and Phylogeny. Belknap Press of Harvard University Press, Cambridge, MA
10. Ambros V, Horvitz HR (1984) Heterochronic mutants of the nematode Caenorhabditis elegans. Science 226: 409–416
11. Slack F, Ruvkun G (1997) Temporal pattern formation by heterochronic genes. Annu Rev Genet 31:611–634
12. Moss EG (2007) Heterochronic genes and the nature of developmental time. Curr Biol 17:R425–434

13. Felix MA, Hill RJ, Schwarz H et al (1999) Pristionchus pacificus, a nematode with only three juvenile stages, displays major heterochronic changes relative to Caenorhabditis elegans. Proc Biol Sci 266:1617–1621
14. Rougvie AE (2005) Intrinsic and extrinsic regulators of developmental timing: from miRNAs to nutritional cues. Development 132:3787–3798
15. Resnick TD, McCulloch KA, Rougvie AE (2010) miRNAs give worms the time of their lives: small RNAs and temporal control in Caenorhabditis elegans. Dev Dyn 239:1477–1489
16. Euling S, Ambros V (1996) Heterochronic genes control cell cycle progress and developmental competence of C. elegans vulva precursor cells. Cell 84:667–676
17. Liu ZC, Ambros V (1989) Heterochronic genes control the stage-specific initiation and expression of the dauer larva developmental program in Caenorhabditis elegans. Genes Dev 3:2039–2049
18. Ambros V (1989) A hierarchy of regulatory genes controls a larva-to-adult developmental switch in C. elegans. Cell 57:49–57
19. Rougvie AE, Ambros V (1995) The heterochronic gene lin-29 encodes a zinc finger protein that controls a terminal differentiation event in Caenorhabditis elegans. Development 121:2491–2500
20. Lee RC, Feinbaum RL, Ambros V (1993) The C. elegans heterochronic gene lin-4 encodes small RNAs with antisense complementarity to lin-14. Cell 75:843–854
21. Reinhart BJ, Slack FJ, Basson M et al (2000) The 21-nucleotide let-7 RNA regulates developmental timing in Caenorhabditis elegans. Nature 403:901–906
22. Feinbaum R, Ambros V (1999) The timing of lin-4 RNA accumulation controls the timing of postembryonic developmental events in Caenorhabditis elegans. Dev Biol 210:87–95
23. Chalfie M, Horvitz HR, Sulston JE (1981) Mutations that lead to reiterations in the cell lineages of C. elegans. Cell 24:59–69
24. Moss EG, Lee RC, Ambros V (1997) The cold shock domain protein LIN-28 controls developmental timing in C. elegans and is regulated by the lin-4 RNA. Cell 88:637–646
25. Abbott AL, Alvarez-Saavedra E, Miska EA et al (2005) The let-7 MicroRNA family members mir-48, mir-84, and mir-241 function together to regulate developmental timing in Caenorhabditis elegans. Dev Cell 9:403–414
26. Bethke A, Fielenbach N, Wang Z et al (2009) Nuclear hormone receptor regulation of microRNAs controls developmental progression. Science 324:95–98
27. Slack FJ, Basson M, Liu Z et al (2000) The lin-41 RBCC gene acts in the C. elegans heterochronic pathway between the let-7 regulatory RNA and the LIN-29 transcription factor. Mol Cell 5:659–669
28. Abrahante JE, Daul AL, Li M et al (2003) The Caenorhabditis elegans hunchback-like gene lin-57/hbl-1 controls developmental time and is regulated by microRNAs. Dev Cell 4:625–637
29. Lin SY, Johnson SM, Abraham M et al (2003) The C elegans hunchback homolog, hbl-1, controls temporal patterning and is a probable microRNA target. Dev Cell 4:639–650
30. Newman AP, Inoue T, Wang M et al (2000) The Caenorhabditis elegans heterochronic gene lin-29 coordinates the vulval-uterine-epidermal connections. Curr Biol 10:1479–1488
31. Bettinger JC, Euling S, Rougvie AE (1997) The terminal differentiation factor LIN-29 is required for proper vulval morphogenesis and egg laying in Caenorhabditis elegans. Development 124:4333–4342
32. Abraham MC, Lu Y, Shaham S (2007) A morphologically conserved nonapoptotic program promotes linker cell death in Caenorhabditis elegans. Dev Cell 12:73–86
33. Singh RN, Sulston JE (1978) Some observations on moulting in Caenorhabditis elegans. Nematologica 24: 63–71
34. Sulston JE, Horvitz HR (1977) Post-embryonic cell lineages of the nematode, Caenorhabditis elegans. Dev Biol 56:110–156
35. Cox GN, Hirsh D (1985) Stage-specific patterns of collagen gene expression during development of Caenorhabditis elegans. Mol Cell Biol 5:363–372
36. Liu Z, Kirch S, Ambros V (1995) The Caenorhabditis elegans heterochronic gene pathway controls stage-specific transcription of collagen genes. Development 121:2471–2478
37. Hayes GD, Frand AR, Ruvkun G (2006) The mir-84 and let-7 paralogous microRNA genes of Caenorhabditis elegans direct the cessation of molting via the conserved nuclear hormone receptors NHR-23 and NHR-25. Development 133:4631–4641
38. Kimble J (1981) Alterations in cell lineage following laser ablation of cells in the somatic gonad of Caenorhabditis elegans. Dev Biol 87:286–300
39. Sternberg PW, Horvitz HR (1986) Pattern formation during vulval development in C. elegans. Cell 44: 761–772
40. Newman AP, White JG, Sternberg PW (1995) The Caenorhabditis elegans lin-12 gene mediates induction of ventral uterine specialization by the anchor cell. Development 121:263–271
41. Newman AP, White JG, Sternberg PW (1996) Morphogenesis of the C. elegans hermaphrodite uterus. Development 122:3617–3626

42. Shemer G, Suissa M, Kolotuev I et al (2004) EFF-1 is sufficient to initiate and execute tissue-specific cell fusion in C. elegans. Curr Biol 14:1587–1591
43. Podbilewicz B, Leikina E, Sapir A et al (2006) The C. elegans developmental fusogen EFF-1 mediates homotypic fusion in heterologous cells and in vivo. Dev Cell 11:471–481
44. Oren-Suissa M, Hall DH, Treinin M et al (2010) The fusogen EFF-1 controls sculpting of mechanosensory dendrites. Science 328:1285–1288
45. Friedlander-Shani L (2010) Heterochronic control of AFF-1-mediated cell-to-cell fusion in C. elegans. MSc Thesis, Technion-Israel Institute of Technology, Haifa
46. Francis R, Waterston RH (1991) Muscle cell attachment in Caenorhabditis elegans. J Cell Biol 114:465–479
47. Waterston RH (1988) Muscle. In: Wood WB (eds) The Nematode Caenorhabditis elegans. Cold Spring Harbor Laboratory Press, New York, NY

Chapter 3
Role of SNAREs in Membrane Fusion

Bhanu P. Jena

Abstract Fusion between opposing cellular membranes is essential for numerous cellular activities such as protein maturation, neurotransmission, hormone secretion, and enzyme release. The universal molecular mechanism of membrane fusion involves Ca^{2+}, and the assembly of a specialized set of proteins present in the opposing membrane bilayers. For example in cell secretion, target membrane proteins at the cell plasma membrane SNAP-25 and syntaxin termed t-SNAREs, and secretory vesicle-associated protein VAMP or v-SNARE, are part of the conserved protein complex involved in fusion of opposing membranes. In the presence of Ca^{2+}, t-SNAREs and v-SNARE in opposing bilayers interact and self-assemble in a ring conformation, to form conducting channels. Such self-assembly of t-/v-SNARE ring occurs only when the respective SNAREs are in association with membrane. The size of the SNARE ring complex is dependent on the curvature of the opposing bilayers. Electron density map and 3-D topography of the SNARE ring complex, suggests the formation of a leak-proof channel measuring 25 Å in ring thickness, and 42 Å in height. The mechanism of membrane-directed SNARE ring complex assembly, and the mathematical prediction of SNARE ring size, has been determined. X-ray diffraction measurements and simulation studies have further advanced that membrane-associated t-SNAREs and v-SNARE overcome repulsive forces to bring the opposing membranes close to within a distance of approximately 2.8 Å. Calcium is then able to bridge the closely apposed bilayers, leading to the release of water from hydrated Ca^{2+} ions as well as the loosely coordinated water at phospholipid head groups, leading to membrane destabilization and fusion.

3.1 Introduction

Membrane fusion is essential for numerous cellular activities, including hormone secretion, enzyme release, and neurotransmission. In live cells, membrane fusion is mediated via a specialized set of proteins present in opposing bilayers. In the past 2 decades, much progress has been made in our understanding of membrane fusion in cells, beginning with the discovery of an *N*-ethylmaleimide-sensitive factor (NSF) [1] and SNARE proteins [2–4], and the determination of their participation in membrane fusion [5–11]. VAMP and syntaxin are both integral membrane proteins, with the soluble SNAP-25 associating with syntaxin. Therefore the understanding of SNARE-induced membrane fusion requires determining the atomic arrangement and interactions between membrane-associated v- and t-SNARE proteins. Ideally, the atomic coordinates of membrane-associated SNARE complex using x-ray crystallography would help to elucidate the chemistry of SNARE-induced membrane

B.P. Jena (✉)
Department of Physiology, Wayne State University School of Medicine, Detroit, MI 48201, USA
e-mail: bjena@med.wayne.edu

T. Dittmar, K.S. Zänker (eds.), *Cell Fusion in Health and Disease*, Advances in Experimental Medicine and Biology 713, DOI 10.1007/978-94-007-0763-4_3, © Springer Science+Business Media B.V. 2011

fusion in cells. So far such structural details at the atomic level of membrane-associated t-/v-SNARE complex has not been possible, primarily due to solubility problems of membrane-associated SNAREs, compounded with the fact that v-SNARE and t-SNAREs need to reside in opposing membranes when they meet, to assemble in a physiologically relevant SNARE complex. The remaining option, the use of nuclear magnetic resonance spectroscopy (NMR) has also been of little help, since the size of t-/v-SNARE ring complex is beyond the optimal limit for NMR studies. Regardless, high-resolution AFM force spectroscopy, and EM electron density map and 3-D topography of the SNARE ring complex has enabled an understanding of the structure and assembly, and also the disassembly of membrane-associated t-/v-SNARE complexes in physiological buffer solution [5–10, 12].

The structure and arrangement of SNAREs associated with lipid bilayers were first determined using AFM [5], almost a decade ago. Electrophysiological measurements of membrane conductance and capacitance, enabled the determination of fusion of v-SNARE-reconstituted liposomes with t-SNARE-reconstituted membrane. Results from these studies demonstrated that t-SNAREs and v-SNARE when present in opposing membrane interact and assemble in a circular array, and in presence of calcium, form conducting channels [5]. The interaction of t-/v-SNARE proteins to form such conducting channels is strictly dependent on the presence of t-SNAREs and v-SNARE in opposing membranes. Simple addition of purified recombinant v-SNARE to a t-SNARE-reconstituted lipid membrane, fails to form the SNARE ring complex, and is without influence on the electrical properties of the membrane [5]. However when v-SNARE vesicles are added to t-SNARE reconstituted membrane, SNAREs assemble in a ring conformation, and in the presence of calcium, establish continuity between the opposing membrane. The establishment of continuity between the opposing t-SNARE and v-SNARE reconstituted bilayers, is reflected in the increase in membrane capacitance and conductance. These results confirm that t- and v-SNAREs are required to reside in opposing membrane, similar to their presence in cells, to allow appropriate t-/v-SNARE interactions leading to membrane fusion [5, 7]. Studies using SNARE-reconstituted liposomes and bilayers [9] further demonstrate a low fusion rate (τ=16 min) between t- and v-SNARE-reconstituted liposomes in the absence of Ca^{2+}. Exposure of t-/v-SNARE liposomes to Ca^{2+} drives vesicle fusion on a near physiological relevant time-scale ($\tau \sim$10 s), demonstrating Ca^{2+} and SNAREs in combination to be the universal fusion machinery in cells [9]. Native and synthetic vesicles exhibit a significant negative surface charge primarily due to the polar phosphate head groups, generating a repulsive force that prevent the aggregation and fusion of opposing vesicles. In cells, SNAREs provide direction and specificity, bring opposing bilayers closer to within a distance of 2–3 Å [9], enabling Ca^{2+} bridging and membrane fusion. The bound Ca^{2+} then leads to the expulsion of water between the bilayers at the bridging site, leading to lipid mixing and membrane fusion. Hence SNAREs, besides bringing opposing bilayers closer, dictate the site and size of the fusion area during cell secretion. The size of the t-/v-SNARE complex is dictated by the curvature of the opposing membranes [7], hence smaller the vesicle, the smaller the channel formed.

A unique set of chemical and physical properties of the Ca^{2+} ion make it ideal for participating in the membrane fusion reaction. Calcium ion exists in its hydrated state within cells. The properties of hydrated calcium have been extensively studied using x-ray diffraction, neutron scattering, in combination with molecular dynamics simulations [13–16]. The molecular dynamic simulations include three-body corrections compared with ab initio quantum mechanics/molecular mechanics molecular dynamics simulations. First principle molecular dynamics has also been used to investigate the structural, vibrational, and energetic properties of $[Ca(H_2O)_n]^{2+}$ clusters, and the hydration shell of the calcium ion [13]. These studies demonstrate that hydrated calcium $[Ca(H_2O)_n]^{2+}$ has more than one shell around the Ca^{2+}, with the first hydration shell having six water molecules in an octahedral arrangement [13]. In studies using light scattering and X-ray diffraction of SNARE-reconstituted liposomes, it has been demonstrated that fusion proceeds only when Ca^{2+} ions are available between the t- and v-SNARE-apposed proteoliposomes [8, 9]. Mixing of t- and v-SNARE proteoliposomes in the absence of Ca^{2+} leads to a diffuse and asymmetric diffractogram in X-ray diffraction studies, a typical characteristic of short range ordering in a liquid system [15]. In contrast, when t-SNARE and v-SNARE proteoliposomes in the presence of Ca^{2+} are mixed, it leads to a more structured diffractogram, with approximately a 12% increase in X-ray scattering intensity, suggesting an increase in the number of contacts between opposing bilayers, established presumably through calcium-phosphate bridges, as previously suggested [8, 9, 16]. The ordering effect of Ca^{2+} on inter-bilayer contacts observed in X-ray studies [9] is in good agreement with light, AFM, and spectroscopic studies, suggesting close apposition of PO-lipid head groups in the presence of Ca^{2+}, followed by formation of Ca^{2+}-PO bridges between the adjacent bilayers [8, 9, 17]. X-ray diffraction studies show that the effect of Ca^{2+} on bilayers orientation and inter-bilayer contacts is most prominent in the area of 3 Å, with additional appearance of a new peak at position 2.8 Å, both of which are within the ionic radius of Ca^{2+} [9]. These studies further suggest that the ionic radius of Ca^{2+} may make it an ideal player in the membrane fusion reaction. Hydrated calcium $[Ca(H_2O)_n]^{2+}$ however, with a hydration shell having six water molecules and measuring ~6 Å would be excluded from the t-/v-SNARE apposed inter-bilayer space, hence calcium has to be present in the buffer solution when t-SNARE vesicles and v-SNARE vesicles meet. Indeed, studies demonstrate that if t- and v-SNARE vesicles are allowed to mix in a calcium-free buffer, there is no fusion following post addition of calcium [8]. How does calcium work? Calcium bridging of apposing bilayers may lead to the release of water from the hydrated Ca^{2+} ion, leading to bilayer destabilization and membrane fusion. Additionally, the binding of calcium to the phosphate head groups of the apposing bilayers may also displace the loosely coordinated water at the PO-lipid head groups, resulting in further dehydration, leading to destabilization of the lipid bilayer and membrane fusion. Recent studies in the laboratory [18], using molecular dynamics simulations in the isobaric-isothermal ensemble to determine whether Ca^{2+} was capable of bridging opposing phospholipid head groups in the early stages of the membrane fusion process, demonstrate indeed this to be the case. Furthermore, the distance between the oxygen atoms of the opposing PO-lipid head groups bridged by calcium was in agreement with the 2.8 Å distance previously determined using X-ray diffraction measurements. The hypothesis that there is loss of coordinated water both from the hydrated calcium ion and oxygen of the phospholipid head groups in opposing bilayers, following calcium bridging, is further demonstrated from the study.

In presence of ATP, the highly stable, membrane-directed, and self-assembled t-/v-SNARE complex, can be disassembled by a soluble ATPase, the *N*-ethylmaleimide-sensitive factor (NSF). Careful examination of the partially disassembled t-/v-SNARE bundles within the complex using AFM, demonstrates a left-handed super coiling of SNAREs. These results demonstrate that t-/v-SNARE disassembly requires the right-handed uncoiling of each SNARE bundle within the ring complex, demonstrating NSF to behave as a right-handed molecular motor [6]. Furthermore, recent studies in the laboratory [19] using circular dichroism (CD) spectroscopy, we report for the first time that both t-SNAREs and v-SNARE and their complexes in buffered suspension, exhibit defined peaks at CD signals of 208 and 222 nm wavelengths, consistent with a higher degree of helical secondary structure. Surprisingly, when incorporated in lipid membrane, both SNAREs and their complexes exhibit reduced folding. NSF, in presence of ATP, disassembles the SNARE complex as reflected from the CD signals demonstrating elimination of a-helices within the structure. These results further demonstrate that NSF-ATP is sufficient for the disassembly of the t-/v-SNARE complex. These studies have provided a molecular understanding of SNARE-induced membrane fusion in cells. Findings from the studies outlined above are described in this chapter.

3.2 Materials and Methods

3.2.1 Preparation of Lipid Bilayer

Lipid bilayers were prepared using brain phosphatidylethanolamine (PE) and phosphatidylcholine (PC), and dioleoylphosphatidylcholine (DOPC), and dioleylphosphatidylserine (DOPS), obtained from Avanti Lipids, Alabaster, AL. A suspension of PE:PC in a ratio of 7:3, and at a concentration of 10 mg/ml was prepared. 100 μl of the lipid suspension was dried under nitrogen gas and resuspended in 50 μl of decane. To prepare membranes reconstituted with VAMP, 625 ng/ml VAMP-2 protein stock was added to the lipid suspension and brushed onto a 200 μm hole in the bilayer preparation cup until a stable bilayer with a capacitance between 100 and 250 pF was formed.

3.2.2 Lipid Membrane on Mica Surface

To prepare lipid membrane on mica for AFM studies, freshly cleaved mica disks were placed in a fluid chamber. One hundred eighty microliters of bilayer bath solution containing 140 mM NaCl, 10 mM HEPES, and 1 mM $CaCl_2$ was placed at the center of the cleaved mica disk. Twenty μl of PC:PS vesicles were added to the above bath solution. The mixture was then allowed to incubate for 60 min at RT, prior to washing (X10), using 100 μl of bath solution/wash. The lipid membrane on mica was then imaged before and following the addition of SNARE proteins and or v-SNARE reconstituted vesicles. Ten microliters of t-SNAREs (10 μg/ml stock) and or v-SNAREs (5 μg/ml stock), was added to the lipid membrane. Similarly, 10 μl of v-SNARE reconstituted vesicles was added to either the lipid membrane alone or lipid membrane containing t-SNAREs.

3.2.3 Atomic Force Microscopy

Atomic Force Microscopy was performed on mica and on lipid membrane. Lipid membrane alone or in the presence of SNAREs and or v-SNARE reconstituted vesicles on mica, were imaged using the Nanoscope IIIa, an AFM from Digital Instruments, Santa Barbara, CA. Images were obtained both in the "contact" and "tapping" mode in fluid. However, all images presented in this manuscript were obtained in the "tapping" mode in fluid, using silicon nitride tips with a spring constant of 0.38 $N{\cdot}m^{-1}$, and an imaging force of <200 pN. Images were obtained at line frequencies of 2 Hz, with 512 lines per image, and constant image gains. Topographical dimensions of SNARE complexes and lipid vesicles were analyzed using the software nanoscopeIIIa4.43r8 supplied by Digital Instruments.

3.2.4 EPC9 Electrophysiological Lipid Bilayer Setup

Electrical measurements of the artificial lipid membrane were performed using a bilayer setup [20–22]. Current verses time traces were recorded using pulse software, an EPC9 amplifier and probe from HEKA (Lambrecht, Germany). Briefly, membranes were formed while holding at 0 mV. Once a bilayer was formed and demonstrated to be in the capacitance limits for a stable bilayer membrane according to the hole diameter, the voltage was switched to –60 mV. A baseline current was established before the addition of proteins or vesicles.

3.2.5 Preparation of Lipid Vesicles and SNARE Protein Reconstitutions

Purified recombinant SNAREs were reconstituted into lipid vesicles using mild sonication. Three hundred microliters of PC:PS, 100 μl ergosterol and 15 μl of nystatin (Sigma Chemical Company, St. Louis, MO.) were dried under nitrogen gas. The lipids were resuspended in 543 μl of 140 mM NaCl, 10 mM HEPES, and 1 mM $CaCl_2$. The suspension was vortexed for 5 min, sonicated for 30 s and aliquoted into 100 μl samples (AVs). Twenty five μl of syntaxin 1A-1 and SNAP-25 (t-SNAREs) at a concentration of 25 μg/ml was added to 100 μl of AVs. The t-SNARE vesicles were frozen and thawed 3 times and sonicated for 5 s before use. Bilayer bath solutions contained 140 mM NaCl and 10 mM HEPES. KCl at a concentration of 300 mM was used as a control for testing vesicle fusion.

3.2.6 Circular Dichroism Spectroscopy

Overall secondary structural content of SNAREs and their complexes, both in suspension and membrane-associated, were determined by CD spectroscopy using an Olis DSM 17 spectrometer. Data were acquired at 25°C with a 0.01 cm path length quartz cuvette (Helma). Spectra were collected over a wavelength range of 185–260 nm using a 1-nm step spacing. In each experiment, 30 scans were averaged per sample for enhanced signal to noise, and data were acquired on duplicate independent samples to ensure reproducibility. SNAREs and their complexes, both in suspension and membrane-associated, were analyzed for the following samples: v-SNARE, t-SNAREs, v-SNARE + t-SNAREs, v-SNARE + t-SNAREs + N-ethylmaleimide sensitive factor (NSF) and v-SNARE + t-SNAREs + NSF + 2.5 mM ATP. All samples had final protein concentrations of 10 μM in 5 mM sodium phosphate buffer at pH 7.5 and were baseline subtracted to eliminate buffer (or liposome in buffer) signal. Data were analyzed using the GLOBALWORKS software (Olis), which incorporates a smoothing function and fit using the CONTINLL algorithm [19].

3.2.7 Wide-Angle X-Ray Diffraction

Ten microliter of a 10 mM lipid vesicle suspension was placed at the center of an X-ray polycarbonate film mounted on an aluminum sample holder and placed in a Rigaku RU2000 rotating anode X-ray diffractometer equipped with automatic data collection unit (DATASCAN) and processing software (JADE). Similarly, X-ray diffraction studies were also performed using t- and v-SNARE reconstituted liposomes, both in the presence and absence of Ca^{2+}. Samples were scanned with a rotating anode, using the nickel-filtered Cu Kα line (λ=1.5418 Å) operating at 40 kV and 150 mA. Diffraction patterns were recorded digitally with scan rate of 3°/min. using a scintillation counter detector. The scattered X-ray intensities were evaluated as a function of scattering angle 2θ and converted into Å units, using the formula d (Å)=λ/2sinθ.

3.3 Discussion

3.3.1 V-SNARE and t-SNAREs Need to Reside in Opposing Membrane to Appropriately Interact and Establish Continuity Between Those Membranes

Purified recombinant t- and v-SNARE proteins, when applied to a lipid membrane, form globular complexes (Fig. 3.1a–d) ranging in size from 30 to 100 nm in diameter and 3 to 15 nm in height when examined using AFM. Section analysis of t-SNARE complexes (Fig. 3.1d) in lipid membrane, prior to (Fig. 3.1b), and following addition of v-SNARE (Fig. 3.1c), demonstrate changes only in the size of the complex. A 5% increase in diameter and 40% increase in height were seen following addition of v-SNARE to the t-SNARE complexes in the lipid membrane. Concomittant studies of conductance changes in the bilayer following reconstitution of SNAREs into phospholipid membranes supported the AFM observations. Addition of t-SNAREs to v-SNARE reconstituted lipid membranes did not alter membrane current (Fig. 3.1e). Similarly, when t-SNAREs were added to the lipid membrane

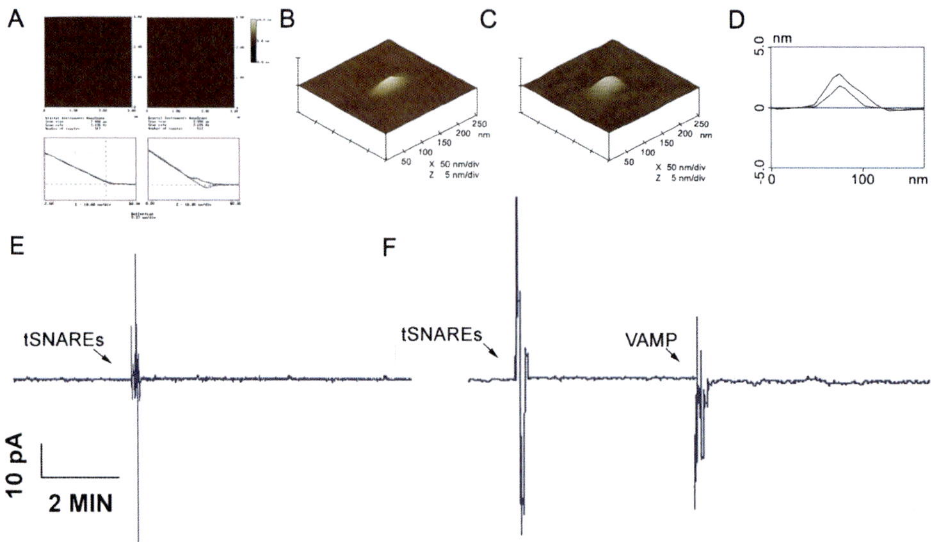

Fig. 3.1 AFM micrographs and force plots of mica and lipid surface and of SNAREs on lipid membrane. AFM performed on freshly cleaved mica (**a**, *left*), and on lipid membrane formed on the same mica surface (**a**, *right*), demonstrating differences in the force vs. distant curves. Note the curvilinear shape exhibited in the force vs. distant curves of the lipid surface in contrast to mica. Three dimensional AFM micrographs of neuronal t-SNAREs deposited on the lipid membrane (**b**), and following the addition of v-SNARE (**c**). Section analysis of the SNARE complex in (**b**) and (**c**) is depicted in (**d**). Note the smaller curve belonging to the t-SNARE complex in (**b**), is markedly enlarged following addition of v-SNARE. Artificial bilayer lipid membranes are nonconducting either in the presence or absence of SNAREs (**e, f**). Current verses time traces of bilayer membranes containing proteins involved in docking and fusion of synaptic vesicles while the membranes are held at –60 mV (current/reference voltage). (**e**) When t-SNAREs are added to the planar lipid bilayer containing the synaptic vesicle protein, VAMP-2, no occurrence of current spike for fusion event at the bilayer membrane is observed (n=7). (**f**) Similarly, no current spike is observed when t-SNAREs (syntaxin 1A-1 and SNAP25) are added to the cis side of a bilayer chamber following with VAMP-2. Increasing the concentration of t-SNAREs and VAMP-2 protein [5]

prior to addition of v-SNARE, no change in the baseline current of the bilayer membrane was demonstrated (Fig. 3.1f). In contrast, when t-SNAREs and v-SNARE in opposing bilayers were exposed to each other, they interact and arrange in circular pattern, forming channel-like structures (Fig. 3.2a–d). These channels are conducting, since some vesicles are seen to have discharged their contents and are therefore flattened (Fig. 3.2b), measuring only 10–15 nm in height as compared to the 40–60 nm size of filled vesicles (Fig. 3.2a). Since the t-/v-SNARE complex lies between the opposing bilayers, the discharged vesicles clearly reveal t-/v-SNAREs forming a rosette pattern with a dimple or channel-like opening at the center (Fig. 3.2b–d). On the contrary, unfused v-SNARE vesicles associated with the t-SNARE reconstituted lipid membrane, exhibit only the vesicle profile (Fig. 3.2a). These studies demonstrate that the t-/v-SNARE arrangement is in a circular array, having a channel-like opening at the center of the complex.

In order to determine if the channel-like structures were capable of establishing continuity between the opposing bilayers, changes in current across the bilayer were examined. T-SNARE vesicles containing the antifungal agent nystatin, and the cholesterol homologue ergosterol, where added to the cis side of the bilayer chamber containing v-SNARE in the bilayer membrane. Nystatin, in the presence of ergosterol, forms a cation-conducting channel in lipid membranes [20–23]. When vesicles containing nystatin and ergosterol incorporate into an ergosterol-free membrane, a current spike can be observed since the nystatin channel collapses as ergosterol diffuses into the lipid membrane

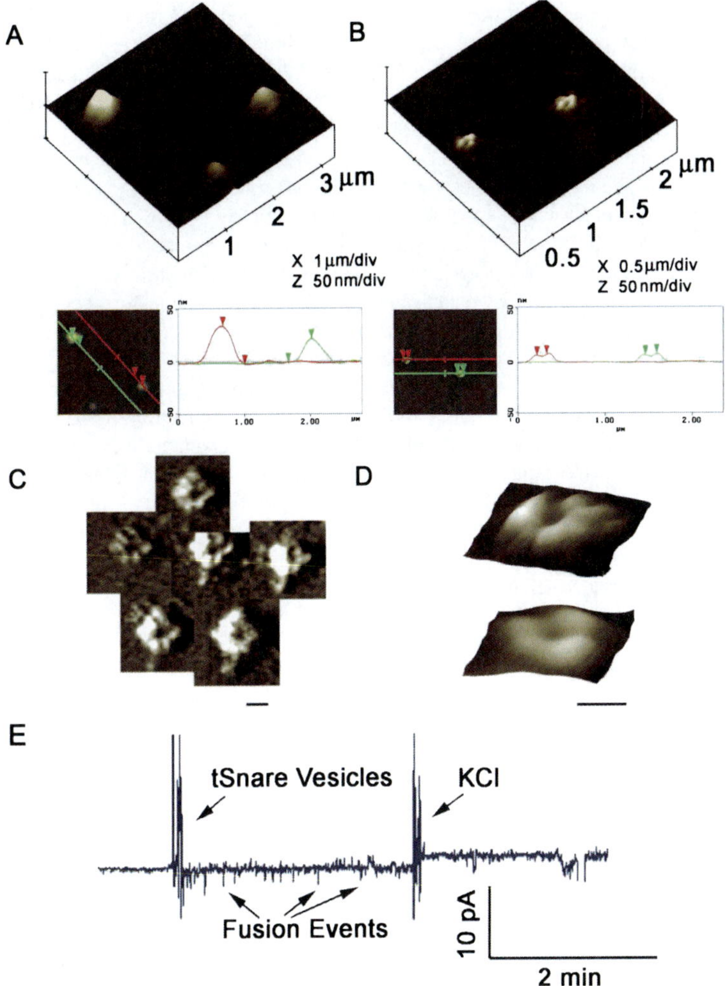

Fig. 3.2 Pore-like structures are formed when t-SNAREs and v-SNARE in opposing bilayers interact. (**a**) Unfused v-SNARE vesicles on t-SNARE reconstituted lipid membrane. (**b**) Dislodgement and/or fusion of v-SNARE-reconstituted vesicles with a t-SNARE-reconstituted lipid membrane, exhibit formation of channel-like structures due to the interaction of v- and t-SNAREs in a circular array. The size of these channels range between 50 and 150 nm (**b–d**). Several 3D AFM amplitude images of SNAREs arranged in a circular array (**c**) and some at higher resolution (**d**), illustrating a channel-like structure at the center is depicted. Scale bar is 100 nm. Recombinant t-SNAREs and v-SNARE in opposing bilayers drive membrane fusion. (**e**) When t-SNARE vesicles were exposed to v-SNARE reconstituted bilayers, vesicles fused. Vesicles containing nystatin/ergosterol and t-SNAREs were added to the cis side of the bilayer chamber. Fusion of t-SNARE containing vesicles with the membrane observed as current spikes that collapse as the nystatin spreads into the bilayer membrane. To determine membrane stability, the transmembrane gradient of KCl was increased, allowing gradient driven fusion of nystatin-associated vesicles [5]

[20–22]. As a positive control, a KCl gradient was established to test the ability of vesicles to fuse at the lipid membrane (410 mM cis: 150 mM trans). The KCl gradient provided a driving force for vesicle incorporation that was independent of the presence of SNARE proteins [21]. When t-SNARE vesicles were exposed to v-SNARE reconstituted bilayers, vesicles fused (Fig. 3.2e). Fusions of t-SNARE containing vesicles with the membrane were observed as current spikes. To verify if the channel-like structures were continuous across the membrane, capacitance and conductance measurements of the membrane were carried out (Fig. 3.3a). Phospholipid vesicles that come in contact with the bilayer membrane do not readily fuse with the membrane. When v-SNARE-reconstituted phospholipid vesicles were added to the cis compartment of the bilayer chamber, a small increase in capacitance and a simultaneous increase in conductance was observed with little or no further increase over a 5 min period. The increase and no further change in conductance or capacitance is consistent with vesicles making contact with the membrane but not fusing (Fig. 3.3b). These vesicles were fusogenic because of a salt (KCl) gradient across the bilayer membrane, inducing fusion of vesicles with the lipid membrane. When t-SNARE vesicles containing nystatin and ergosterol are added to the cis side of the bilayer chamber, an initial increase in capacitance and conductance occurred followed by a stepwise increase in both membrane capacitance and conductance (Fig. 3.3c) along with several fusion events, observed as current spikes in separate recordings (Fig. 3.2e). The stepwise increase in capacitance demonstrates that the docked t-SNARE vesicles are continuous with the bilayer membrane. The simultaneous increase in membrane conductance is a reflection of the vesicle-associated nystatin channels that are conducting through SNARE-induced channels formed, allowing conductance of ions from cis to the trans compartment of the bilayer membrane. SNARE induced fusion occurrs at an average rate of four t-SNARE vesicle incorporations every 5 min into the v-SNARE reconstituted bilayer without osmotic pressure, compared to 6 vesicles using a KCl gradient (n=7). These studies demonstrate that when opposing bilayers meet, SNAREs arrange in a ring pattern results in the formation of a conducting channel [5].

3.3.2 Membrane Curvature Dictate the Size of the SNARE Ring Complex

SNARE-ring complexes ranging in size from approximately 15 to 300 nm in diameter are formed when t-SNARE-reconstituted and v-SNARE-reconstituted lipid vesicles meet. Since vesicle curvature would dictate the contact area between opposing vesicles, this broad spectrum of SNARE complexes observed, may be due to the interaction between SNARE-reconstituted vesicles of different size. To test this hypothesis, t-SNARE- and v-SNARE-reconstituted liposomes (proteoliposomes) of distinct diameters were used [7]. Lipid vesicles of different sizes used in the study were isolated using published extrusion method [9]. The size of each vesicle population was further assessed using the AFM (Fig. 3.4). AFM section analysis demonstrates the presence of small 40–50 nm-in diameter vesicles isolated using a 50 nm extruder filter (Fig. 3.4a, b). Similarly, representative samples of large vesicles measuring 150–200 and 800–1,000 nm were obtained using different size filters in the extruder. Such large vesicles are shown in the AFM micrograph (Fig. 3.4c, d). Analysis of vesicle size using photon correlation spectroscopy, further confirmed the uniformity in the size of vesicles within each vesicle population. The morphology and size of the SNARE complex formed by the interaction of t-SNARE- and v-SNARE-reconstituted vesicles of different diameter were examined using the AFM (Fig. 3.5). In each case, the t-SNARE and v-SNARE proteins in opposing proteoliposomes, interact and self-assemble in a circular pattern, forming channel-like structures. The interaction and arrangement of SNAREs in a characteristic ring pattern were observed for all populations of proteoliposomes examined (Fig. 3.5a–d). However, the size of the SNARE complex was determined to be dictated by the diameter of the proteoliposomes used (Fig. 3.5) [7]. When small (~50 nm) t-SNARE- and v-SNARE-reconstituted vesicles were allowed to interact, SNARE-ring complexes of ~20 nm in diameter were generated (Fig. 3.5a, b) [7]. With increase in the diameter of proteoliposomes, larger t-/v-SNARE complexes were formed (Fig. 3.5c, d). A strong linear relationship between size of the SNARE complex and vesicle diameter is demonstrated from these studies (Fig. 3.6) [7]. The experimental data fit well with the high correlation coefficient, R^2=0.9725 between vesicle diameter and SNARE-complex size (Fig. 3.6).

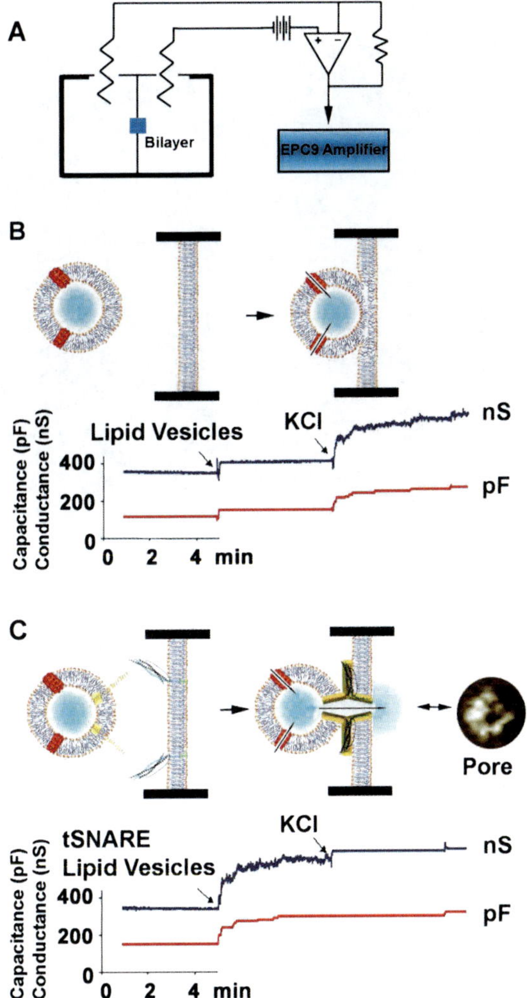

Fig. 3.3 Opposing bilayers containing t- and v-SNAREs respectively, interact in a circular array to form conducting pores. (**a**) Schematic diagram of the bilayer-electrophysiology setup. (**b**) Lipid vesicle containing nystatin channels (in *red*) and both vesicles and membrane bilayer without SNAREs, demonstrate no significant changes in capacitance and conductance. Initial increase in conductance and capacitance are due to vesicle-membrane attachment. To demonstrate membrane stability (both bilayer membrane and vesicles), the transmembrane gradient of KCl was increased to allow gradient driven fusion and a concomitance increase of conductance and capacitance. (**c**) When t-SNARE vesicles were added to a v-SNARE membrane support, the SNAREs in opposing bilayers arranged in a ring pattern, forming pores (as seen in the AFM micrograph on the extreme right) and there were seen stepwise increases in capacitance and conductance (–60 mV holding potential). Docking and fusion of the vesicle at the bilayer membrane, opens vesicle-associated nystatin channels and SNARE-induced pore formation, allowing conductance of ions from cis to the trans side of the bilayer membrane. Then further addition of KCl to induce gradient driven fusion, resulted in little or no further increase in conductance and capacitance, demonstrating docked vesicles have already fused [5]

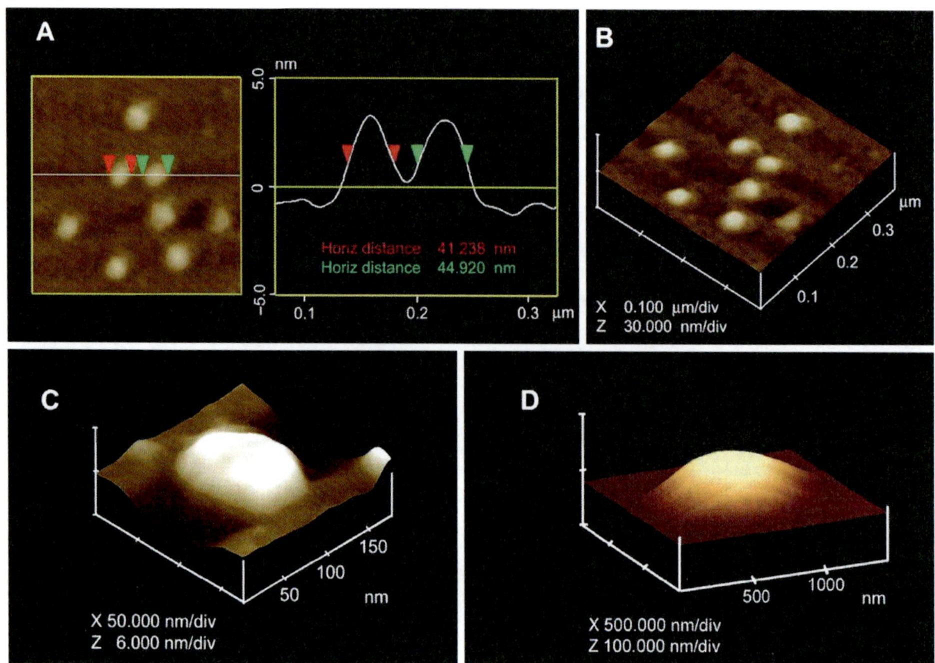

Fig. 3.4 AFM micrograph of t-SNARE and v-SNARE reconstituted liposomes of different sizes. Note the ~40–50 nm vesicles (**a**, **b**), the ~150 nm (**c**) and ~800 nm vesicle (**d**) [7]

3.3.3 Disassembly of the SNARE Complex

Studies demonstrate that the soluble *N-ethylmaleimide-sensitive factor* (NSF) an ATPase, disassembles the t-/v-SNARE complex in presence of ATP [10]. This study was also the first conformation by direct physical observation, that NSF-ATP alone can lead to SNARE complex disassembly. In this study, using purified recombinant NSF, and t- and v-SNARE-reconstituted liposomes, the disassembly of the t-/v-SNARE complex was examined. Lipid vesicles ranging in size from 0.2–2 μm, were reconstituted with either t-SNAREs or v-SNARE. Kinetics of association and dissociation of t-SNARE- and v-SNARE-reconstituted liposomes in solution, in the presence or absence of NSF, ATP, and AMP-PNP (the non-hydrolyzable ATP analogue), were monitored by right angle light scattering (Fig. 3.7a, b). Addition of NSF and ATP to the t/v-SNARE-vesicle mixture led to a rapid and significant increase in intensity of light scattering (Fig. 3.7a, b), suggesting rapid disassembly of the SNARE complex and dissociation of vesicles. Dissociation of t-/v-SNARE vesicles occurs on a logarithmic scale that can be expressed by first order equation, with rate constant $k=1.1\ s^{-1}$ (Fig. 3.7b). To determine whether NSF-induced dissociation of t- and v-SNARE vesicles is energy driven, experiments were performed in the presence and absence of ATP and AMP-PNP. No significant change with NSF alone, or in presence of NSF-AMP-PNP, was observed (Fig. 3.7c). These results demonstrate that t-/v-SNARE disassembly is an enzymatic and energy-driven process.

To further confirm the ability of NSF-ATP in the disassembly of the t-/v-SNARE complex, immunochemical studies were performed. It has been demonstrated that v-SNARE and t-SNAREs form an SDS-resistant complex [24]. NSF binds to SNAREs and forms a stable complex when locked in the ATP-bound state (ATP-NSF). Thus, in the presence of ATP+EDTA, VAMP antibody has been demonstrated to be able to immunoprecipitate this stable NSF-SNARE complex [24]. Therefore, in

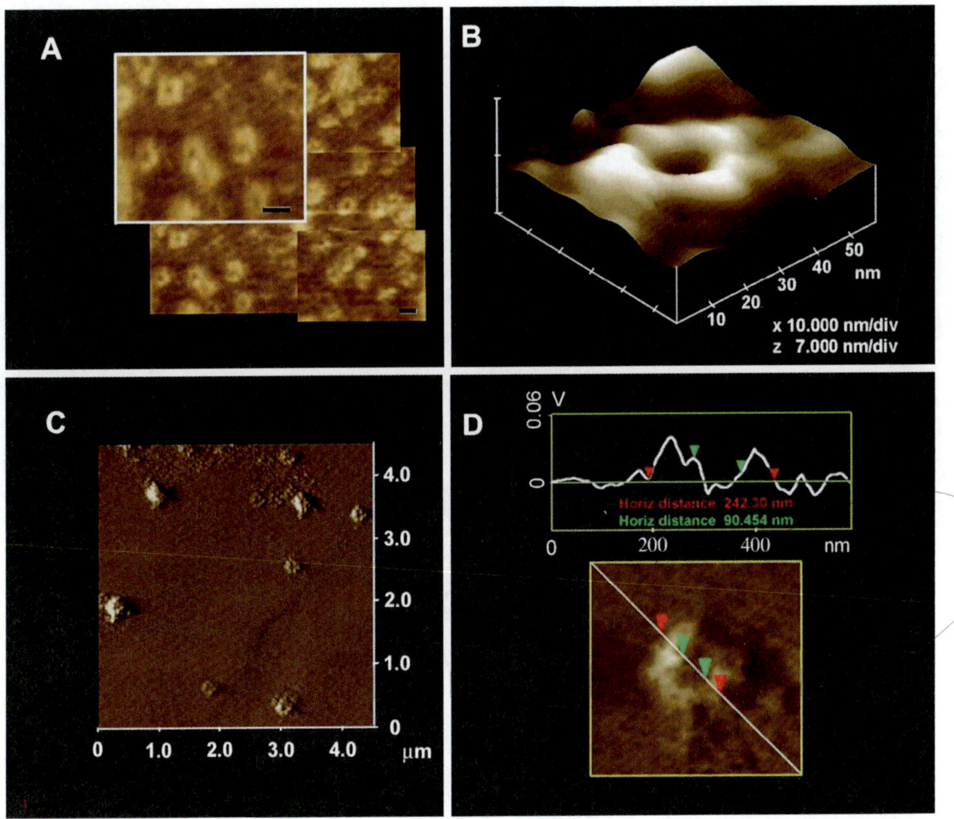

Fig. 3.5 Representative AFM micrograph of t-/v-SNARE complexes formed when small (a, b) or large (c, d) t-SNARE and v-SNARE reconstituted vesicles interact with each other. Note the formation of different size SNARE complexes, which are arranged in a ring pattern. *Bar*=20 nm. AFM section analysis (**d**) shows the size of a large SNARE complex [7]

the present study, when t- and v-SNARE vesicles were mixed in the presence or absence of ATP, NSF, NSF+ATP, or NSF+AMP-PNP, and resolved using SDS-PAGE followed by immunoblots using syntaxin-1 specific antibody, t-/v-SNARE disassembly was found to be complete only in the presence of NSF-ATP (Fig. 3.7d, e). To further confirm these findings (Fig. 3.7), direct observation of the t-/v-SNARE complex disassembly was assessed using AFM. When purified recombinant t-SNAREs and v-SNARE in opposing bilayers interact and self-assemble to form supramolecular ring complexes, they disassembled when exposed to recombinant NSF and ATP, as observed at nm resolution using AFM (Fig. 3.8). Since SNARE ring complex requires v-SNARE and t-SNAREs to be membrane-associated, suggested that NSF may require the t-/v-SNARE complex to be arranged in a specific configuration or pattern, for it to bind and disassemble the complex in presence of ATP. To test this hypothesis, t-SNAREs followed by v-SNARE, NSF and ATP were added to a lipid membrane, and continuously imaged in buffer by AFM (Fig. 3.9). Results from this study demonstrate that both SNARE complexes either in presence or absence of membrane disassemble [10]. Furthermore, close examination of the NSF-ATP-induced disassembled SNARE complex by AFM, demonstrates NSF to function as a right-handed molecular motor [6].

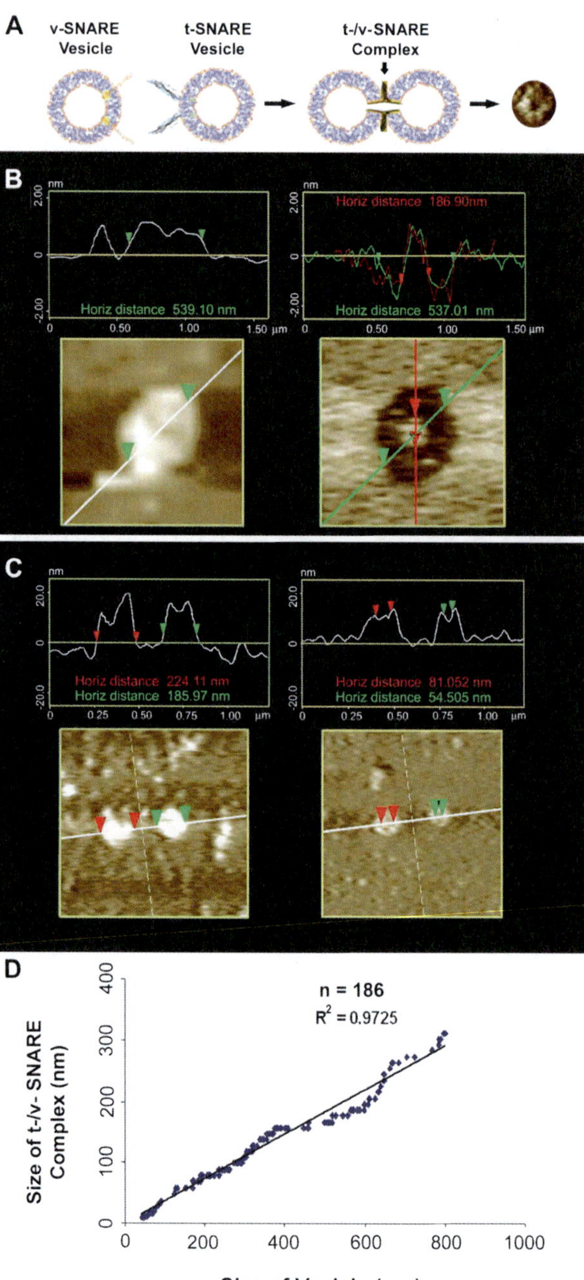

Fig. 3.6 SNARE complex is directly proportional to vesicle diameter. Schematic diagram depicting the interaction of t-SNARE and v-SNARE reconstituted vesicles. At the extreme right, is a single t-/v-SNARE complex imaged by AFM (**a**). AFM images of vesicles before and after their removal by the AFM cantilever tip, exposing the t-/v-SNARE complex (**b**). Interacting t-SNARE- and v-SNARE-vesicles imaged by AFM at low (<200 pN) and high forces (300–500 pN). Note, at low imaging forces, only the vesicle profile is imaged (*left* **c**). However at higher forces, the soft vesicle is flattened, allowing the SNARE complex to be imaged (*right* **c**). Plot of vesicle diameter vs. size of the SNARE complex. Note the high correlation coefficient (R^2=0.9725) between vesicle diameter and the size of the SNARE complex (**d**) [7]

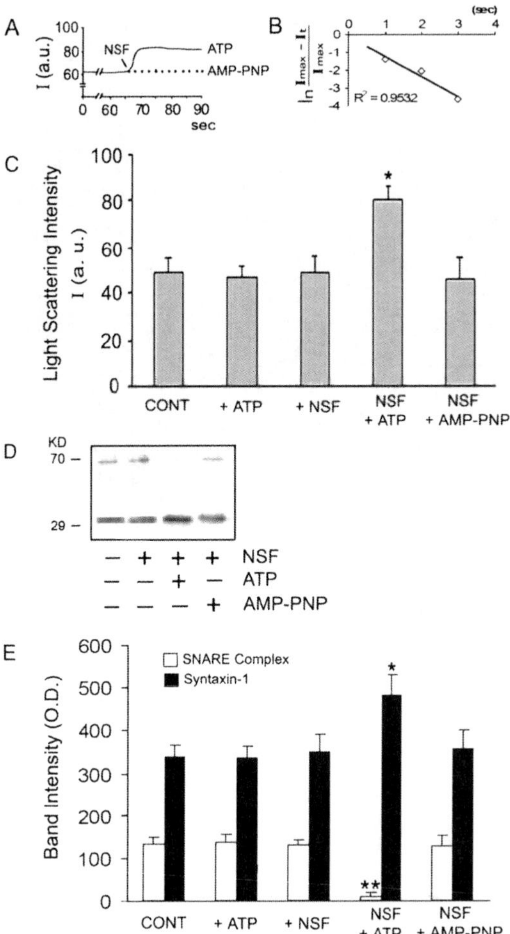

Fig. 3.7 NSF-ATP induced dissociation of t-SNARE and v-SNARE associated liposomes. (**a**) Real-time light scattering profiles of interacting t-SNARE and v-SNARE vesicles in solution in the presence and absence of NSF (depicted by *arrow*). In presence of ATP, NSF rapidly disassembles the SNARE complex and dissociates SNARE-vesicles represented as a rapid increase in light scattering. No change in light scattering is observed when ATP is replaced with the non-hydrolyzable analog AMP-PNP. (**b**) Kinetics of NSF-induced dissociation. The graph depicts first-order kinetics of vesicles dissociation elicited by NSF-ATP. (**c**) NSF requires ATP to dissociate vesicles. NSF in the presence of ATP dissociates vesicles ($p<0.05$, $n=4$, Student's *t*-test). However, NSF alone or NSF in the presence of AMP-PNP had no effect on the light scattering properties of SNARE-associated vesicle ($p>0.05$, $n=4$, Student's *t*-test). (**d**) When t- and v-SNARE vesicles are mixed in the presence or absence of ATP, NSF, NSF+ATP, or NSF+AMP-PNP, and resolved by SDS-PAGE followed by immunoblots using syntaxin-1 specific antibody, t-/v-SNARE disassembly was found to be complete only in the presence of NSF-ATP (**e**). Densitometric scan of the bands reveals significant changes in SNARE complex and syntaxin-1 reactivity only when NSF and ATP were included in reaction mixture ($p<0.05$, $n=3$; and $p<0.01$, $n=3$, Student's *t*-test) [10]

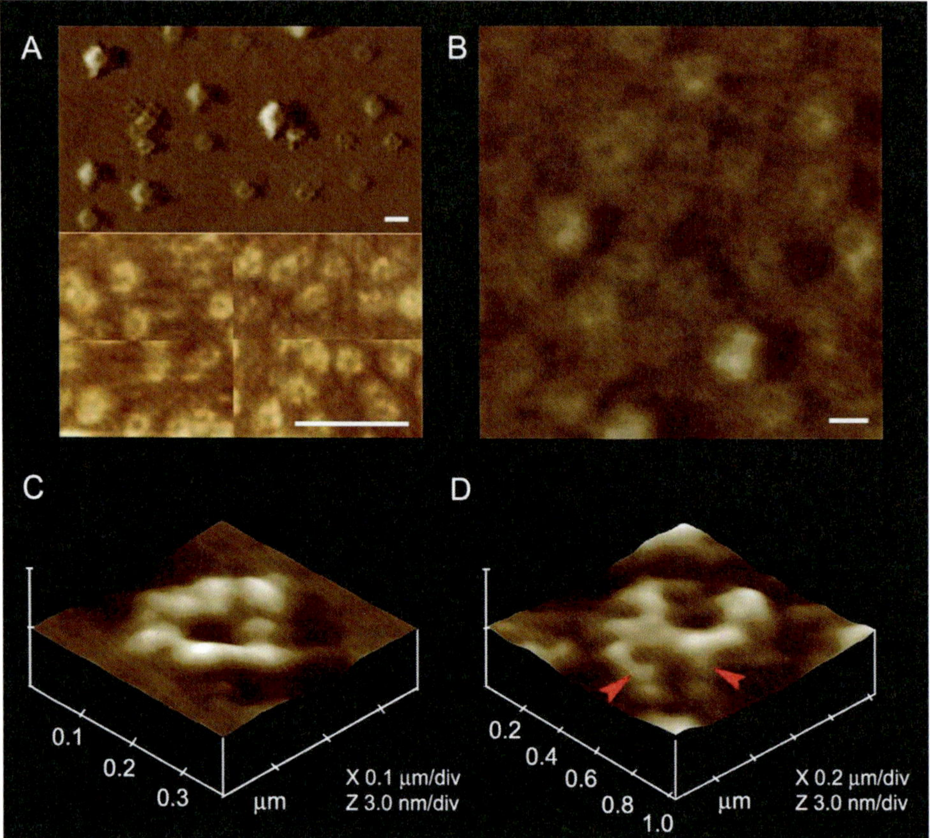

Fig. 3.8 AFM micrographs of NSF-ATP induced disassembly of the t-/v-SNARE ring complex. Representative AFM micrograph of t-/v-SNARE complexes formed when large (*top panel* **a**) or small (*bottom panel* **a**) t-/v-SNARE ring complexes are formed due to the interaction of large and small v-SNARE reconstituted vesicles interact with a t-SNARE reconstituted lipid membrane. *Bar* = 250 nm. (**b**) Disassembly of large t-/v-SNARE complex. *Bar* = 250 nm. (**c**) High resolution of a t-/v-SNARE ring complex, and a disassembled one (**d**) [10]

3.3.4 CD Spectroscopy Confirm the Requirement of Membrane for Appropriate t-/v-SNARE Complex Assembly, and that NSF-ATP Alone Can Mediated SNARE Disassembly

The overall secondary structural content of full-length neuronal v-SNARE and t-SNAREs, and the t-/v-SNARE complex, both in suspension and membrane-associated, were determined by CD spectroscopy using an Olis DSM 17 spectrometer [19]. Circular dichroism spectroscopy reveals that v-SNARE in buffered suspension (Fig. 3.10ai), when incorporated into liposomes (Fig. 3.10bi), exhibit reduced folding (Table 3.1). This loss of secondary structure following incorporation of full-length v-SNARE in membrane may be a result of self-association of the hydrophobic regions of the protein in absence of membrane. When incorporated into liposomes, v-SNARE may freely unfold without the artifactual induction of secondary structure, as reflective of the lack in CD signals at 208 and 222 nm, distinct for a-helical content. The t-SNAREs (Fig. 3.10aii, bii), shows clearly defined peaks

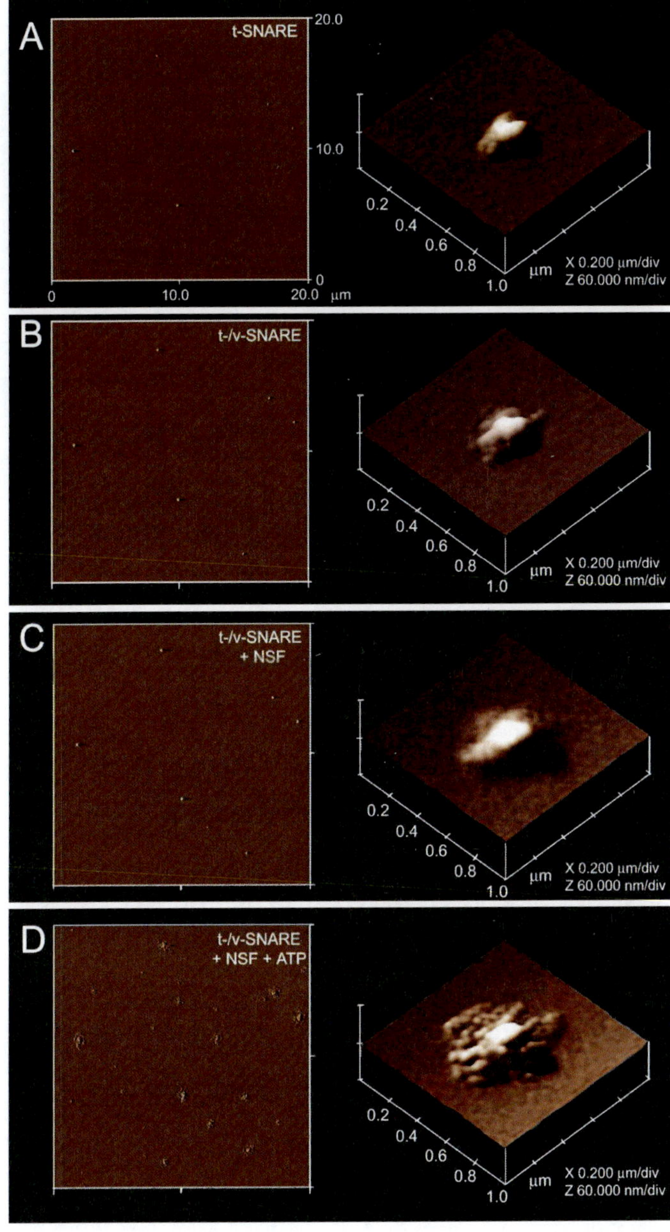

Fig. 3.9 AFM micrographs of NSF-ATP induced disassembly of the t-/v-SNARE complex formed when v-SNARE is added to a t-SNARE reconstituted lipid membrane. The *left panel* **a–d** shows at low resolution, the sequential AFM micrographs of one of ten representative experiments, where v-SNARE is added to **a** t-SNARE reconstituted lipid membrane, followed by NSF and then ATP. Note the dramatic disassembly of the SNARE complexes in **d**. The *right panel* shows at higher resolution, the disassembly of one of such SNARE complexes [10]

at both these wavelengths, consistent with a higher degree of helical secondary structures formed both in buffered suspension and in membrane, at ca. 66 and 20%, respectively (Table 3.1). Again, the membrane-associated SNARE exhibits less helical content than when in suspension. Similarly, there appears to be a dramatic difference in the CD signal observed in t-/v-SNARE complexes in suspension, and those complexes that are formed when membrane-associated SNAREs interact

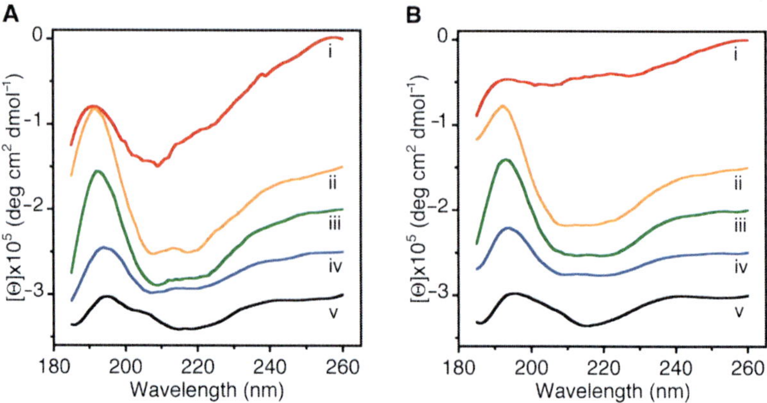

Fig. 3.10 Circular dichroism data reflecting structural changes to SNAREs, both in suspension and in association with membrane. Structural changes, following the assembly and disassembly of the t-/v-SNARE complex is further shown. (**a**) CD spectra of purified full-length SNARE proteins in suspension and (**b**) in membrane-associated; their assembly and (NSF–ATP)-induced disassembly is demonstrated. (i) v-SNARE; (ii) t-SNAREs; (iii) t-/v-SNARE complex; (iv) t-/v-SNARE + NSF and (v) t-/v-SNARE + NSF + 2.5 mM ATP, is shown. CD spectra were recorded at 25°C in 5 mM sodium phosphate buffer (pH 7.5), at a protein concentration of 10 mM. In each experiment, 30 scans were averaged per sample for enhanced signal to noise, and data were acquired on duplicate independent samples to ensure reproducibility [19]

(Fig. 3.10aiii, biii). Interestingly, there is no increase of secondary structure upon complex formation. Rather, the CD spectra of the complexes are identical to a combination of individual spectra. Moreover, membrane associated t-/v-SNAREs are less folded than the purified SNARE complex. This data supports previous AFM results that lipid is required for proper arrangement of the SNARE proteins in membrane fusion. Addition of NSF to the t-/v-SNARE complex results in an increase in the unordered fraction (Fig. 3.10aiv, biv and Table 3.1), which may be attributed to an overall disordered secondary structure of the NSF, and not necessarily unfolding of the t-/v-SNARE complex. In contrast, activation of NSF by the addition of ATP almost completely abolishes all α-helical content within the multi-protein complex (Fig. 3.10av, bv). This direct observation of the helical unfolding of the SNARE complex using CD spectroscopy under physiologically relevant conditions (i.e. in membrane-associated SNAREs), confirms earlier AFM reports on NSF–ATP-induced t-/v-SNARE complex disassembly [10]. In further agreement with previously reported studies using the AFM, the consequence of ATP addition to the t-/v-SNARE–NSF complex is disassembly, regardless of whether

Table 3.1 Secondary structural fit parameters of SNARE complex formation and dissociation [19]

	Suspension (100 × f[a])					Membrane-associated (100 × f)				
Protein[b]	α	β	O	U	Fit[c]	α	β	O	U	Fit
v-SNARE	4	36	18	43	0.19	0	30	32	38	0.21
t-SNAREs	66	34	0	0	0.02	20	15	21	44	0.84
v-/t-SNAREs	48	52	0	0	0.02	20	19	56	5	0.38
v-/t-SNAREs+NSF	20	25	0	55	0.07	18	6	8	68	0.2
v-/t-SNAREs+NSF+ATP	3	39	18	40	0.22	1	27	34	38	0.23

[a]Abbreviations used: *f*, fraction of residues is a given conformational class; *α*, α-helix; *β*, β-sheet; *O*, other (sum of turns, distorted helix, distorted sheet); *U*, unordered.
[b]Protein constructs: *v-SNARE* (VAMP2); *t-SNAREs* (SNAP-25 + syntaxin 1A); *NSF*, N-ethylmaleimide Sensitive Factor. *ATP*, adenosine triphosphate.
[c]Fit: goodness of fit parameter expressed as Normalized Spectral Fit Standard Deviation (nm).

the t-/v-SNARE+NSF complex is membrane-associated or in buffered suspension. In earlier AFM studies, 0.16–0.2 mg/ml of SNARE proteins were used, as opposed to the 800–1,000 mg/ml protein concentration required for the current CD studies. To determine if t-SNARE and v- SNARE interact differently at higher protein concentrations, both membrane-associated and in-suspension v- and t-SNARE complexes used in CD studies, were imaged using the AFM. In confirmation to previously reported AFM studies, results from the CD study demonstrated the formation of t-/v-SNARE ring complexes, only when t-SNARE-liposomes are exposed to v-SNARE-liposomes. Hence, higher SNARE protein concentrations are without influence on the membrane-directed self-assembly of the SNARE complex [19]. In summary, the CD results demonstrate that v-SNARE in suspension, when incorporated into liposomes, exhibits reduced folding. Similarly, t-SNAREs which exhibit clearly defined peaks at CD signals of 208 and 222 nm wavelengths, consistent with a higher degree of helical secondary structure in both the soluble and liposome-associated forms, exhibit reduced folding when membrane associated. ATP-induced activation of NSF bound to the t-/v-SNARE complex, results in disassembly of the SNARE complex, eliminating all α-helices within the structure. In addition, these studies are a further confirmation of earlier reports [10] that NSF–ATP is sufficient for the disassembly of the t-/v-SNARE complex.

3.3.5 SNAREs Bring Opposing Bilayers Closer, Enabling Calcium Bridging and Membrane Fusion

Diffraction patterns of non-reconstituted vesicles and t- and v-SNARE-reconstituted vesicles in the absence and presence of 5 mM Ca^{2+} are shown for comparison in Fig. 3.11. To our knowledge, these are the first recorded wide-angle diffractograms of unilamellar (single bilayer) vesicles in the 2–4 Å diffraction range. They have broad pattern spanning 2θ ranges approximately 23–48° or d values of 3.9–1.9 Å with sharp drop off intensity on either sides of the range. Relatively, broad feature of diffractogram indicate multitude of contacts between atoms of one vesicle as well between different vesicles during collision. However, two broad peaks are visible on the diffractogram, the stronger one at 3.1 Å and a weaker one at 1.9 Å. They indicate that the greatest number of contacts between them have these two distances. Addition of Ca^{2+} or incorporation of SNAREs at the vesicles membrane or both, influence both peaks within the 2.1–3.3 Å intensity range (Fig. 3.11). However, the influence of Ca^{2+}, SNAREs or both is more visible on peak positioned at 3.1 Å in form of an increased Imax of arbitrary units and 2θ. This increase of Imax at the 3.1 Å can be explained in terms of increased vesicle pairing and/or a decrease in distance between apposed vesicles. Incorporation of t- and v-SNARE proteins at the vesicle membrane allows for tight vesicle-vesicle interaction, demonstrated again as an Imax shifts

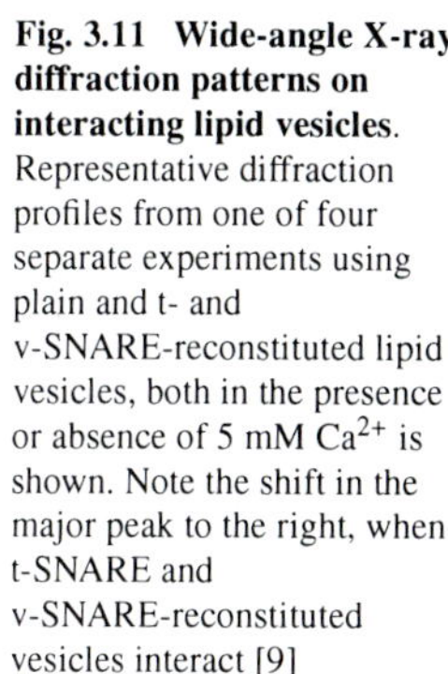

Fig. 3.11 Wide-angle X-ray diffraction patterns on interacting lipid vesicles. Representative diffraction profiles from one of four separate experiments using plain and t- and v-SNARE-reconstituted lipid vesicles, both in the presence or absence of 5 mM Ca^{2+} is shown. Note the shift in the major peak to the right, when t-SNARE and v-SNARE-reconstituted vesicles interact [9]

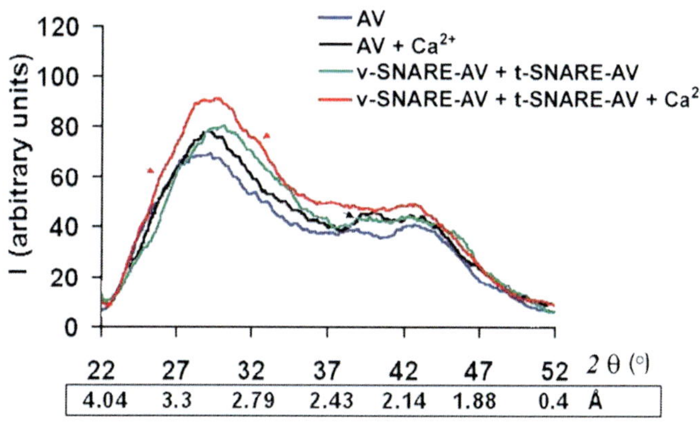

to 30.5° or 2.9 Å from 3.1 Å. Ca^{2+} and SNAREs work in manner that induces a much higher increase of peak intensity with appearance of shoulders on both sides of the peak at 2.8 and 3.4 Å (Fig. 3.11). This indicates an increase in number of vesicle contact points at a constant distance between them. Vesicles containing either t- or v-SNAREs have little effect on the X-ray scattering patterns. Only as discussed above, when t-SNARE and v-SNARE reconstituted-vesicles were brought together, we did detect change in the X-ray diffraction patterns. Since exposure of t-SNARE vesicle and v-SNARE vesicle mixture to Ca^{2+} results in maximum increase in a.u. and 2θ using X-ray diffraction, the effect of Ca^{2+} on fusion and aggregation of t-/v-SNARE vesicles were examined using light scattering, light microscopy, AFM, fluorescent dequenching and electrical measurements of fusion [9].

In recent studies [12], using high-resolution electron microscopy, the electron density maps and 3-D topography of the membrane-directed SNARE ring complex was determined at nm resolution (Fig. 3.12). Similar to the t-/v-SNARE ring complex formed when 50 nm v-SNARE liposomes meet

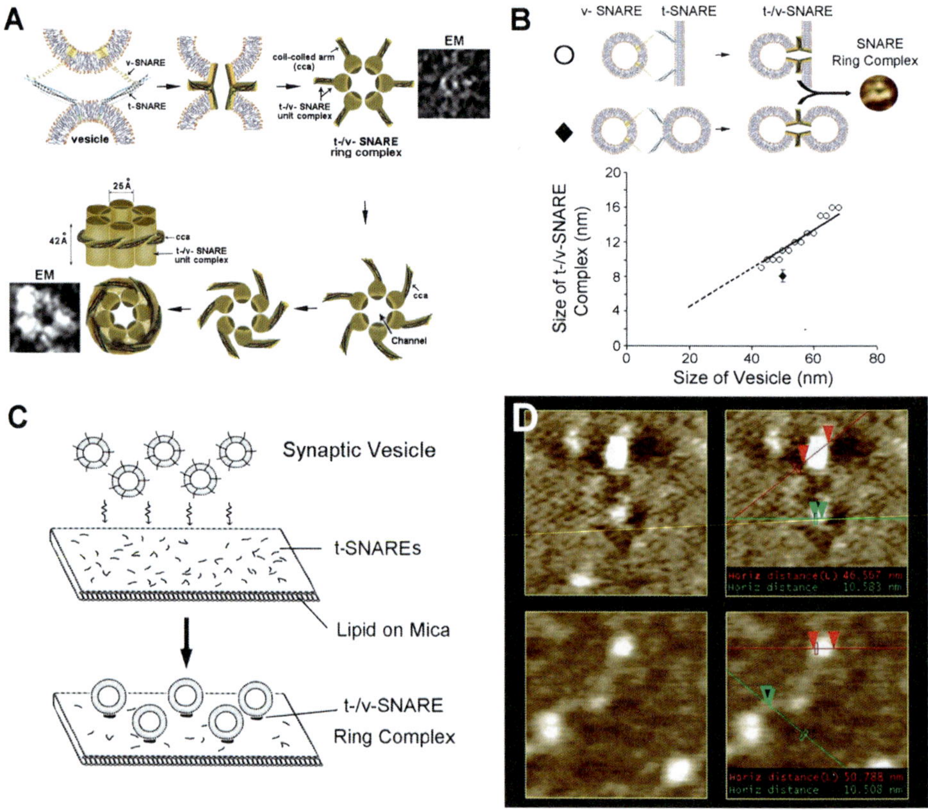

Fig. 3.12 The possible establishment of a leak-proof SNARE ring complex channel is demonstrated. (**a**) Size of the t-/v-SNARE ring complex is directly proportional to the size of the SNARE-associated vesicle (**c**). Different sizes of v-SNARE-associated vesicles, when interact with t-SNARE-associated membrane (○), demonstrate the SNARE ring size to be directly proportional to the vesicle size. When a 50 nm in diameter v-SNARE-reconstituted vesicle interacts with a t-SNARE-reconstituted membrane, an 11 nm in diameter t-/v-SNARE ring complex is formed. Similarly, the present study demonstrates that when a 50 nm in diameter v-SNARE-reconstituted vesicle, interacts with a 50 nm in diameter t-SNARE-reconstituted vesicle, a 8 nm in diameter t-/v-SNARE ring complex is established (♦). Analogous to the 11 nm in diameter t-/v-SNARE ring complexes formed when 50 nm v-SNARE vesicles meet a t-SNARE-reconstituted planer membrane (**b**), approximately 11 nm in diameter t-/v-SNARE ring complexes are formed when 50 nm in diameter synaptic vesicles meets a t-SNARE-reconstituted planer membrane (**c**, **d**) [12]

a t-SNARE-reconstituted planer membrane, SNARE rings are also formed when 50 nm in diameter isolated synaptic vesicles meet a t-SNARE-reconstituted planer lipid membrane. Furthermore, the mathematical prediction of the SNARE ring complex size with reasonable accuracy, and the possible mechanism of membrane-directed t-/v-SNARE ring complex assembly, was determined from the study. Using both lipososome-reconstituted recombinant t-/v-SNARE proteins, and native v-SNARE present in isolated synaptic vesicle membrane, the membrane-directed molecular assembly of the neuronal SNARE complex was determined for the first time and its size mathematically predicted. These results provide a new molecular understanding of the universal machinery and mechanism of membrane fusion in cells, having fundamental implications in human health and disease. The above mention studies and their findings provide a molecular understanding of SNARE-induced membrane fusion in cells.

Acknowledgments The author thanks the many students and collaborators who have participated in the various studies discussed in this article. Support from the National Institutes of Health (USA), the National Science Foundation (USA), and Wayne State University, is greatly appreciated.

References

1. Malhotra V, Orci L, Glick BS et al (1988) Role of an N-ethylmaleimide-sensitive transport component in promoting fusion of transport vesicles with cisternae of the Golgi stack. Cell 54:221–227
2. Bennett MK, Calakos N, Scheller RH (1992) Syntaxin: a synaptic protein implicated in docking of synaptic vesicles at presynaptic active zones. Science 257:255–259
3. Oyler GA, Higgins GA, Hart RA et al (1989) The identification of a novel synaptosomal-associated protein, SNAP-25, differentially expressed by neuronal subpopulations. J Cell Biol 109:3039–3052
4. Trimble WS, Cowan DM, Scheller RH (1988) VAMP-1: a synaptic vesicle-associated integral membrane protein. Proc Natl Acad Sci USA 85:4538–4542
5. Cho SJ, Kelly M, Rognlien KT et al (2002) SNAREs in opposing bilayers interact in a circular array to form conducting pores. Biophys J 83:2522–2527
6. Cho WJ, Jena BP (2007) N-ethylmaleimide-Sensitive Factor is a Right-Handed Molecular Motor. J Biomed Nanotech 3:209–211
7. Cho WJ, Jeremic A, Jena BP (2005) Size of supramolecular SNARE complex: membrane-directed self-assembly. J Am Chem Soc 127:10156–10157
8. Jeremic A, Cho WJ, Jena BP (2004) Membrane fusion: what may transpire at the atomic level. J Biol Phys Chem 4:139–142
9. Jeremic A, Kelly M, Cho JA et al (2004) Calcium drives fusion of SNARE-apposed bilayers. Cell Biol Int 28:19–31
10. Jeremic A, Quinn AS, Cho WJ et al (2006) Energy-dependent disassembly of self-assembled SNARE complex: observation at nanometer resolution using atomic force microscopy. J Am Chem Soc 128:26–27
11. Weber T, Zemelman BV, McNew JA et al (1998) SNAREpins: minimal machinery for membrane fusion. Cell 92:759–772
12. Cho WJ, Lee JS, Ren G et al (2011) Membrane-directed molecular assembly of the neuronal SNARE complex. J Cell Mol Med 15:31–37
13. Bako I, Hutter J, Palinkas G (2002) Car–Parrinello molecular dynamics simulation of the hydrated calcium ion. J Chem Phys 117:9838–9843
14. Chialvo AA, Simonson JM (2003) The structure of CaCl2 aqueous solutions over a wide range of concentration. Interpretation of diffraction experiments via molecular simulation. J Chem Phys 119:8052–8061
15. McIntosh TJ (2000) Short-range interactions between lipid bilayers measured by X-ray diffraction. Curr Opin Struct Biol 10:481–485
16. Portis A, Newton C, Pangborn W et al (1979) Studies on the mechanism of membrane fusion: evidence for an intermembrane Ca^{2+}-phospholipid complex, synergism with Mg^{2+}, and inhibition by spectrin. Biochemistry 18:780–790
17. Laroche G, Dufourc EJ, Dufourcq J et al (1991) Structure and dynamics of dimyristoylphosphatidic acid/calcium complexes by 2H NMR, infrared, spectroscopies and small-angle x-ray diffraction. Biochemistry 30:3105–3114
18. Potoff JJ, Issa Z, Manke CW, Jr et al (2008) Ca^{2+}-dimethylphosphate complex formation: providing insight into Ca^{2+}-mediated local dehydration and membrane fusion in cells. Cell Biol Int 32:361–366
19. Cook JD, Cho WJ, Stemmler TL et al (2008) Circular dichroism (CD) spectroscopy of the assembly and disassembly of SNAREs: The proteins involved in membrane fusion in cells. Chem Phy Lett 462:6–9

20. Cohen FS, Niles WD (1993) Reconstituting channels into planar membranes: a conceptual framework and methods for fusing vesicles to planar bilayer phospholipid membranes. Methods Enzymol 220:50–68
21. Kelly ML, Woodbury DJ (1996) Ion channels from synaptic vesicle membrane fragments reconstituted into lipid bilayers. Biophys J 70:2593–2599
22. Woodbury DJ (1999) Nystatin/ergosterol method for reconstituting ion channels into planar lipid bilayers. Methods Enzymol 294:319–339
23. Woodbury DJ, Miller C (1990) Nystatin-induced liposome fusion. A versatile approach to ion channel reconstitution into planar bilayers. Biophys J 58:833–839
24. Jeong EH, Webster P, Khuong CQ et al (1998) The native membrane fusion machinery in cells. Cell Biol Int 22:657–670

Chapter 4
Molecular and Cellular Mechanisms of Mammalian Cell Fusion

Xiaofeng Zhou and Jeffrey L. Platt

Abstract The fusion of one cell with another occurs in development, injury and disease. Despite the diversity of fusion events, five steps in sequence appear common. These steps include programming fusion-competent status, chemotaxis, membrane adhesion, membrane fusion, and post-fusion resetting. Recent advances in the field start to reveal the molecules involved in each step. This review focuses on some key molecules and cellular events of cell fusion in mammals. Increasing evidence demonstrates that membrane lipid rafts, adhesion proteins and actin rearrangement are critical in the final step of membrane fusion. Here we propose a new model for the formation and expansion of membrane fusion pores based on recent observations on myotube formation. In this model, membrane lipid rafts first recruit adhesion molecules and align with opposing membranes, with the help of a cortical actin "wall" as a rigid supportive platform. Second, the membrane adhesion proteins interact with each other and trigger actin rearrangement, which leads to rapid dispersion of lipid rafts and flow of a highly fluidic phospholipid bilayer into the site. Finally, the opposing phospholipid bilayers are then pushed into direct contact leading to the formation of fusion pores by the force generated through actin polymerization. The actin polymerization generated force also drives the expansion of the fusion pores. However, several key questions about the process of cell fusion still remain to be explored. The understanding of the mechanisms of cell fusion may provide new opportunities in correcting development disorders or regenerating damaged tissues by inhibiting or promoting molecular events associated with fusion.

4.1 Introduction

Many anatomic structures and processes of mammalian development depend on fusion of one cell with another. Although fusion of cells was first recognized from anatomic features [1], anatomy alone provides poor measure of the relative contribution of cell–cell fusion to the anatomic structure, development and physiology of tissues. Cell fusion can be underestimated if nuclei of multinucleated cells are fused or shed, or overestimated if mitosis proceeds without cytokinesis [2]. Accordingly, the impact of cell fusion on structure and function of tissues may be best evaluated by examining the expression or function of molecules that regulate the process. This chapter will review some molecules

J.L. Platt (✉)
Departments of Surgery and Microbiology and Immunology, University of Michigan, Ann Arbor, MI 48109-2200, USA
e-mail: plattjl@umich.edu

T. Dittmar, K.S. Zänker (eds.), *Cell Fusion in Health and Disease*, Advances in Experimental Medicine and Biology 713, DOI 10.1007/978-94-007-0763-4_4, © Springer Science+Business Media B.V. 2011

thought to be important for cell fusion in mammals. We focus not only on the molecules involved in fusion of one cell with another, but also the molecules that prepare cells for fusion or make fusion more or less likely.

The types of cell fusion best understood in mammals include the fusion of sperm and egg [3], fusion of cytotrophoblast cells to form syncytiotrophoblast [4], fusion of myoblasts to form myotubes [5], and fusion of macrophages to form multi-nucleated giant cells [6]. Recently, fusion of bone marrow derived stem cells with parenchymal cells in the course of tissue repair has also been appreciated as a mechanism of regeneration and apparent adult stem cell plasticity [2, 7, 8].

Certain processes or steps occurring in sequence appear common to cell fusion in disparate systems. The first step involves "programming" of cells to make them competent for fusion. The second step involves movement of fusion-competent cells toward each other by releasing or responding to attractants. In the third step, recognition and attachment of the cells bring plasma membranes into apposition. The fourth step involves mixing of lipids in apposing membranes leading to formation and expansion of fusion pores. The fifth step involves physiologic adaptation of the multinucleated cells either to undergo or to prevent (in the case of sperm-egg fusion) further rounds of fusion.

Although the mechanisms of the steps above are not fully understood, each of the fusion steps appears tightly regulated at the molecular level. Gene knockout, RNA interference, overexpression of wild-type or dominant negative protein forms, and antibody neutralization have been used to identify molecules important for cell fusion. Some molecules seem to be specific for certain cell types, while others are important in many fusion systems. Table 4.1 lists some molecules involved in various steps of cell fusion.

In the sections that follow, we discuss the molecules thought to govern cell fusion. Since some molecules are involved in multiple steps they are mentioned repeatedly. In the section of fusion

Table 4.1 A partial list of molecules involved in each step of cell–cell fusion in mammals

Priming	Chemotaxis	Adhesion	Fusion	Post-fusion
Cytokines:	Chemoattractants:	Ig domain proteins:	Actin network:	Myoferlin
IL-4	HGF	Nephrin	Actin	Bcl-2
IFN-γ	SDF1	Cdo	Myosin	c-Flip
RANKL	IL-4	Neogenin	Arp2/3	
M-CSF	CXCR4	NCAM	SCAR/WAVE	
DAP12	MOR23	Izumo	WASP	
Phosphatidylserine	MCP-1	CD47	Rac1	
Calcium	Progesterone	SIRP-α	Cdc42	
Caspase 8	Integrins:	Cadherins:	Dock1	
Caspase 9	α7β1	N-cadherin	Brag2	
Calpain:	αvβ3	M-cadherin	Trio	
m-calpain	V-Atp6v0d2	E-cadherin	Nap1	
Calpain 3	Actin regulators:	Cadherin-11	Lipid rafts:	
Glis3	WAVE2	Tetraspanins:	Cholesterol	
	WASP	CD9	Sphingolipids	
	ROCK-I	CD81	Syncytin-1	
	MR	CD63	Syncytin-2	
	MMP-9	Integrin β1		
		GPI-anchored proteins		
		ADAM		
		DC-STAMP		

pore formation and expansion (Section 4.5.3), we propose a new model of the events leading to cell membrane fusion based on recent observations.

4.2 Programming Cellular Competence for Fusion

Most cells do not fuse spontaneously. The lipid bilayers of cell membranes maintain the integrity of individual cells unless they change in ways that overcome barriers to fusion. The process by which cells acquire the ability to fuse is called "priming" and the primed cell is called a "competent cell". The priming process might include (i) expression of fusion machinery, such as adhesion molecules; (ii) alteration of cell membrane lipid composition, for instance translocation of inner-leaflet lipids; and (iii) loss of inhibitory state, including degradation of extracellular matrix allowing cell migration and cell–cell adhesion. Various molecules program fusion-competent status in different cell fusion systems.

4.2.1 Cytokines

Cytokines can make monocytes and macrophages competent for fusion [9]. Cytokines are small secreted proteins that act on specific cell membrane receptors initiating signaling pathways to alter gene transcription.

Interleukin (IL)-4, IL-13, IL-3, IL-17A, and interferon (IFN)-γ induce monocytes and macrophages to form multinucleated giant cells [10–15]. For examples, McInnes and Rennick [14] found that culturing mouse monocytes/macrophages with IL-4 induces the formation of multinucleated giant cells, and Chensue et al. [18] demonstrated that anti-IL-4 antibodies inhibit or reduce the formation of multinucleated giant cells. Weinberg et al. [16] generated multinucleated giant cells after culture of macrophages with IFN-γ, while anti-IFN-γ antibodies block the fusion [16–18].

How IL-4 induces macrophage fusion-competent status has been elucidated in part. Stimulation of IL-4 receptor α chain (IL4Rα) by IL-4 or IL-13 initiates phosphorylation and activation of the transcription factor STAT6. *STAT6*-knockout macrophages do not form multinucleated giant cells in the presence of IL-4 in vitro [19, 20]. The IL-4/STAT6 axis turns on the expression of E-cadherin, dendritic cell-specific transmembrane protein (DC-STAMP) and matrix metalloproteinase 9 (MMP-9) [20–23]. MMP-9 can digest structural components of the extracellular matrix and cellular surface facilitating cell migration and adhesion [22]. E-cadherin mediates cell adhesion to bring membranes of cells into close proximity, while DC-STAMP acts by an unknown mechanism to promote cell membrane fusion [24, 25]. IL-4 also upregulates the mannose receptor [26], which may promote fusion of macrophages [6, 27, 28]. IL-4 only causes macrophages competent for fusion but not other cell types, indicating that the presence of macrophage-specific proteins must be essential for IL-4 to prime macrophages.

Although both osteoclasts (multinucleated bone resorbing cells) and multinucleated giant cells originate from a common progenitor, formation of osteoclasts requires different cytokines than formation of multi-nucleated giant cells. Stimulation of receptor activator for nuclear factor κB ligand (RANKL) and macrophage colony stimulating factor (M-CSF) induces the commitment of osteoclast differentiation and fusion-competence of macrophages [29, 30]. RANKL stimulates the TRAF6/NF-κB and c-Fos pathways, which induce the expression of nuclear factor of activated T cells, cytoplasmic 1 (NFATc1) [31]. NFATc1 is a master transcription factor that not only regulates osteoclast-specifying genes, but also upregulates the genes that promote cell fusion, such as DC-STAMP and the d2 isoform of vacuolar ATPase V0 domain (Atp6v0d2) [32–34]. Osteoblasts and their precursors appear to be the main regulators of osteoclast formation. Osteoblasts express M-CSF and RANKL to promote osteoclastogenesis, as well as osteoprotegerin that inhibits osteoclastogenesis. Thus osteoblasts control the balance between bone construction and absorption [35, 36].

4.2.2 DNAX Activating Protein 12

DNAX activating protein 12 (DAP12) programs fusion-competence for cells that become osteoclasts and multinucleated giant cells [37–40]. DAP12 is a signaling adaptor that associates with a number of cell surface receptors including the triggering receptor expressed in myeloid cells 2 (TREM-2) [41]. Binding of an unknown ligand to TREM-2 triggers the phosphorylation of DAP12, which in turn forms docking sites for the protein tyrosine kinase Syk and the p85 subunit of phosphoinositide (PI) 3-kinase [40]. DAP12/Syk complex increases the expression of DC-STAMP, MMP-9 and E-cadherin, all essential for the formation of multinucleated giant cells in an IL-4 independent manner [37]. In addition, binding of M-CSF to its receptor, c-Fms, also activates DAP12/Syk axis [42]. In osteoclast precursor cells, DAP12/Syk activates phospholipase Cγ (PLCγ) to regulate the expression of the master transcription factor NFATc1 [31].

4.2.3 Phosphatidylserine

Modification of the distribution of lipids in the inner and outer leaflets of plasma membranes is a key event in preparing plasma membranes in formation of myotubes [43], syncytiotrophoblast [44], fertilization of eggs [45] and multinucleated giant cell formation [46]. Phosphatidylserine, a negatively charged amine-containing phospholipid normally of the inner-leaflet (the cytosolic side) of cell membranes, translocates to the outer-leaflet before cell fusion. Interestingly, redistribution of phosphatidylserine also occurs in early apoptosis [47]. Whether plasma membrane fusion and the early apoptosis share a common molecular mechanism is debated [48, 49].

Exposing phosphatidylserine at the outer surface of the lipid bilayer is believed to be essential for sperm “capacitation” [50], that is a membrane destabilization process in sperm that allows binding and penetrating the zona pellucida (outer coating of the egg). Capacitation takes place in the female reproductive tract and can be mimicked in vitro. Plasma membrane architecture changes through phosphatidylserine and phosphatidylethanolamine efflux in capacitation have profound implications for fertilization [50]. Surface exposure of phosphatidylserine and phosphatidylethanolamine is limited to the plasma membrane of the apical head (the anterior acrosomal region) of sperm [51]. Phosphatidylserine and phosphatidylethanolamine appear to attract cholesterol from the post-acrosomal region to the apical acrosomal region [52]. Next, complexes of cholesterol and filipin are depleted in an albumin-dependant manner and this depletion increases membrane fluidity which in turn may facilitate membrane fusion. The increase in sperm membrane fluidity after phosphatidylserine and phosphatidylethanolamine exposure can be detected by fluorescence labeled membrane lipids in boar sperm [53]. Lowered levels of cholesterol may also promote lipid raft dynamics, weaken the binding of surface proteins, and alter steady-state intracellular ion concentrations [50].

Bicarbonate triggers phosphatidylserine exposure and reorganization of plasma membrane of sperm [53]. A rapid increase in scrambling of boar sperm membrane lipids was observed within 100 s of exposure to bicarbonate in vitro. Bicarbonate first stimulates sAC, a major adenylyl cyclase isoform in spermatozoa, increasing cyclic AMP [54, 55]. A sudden increase in bicarbonate concentration may be experienced by sperm in the female reproductive tract, as bicarbonate concentration in the epididymis, where mature sperm are stored, is much lower [56]. The bicarbonate-induced membrane change is inhibited by H89, a protein kinase A inhibitor, suggesting that cAMP-dependent protein kinase A signaling controls phosphatidylserine exposure [55]. Redistribution of phosphatidylserine in sperm is not associated with apoptosis as measured by DNA degeneration or mitochondrial function. Furthermore, the process of phosphatidylserine efflux is caspase-independent as it is not affected by caspase inhibitors [57].

Transposition of phosphatidylserine to the surface in macrophages is also an early step in multinucleated giant cell formation [46, 58]. When the outer-leaflet phosphatidylserine is masked by annexin V or phosphatidylserine-containing liposomes, IL-4 induced macrophage fusion is blocked [46]. What controls the exposure of phosphatidylserine in macrophages is not known, but the P2X7 purinoceptor has been implicated since activation of P2X7 receptors exposes phosphatidylserine on the cell surface [9, 59]. Cell surface phosphatidylserine may be recognized by the scavenger receptor CD36 that is present at sites of contact between macrophages [46]. Monoclonal antibodies against CD36 suppress fusion of phosphatidylserine-containing liposomes and macrophages [46]. How CD36 recognizes phosphatidylserine is not yet known.

Exposure of phosphatidylserine on myoblasts facilitates myotube formation and development of skeletal muscle [60]. In cultured C2C12 and H9C2 myoblasts, transient expression of phosphatidylserine at sites of cell–cell contact is detected prior to the fusion of individual myoblasts into multinucleated myotubes [61]. Myotube formation in vitro is inhibited by annexin V. The transposition of phosphatidylserine in myoblasts differs from the mechanism associated with apoptosis, because the multi-caspase inhibitor zVAD(OMe)-fmk, which prevents apoptosis, does not prevent myotube formation [61].

Efflux of phosphatidylserine also occurs during differentiation of the choriocarcinoma cell line BeWo and formation of villous cytotrophoblast [62, 63]. Antibodies against phosphatidylserine prevent fusion of BeWo cells and inhibit differentiation of isolated first trimester villous cytotrophoblasts [44, 64]. Protein kinase A may induce transposition of phosphatidylserine, since the protein kinase A inhibitor H89 prevents forskolin stimulated BeWo cell fusion [49]. ATP-dependent phospholipid floppases, most of which are members of the ABC transporters, are also critical for phosphatidylserine efflux [65]. The floppase inhibitor vanadate strongly suppresses phosphatidylserine exposure and fusion of BeWo cells. Whether caspase 8 can trigger phosphatidylserine exposure is unclear [48, 49].

4.2.4 Calcium

After capacitated mammalian sperm encounter the zona pellucida of an egg, the acrosome reaction primes sperm for fusion. The acrosome reaction is triggered by influx of Ca^{2+} [66]. The acrosome is a secretory vesicle containing zona-digesting enzymes in the anterior portion of the sperm head. During the acrosome reaction, the outer acrosomal membrane fuses with the sperm plasma membrane in multiple locations generating many openings, through which acrosomal contents are released [67]. The hybrid membrane of plasma and outer acrosome at the sperm head is worn down during penetration of the zona pellucida, and thus the inner region of the acrosomal membrane is exposed and fused with sperm plasma membrane at the equator. The conjunction of the inner acrosomal membrane to the plasma membrane facilitates relocation of the sperm-egg fusion proteins, such as PH-20, Izumo and FLOT2, into the plasma membrane [68–71].

Contact between sperm and zona pellucida induces Ca^{2+} influx in sperm [66]. The glycoprotein ZP3 in the zona pellucida stimulates the depolarization of sperm membrane potential, causing Ca^{2+} influx and the acrosome reaction [72, 73]. ZP3 stimulation requires heterotrimeric G protein [74, 75] and phospholipase Cd4 [76, 77]. How ZP3 regulates the G protein and Cd4 is still not clear.

After initial stimulation, Ca^{2+} flux into the acrosome is sustained by canonical transient receptor potential channels which activate the PI3 kinase pathway [66]. Protein kinase Akt (also known as protein kinase B) and the atypical protein kinase C, PKC, act downstream of PI3K [78]. Recent studies have linked Akt and PKC with the activation of soluble N-ethylmaleimide-sensitive factor attachment protein receptors (SNAREs) [79], which are essential for the fusion between the outer acrosomal membrane and sperm plasma membrane [80, 81].

4.2.5 Proteases

Two families of proteases, the initiator caspases and calpains, have been implicated in priming cells for fusion. Caspases are usually associated with apoptosis, but increasing evidence suggests that they also promote cell differentiation [82]. Activation of caspase 8 may trigger differentiation of cytotrophoblasts and priming for fusion [48, 49, 83]. Activated Caspase 8 cleaves the sub-membranous cytoskeletal protein α-fodrin [84]. The degradation of the fodrin network may affect membrane curvature and thus facilitate fusion [85]. Active caspase 8 has been implied in externalization of phosphatidylserine [86, 87]. However, only 0.24% villous cytotrophoblasts contain activated caspase 8 and inhibition of caspase 8 activity only marginally inhibits BeWo cell fusion [49, 84]. Nor has the fate of caspase 8 active cytotrophoblasts been determined.

Caspase 9, the initiator caspase in the mitochondrial death pathway, and caspase 3 regulate myoblast differentiation and fusion [88, 89]. Reduction of caspase 9 activity, using either a shRNA or the caspase inhibitor Bcl-xL, prevents caspase 3 activation and inhibits myoblast fusion [88]. Inhibition of caspase 3 causes a dramatic decrease in myotube and myofiber formation [89]. Caspase 3 acts on mammalian sterile twenty-like kinase 1 (MST1), which activates downstream members of the MAPK cascade, such as MKK6 and p38γ, that effectively promote skeletal muscle differentiation [89].

Calpains, Ca^{2+}-activated intracellular cysteine proteases, facilitate myoblast fusion among other physiological processes [90]. The calpain family includes two ubiquitously expressed members (μ- and m-calpain) and several calpain homologs, including a muscle specific calpain, calpain 3. While μ-calpain does not seem important in myogenesis [91, 92], m-calpain is essential for the fusion of myoblasts to myotubes [93]. Knockdown of calpain-2, the catalytic subunit of m-calpain, by its RNA interference (RNAi) in the skeletal myoblast cell line C2C12, strongly inhibits their fusion to myotubes. Cells with deficiency of calpain-2 also exhibit abnormal actin cytoskeleton and impaired cell migration and adhesion [93].

Contrary to the pro-fusion effect of m-calpain, calpain 3 seems to prevent excessive fusion in myotubes. Myotubes with more nuclei are generated by myoblasts deficient in calpain 3 than those generated from wild type myoblasts, reflecting increased numbers of fusion events [94, 95]. This phenotype is associated with abnormal accumulation of two calpain 3 substrates, β-catenin and M-cadherin, on the myotube membrane [95]. Since the levels of M-cadherin determine the frequency of fusion events in vitro [96], calpain 3 may control the scale of myoblast fusion by limiting the levels of membrane associated M-cadherin.

4.2.6 Glis3

Glis3 is a Krüppel-like zinc finger transcription factor homologous to *Drosophila* protein Lame Duck (Lmd) [97]. Lmd is essential for the specification and differentiation of fusion-competent myoblasts [98–101]. Lmd regulates the expression of myocyte enhancer factor 2 (mef2) and *Drosophila* myoblast specific marker Stick-and-stone (Sns) [99]. However, Glis3 does not seem indispensible for mammalian myogenesis, as Glis3 mutant mice have normal muscle formation, but die shortly after birth from neonatal diabetes [102, 103]. Since Glis1 to 3 proteins share a highly conserved five-C2H2-type zinc finger domain with members of the Gli and Zic subfamilies, functional redundancy among these proteins may compensate for the loss of Glis3.

4.3 Chemoattractant–Receptor Interactions and Cell Migration

Cells made competent for fusion next migrate towards the cells with which they will fuse. Sperm, myoblasts, macrophages and progenitor cells find their fusion partners through chemoattract cues. For example, mononuclear myoblasts respond to chemoattratants by migrating toward each other or toward a nascent myotube. In many cases, chemokines and their receptors are used to guide the migration of fusion-committed cells and migration depends on rearrangement of the actin cytoskeleton.

4.3.1 Secretion and Function of Chemoattractants During Myogenesis

Secreted molecules, including growth factors and chemokines, are central to the fusion of myoblasts to myotubes in development of normal size myofibers and in regeneration of damaged muscle. Such molecules include hepatocyte growth factor/scatter factor [104, 105], IL-4 [15, 106], fibroblast growth factor-2 and 4 [107–109], insulin-like growth factor 1 and 2 [107, 110], platelet-derived growth factor [109], transforming growth factor beta-1 [109], vascular endothelial growth factor [111], leukemia inhibitory factor [109], alpha-chemokine stromal-derived factor 1 (SDF-1) [112], tumor necrosis factor alpha [113, 114], epidermal growth factor [115], thrombospondin-1 [116], regulated on activation normal T cell expressed and secreted (RANTES) [105], fibronectin [117] and laminins [115, 118].

4.3.1.1 Hepatocyte Growth Factor

Hepatocyte growth factor recruits satellite cells (myogenic stem cells) for muscle regeneration [119]. Hepatocyte growth factor is released from injured muscle and the quiescent satellite cells express the hepatocyte growth factor receptor Met. Binding hepatocyte growth factor to Met causes the receptor autophosphorylation which further activates the phosphatidylinositol 3- kinase (PI3K) and Ras-ERK mitogen-activated protein kinase cascade [120]. How interaction of hepatocyte growth factor with Met initiates actin cytoskeleton rearrangement for chemotactic migration is unknown.

4.3.1.2 SDF-1/CXCR4 Axis

The chemokine SDF-1 (also known as CXCL12) and its receptor CXCR4 regulate migration of myoblasts and other progenitor cells [112]. CXCR4 or SDF-1 deficiency impaires limb myogenesis [121]. In response to interaction with its agonist, SDF-1, CXCR4 forms complexes with CD164 [122]. Enhancing or reducing CD164 levels in C2C12 myoblasts promotes or inhibits migration of myoblasts respectively [123]. CXCR4 deficiency also impairs myelopoiesis, cardiogenesis, angiogenesis, neurogenesis and germ cell migration and development, presumably due to the impaired recruitment of progenitor cells [124].

The SDF-1/CXCR4 axis in hematopoietic stem cells is of special interest, because hematopoietic stem cells can regenerate several non-hematopoietic tissues by cell fusion [7, 8]. Normally, interaction of SDF-1 and CXCR4 retains hematopoietic stem cells in bone marrow [125]. However, under conditions of injury such as myocardial infarction, stroke, muscle ischemia, acute renal failure, toxic liver damage or total body irradiation, SDF-1 is expressed in situ [126–131] and released to recruit cells of hematopoietic lineage to fuse with the injured cells [132, 133]. Consistent with this concept, conditions of inflammation causes hematopoietic stem cells to fuse with intestinal epithelial cells, cardiomyocytes, skeletal muscle, hepatocytes, and Purkinje neurons [134–136].

4.3.1.3 IL-4

Newly formed myotubes secret IL-4 to recruit myoblasts for fusion [15]. Human myoblasts migrate toward IL-4 gradients in chemotaxis assays [106]. IL-4 receptor deficient (IL4R$\alpha^{-/-}$) myoblasts fuse infrequently with myotubes, while IL4R$\alpha^{-/-}$ myotubes that can secrete IL-4 recruit wild-type myoblasts as normal [15]. The induction or activation of pro-migratory components, such as urokinase plasminogen activator (u-PA), its receptor (u-PAR), β3 and β1 integrin subunits, by IL-4 treatment in vitro may facilitate the migration of myoblasts [106]. In addition, IL-4 may promote fusion by increasing expression of the mannose receptor in myoblasts [28]. Myoblasts lacking mannose receptor fail to migrate along a gradient of conditioned medium collected from cultures of nascent myotube.

The secretion of IL-4 from nascent myotubes appears to be regulated by nuclear factor of activated T cells, cytoplasmic 2 (NFATc2), because NFATc2 deficient myoblasts do not produce IL-4 [15, 137]. Interestingly, growth hormone is also essential for myoblast and myotube fusion in a way that depends on NFATc2, but not on IL-4 [138].

4.3.2 Mouse Odorant Receptor 23

Odorant receptors are G protein-coupled receptors usually expressed in the olfactory epithelium to detect smell. Interestingly, mouse odorant receptor 23 (MOR23) is also found in male germ cells and skeletal muscle regulating migration of sperm and mouse myoblasts [139, 140]. Both sperm and myoblasts migrate toward a cognate ligand of MOR23 in vitro, and overexpression of MOR23 accelerates the migration [139, 140]. Knockdown MOR23 by siRNA significantly reduces migration toward myotubes [139]. The natural MOR23 ligand is not yet identified.

4.3.3 Monocyte Chemoattractant Protein-1

Monocyte chemoattractant protein-1 (MCP-1), also known as Chemokine (C–C motif) ligand 2 (Ccl2), belongs to the family of CC-type chemokines and is critical for the recruitment and activation of monocytes during acute inflammation [141]. MCP-1 is secreted by many types of cells and serves as an agonist for CCR2, found mainly on cells involved in immune responses. MCP-1 may promote macrophage fusion by stimulating production of MMP-9, which degrades extracellular matrix [23]. The importance of MCP-1 for the migration of macrophage may not be as much as previously thought, because in MCP-1 null mice, macrophages are recruited to sites of implanted material as normal, but fail to form multinucleated foreign body giant cells [142]. When cultured with IL-4, MCP-1-null macrophages exhibit reduced fusion and the fused cells contain fewer nuclei [143]. Deficiency of MCP-1 in osteoclast precursor cells results in decreased expression of DC-STAMP, a cell surface protein critical for cell fusion [144]. DC-STAMP is a putative seven-transmembrane protein with a structure similar to chemokine receptors, but interaction between DC-STAMP and MCP-1 has not been proven [25, 144].

4.3.4 Progesterone

Progesterone is a C-21 steroid produced in the ovaries, the adrenal glands, and the placenta. A concentration gradient of progesterone stimulates sperm to approach the egg. Small amounts of progesterone are found in the cumulus matrix surrounding mammalian oocytes and may induce the oscillation of the concentration of Ca^{2+} in sperm [145, 146]. The Ca^{2+} concentration oscillation cycles alternate and synchronize the sperm flagellar beat mode [145]. Teves et al. [148] identified a number of signaling pathways involved in sperm chemotaxis towards progesterone. Progesterone and its carrier protein corticosteroid-binding globulin first activate the tmAC-cAMP-PKA pathway and then trigger protein tyrosine phosphorylation at equatorial band and flagellum, and calcium mobilization through IP3R and SOC channels. The sGC-cGMP-PKG cascade is activated last [147, 148]. At pharmacological concentrations, progesterone quickly causes influx of Ca^{2+} in sperm and leads acrosome reaction [149]. At the low baseline concentration of progesterone in the cumulus matrix of the egg, progesterone probably acts as chemoattractant rather than a trigger of acrosome reaction.

4.3.5 Integrins

Integrins are type I transmembrane αβ heterodimers that mediate cell–cell or cell- extracellular matrices adhesion. Integrins promote cell migration, membrane fusion and other biological processes [150].

Integrin heterodimer α7β1 supports the migration of myoblasts. Satellite cells (myoblasts) migrate on the surface of basal lamina toward injured muscle sites [151]. The major component of muscle basal lamina is laminin. Both α6β1 and α7β1 integrins are the receptors for laminin [152, 153]; however, depletion of α6 does not decrease cell motility [154]. Antibodies against α7 or β1 integrins impair migration of myoblasts in vitro [154]. Deletion of the *α7* gene causes muscular dystrophy [155], while conditional knock out of β1 integrin results in under-development of myofibers [156]. Overexpression of α7 integrin in human 293 cells, which have little motility on laminin-1 surfaces, increases their motility 8–10 fold [157]. The expression of both α7 and β1 integrins requires Acheron, a RNA binding protein required for myoblast differentiation [158]. In the presence of growth factors, myoblasts remain undifferentiated and express low levels of Acheron in myoblasts, but in the absence of growth factors, Acheron accumulates [159].

Integrins connect to and regulate the actin cytoskeleton by tethering structural and signaling proteins [160]. Integrin cytoplasmic domain-associated protein-1 (ICAP-1) binds the cytoplasmic tail of β1 integrin subunit and the actin cytoskeletal regulator ROCK-I kinase [161]. ICAP-1 and ROCK-I have been co-immune-precipitated together in C2C12 myoblasts and changes in the levels of ICAP-1 or ROCK-I change myoblasts' motility [161].

Migration and adhesion of osteoclast precursors depends on integrin αvβ3 [162–164]. Interaction between integrin αvβ3 and extracellular matrix proteins activates tyrosine kinases Src and Pyk2 triggered phosphorylation cascades [164, 165]. Phosphorylation of the Y402 site in Pyk2 stimulates osteoclast precursors to migrate toward M-CSF [164]. RANKL stimulates the expression of the atypical RhoGTPase Wrich1 which binds the cytoplasmic domain of integrin β3 and inhibits the phosphorylation of Pyk2 and paxillin [166]. Thus Wrich1 stops the migration of osteoclast precursors and allows their fusion with each other by suppressing integrin β3 signaling [166, 167].

4.3.6 The d2 Isoform of Vacuolar ATPase V_0 Domain

The d2 isoform of vacuolar ATPase V_0 domain (V-Atp6v0d2), a subunit of vacuolar ATPase, promotes fusion of osteoclast precursor cells [34]. Atp6v0d2 is highly expressed during osteoclast differentiation, and is stimulated by RANKL through the transcription factor NFATc1 [32]. Atp6v0d2 deficiency causes decreased fusion of osteoclast precursors leading to increased bone mass [34]. The cellular localization and function of Atp6v0d2 have not been elucidated. Atp6v0d2 has been recently found to bind adhesion-regulating molecule 1 (Adrm1) [168]. Adrm1 is a proteasome subunit that functions as an ubiquitin receptor in cytoplasm [169]. Knockdown of Adrm1 impairs preosteoclast migration and fusion, but does not affect adhesion of preosteoclasts to substratum [168], suggesting that Atp6v0d2/Adrm1 complexes encourage cell migration that facilitates osteoclast formation.

4.3.7 Actin Cytoskeleton Regulators

Cell migration requires rearrangement of actin cytoskeleton to achieve cell protrusion (lamellipodia and filopodia), adhesion (lamella and stress fibers) and shape change (cortex) [170]. Treating C2C12 myoblasts with latrunculin B or cytochalasin D, which inhibit actin polymerization, impairs actin-based behaviors, such as lamellipodia and filopodia formation and cell migration [171]. The regulation of cytoskeleton during cell membrane fusion will be further discussed in Section 4.5.1.

WAVE2 is a subunit of the SCAR/WAVE actin-nucleating complex. Inhibition of WAVE2 in C2C12 myoblast cells by expression of a dominant-negative protein dramatically decreases lamellipodia formation and cell motility [172]. Although the SCAR/WAVE complex is known as a downstream target of Rac signal, deficiency of Rac does not change migration of myoblasts or macrophages [173–175], suggesting that the SCAR/WAVE complex may be activated by a yet unknown factor, rather than Rac, to promote cell migration. In parallel, activation of N-WASP, a direct target of both Cdc42 and PI(4,5)P2, induces actin polymerization [176]. Dominant-negative N-WASP inhibits lamellipodia formation and cell motility [172].

Whether RhoA directly regulates myoblast migration has not yet been examined, but a downstream effector of RhoA, ROCK-I, has been found to be important for myoblast migration. ROCK-I translocates to the plasma membrane by binding the ICAP-1/β1 integrin complex and stimulates formation of stress fibers and focal adhesions [161]. ICAP-1 knockdown or inhibition of ROCK-I activity reduces focal adhesion density and cell motility [161].

4.3.8 Mannose Receptor

The mannose receptor has been long implicated in formation of multinucleated giant cells [27] and osteoclasts [177]. More recently, the mannose receptor was found to facilitate the recruitment of myoblasts during myogenesis [28]. The mannose receptor is a type 1 transmembrane protein that binds a variety of soluble and cell surface glycoproteins that have mannose, fucose, N-acetylglucosamine, and glucose residues [178]. The mannose receptor may trigger innate immunity to microorganisms expressing oligosaccharides with terminal mannose residues and may clear serum glycoprotein with terminal mannose substitutions.

During its induction of giant cell formation, IL-4 heightens the expression of the mannose receptor [26] and competitive inhibitors of the mannose receptor block fusion of macrophages [27]. However, macrophages from mannose receptor knockout mice can form normal giant cells after culture with IL-4 [6]. Thus, the mannose receptor is not essential for fusion. Nevertheless, myoblasts from mannose receptor knockout mice migrate more slowly and with less direction toward a gradient of myofiber conditioned medium. How exactly the mannose receptor promotes cell fusion is yet uncertain but the receptor has been proposed to facilitate cell motility via clearance of collagen in extra-cellular matrix [28].

4.3.9 Matrix Metalloproteinases

Matrix metalloproteinases (MMPs) may be important for migration of macrophages [22, 179, 180] and are essential for cell membrane fusion [23]. Matrix metalloproteinases are zinc-dependent endopeptidases with a large range of substrates, including extracellular matrix, growth factors, chemokines, and cell-surface proteins. MMP-9 is highly expressed during the formation of foreign body giant cells [23]. Deficiency of MMP-9 blocks membrane fusion and greatly reduces formation of foreign body giant cells. MMP-9 may degrade extra-cellular matrix, promoting interaction with cell membranes or may activate signal molecules necessary for cell fusion [23].

4.4 Membrane Recognition and Adhesion

Cell fusion requires cell contact. A range of adhesion proteins mediates recognition and adhesion of apposed cell membranes. Those proteins belong to the immunoglobulin super family, cadherins, tetraspanins, integrins or GPI-anchored proteins.

4.4.1 Immunoglobulin Super Family

A number of immunoglobulin (Ig) super family proteins participate in the fusion of mammalian cells. These proteins include Nephrin, Izumo, Cluster of Differentiation (CD) 47 and signal regulatory protein α (SIRPα). Proteins of the Ig super family contain domains of ~70–110 amino acids which form a sandwich-like structure consisting of two sheets of anti-parallel beta strands and a highly conserved disulfide bond [181]. Ig super family members can interact with many membrane surface proteins including homotypic and heterotypic members of the Ig superfamily, integrins and cadherins [182]. The number of Ig domains of the interacting proteins may determine the distance between the opposing membranes.

4.4.1.1 Orthologs of *Drosophila* Ig Super Family Proteins

The Ig proteins involved in myoblast adhesion and eventually in fusion were revealed by genetic studies in *Drosophila* [5]. Myoblast recognition and adhesion in *Drosophila* requires four Ig super family proteins, of which Kirre (also known as Dumbfounded) and Roughest are specifically localized on membrane surface of muscle founder cells (FC), while Sticks and Stones (Sns) and Hibris are specifically expressed in fusion competent myoblasts (FCM) [5]. These Ig super family proteins tether cells and trigger actin polymerization and reorganization.

Nephrin is a mammalian ortholog for *Drosophila* Sns. Like Sns, Nephrin promotes fusion of myoblasts with nascent myotubes [183]. Nephrin is expressed during mouse skeletal muscle development and in diseased or injured adult muscle. Myoblasts isolated from *nephrin* knockout mice can initiate fusion to form nascent myotubes, but fusion of additional myoblasts is impaired. Hence, myotubes in *nephrin* knockout mice are much smaller than those in wild-type mice. Mouse myoblasts lacking *nephrin* fuse poorly to human myotubes in vitro, while *nephrin* null and human hybrid nascent myotubes successfully recruit and fuse with human myoblasts. These results suggest that Nephrin is essential for myoblasts to fuse with nascent myotubes, playing a conserved role as that of *Drosophila* Sns in secondary myoblast fusion [183].

Neph1, Neph2 and Neph3 are mammalian orthologs of Kirre/Roughest. Neph1 deficient mice do have normal muscle development [184], perhaps indicating the compensation of Neph2 and Neph3. Murine Nephrin interacts with itself, Neph1 or Neph2 in a specialized cell–cell adhesion complex in the kidney [185]. The interaction of Nephrin-Neph1 induces actin polymerization at plasma membrane junctions [185].

4.4.1.2 Other Mammalian Ig Proteins Involved in Myoblast Fusion

Some mammalian Ig super family members whose *Drosophila* counterparts do not seem to be involved in cell fusion also promote myoblast fusion [186]. Cdo is a cell surface receptor with Ig and FnIII repeats that binds in *cis* to the ectodomain of N-cadherin in myoblast [187]. The intracellular region of Cdo is associated with Cdc42 and p38α/β MAP kinase through scaffold proteins [188]. Binding of Cdo to N-cadherin activates Cdc42 and p38α/β which stimulates MyoD-dependent, muscle-specific gene expression [189]. Although primary $Cdo^{-/-}$ myoblasts express low levels of

MyoD targeting genes and reduce the fusion of myoblasts to myotubes in vitro [190], Cdo null mice display only a mild delay in skeletal muscle development [191]. Thus, other pathways might compensate to affect MyoD-dependent myoblast differentiation in vivo.

Neogenin is a member of the Ig super family with Ig and FnIII repeats that may be important for myoblast fusion. Neogenin and its ligand, Netrin, are expressed in developing muscle [192]. Interaction of neogenin with netrin activates focal adhesion kinase (FAK) in cultured myoblasts [193]. Myoblasts depleted of neogenin or FAK fuse poorly with myotubes [192, 194]. Thus, the netrin-neogenin-FAK signaling facilitates myoblast fusion and muscle development.

Neural cell adhesion molecule (NCAM), another Ig super family protein, promotes myogenesis by mediating homophilic interactions between myoblasts and myotubes [186]. The GPI-anchored 125-kDa NCAM isoform contains a muscle-specific domain in the extracellular region [195]. This isoform associates with lipid rafts through its GPI-anchor [196]. Overexpression of 125-kDa NCAM in C2 myoblast cells [197] or in skeleton muscle [198] enhances myotube formation. However, neither *NCAM*-null mice nor their myoblasts display obvious defects in muscle development or in myoblast fusion [199]. Thus, NCAM may promote fusion by enhancing cell–cell adhesion, but other proteins may compensate for absence of NCAM.

4.4.1.3 Izumo

Izumo is a sperm Ig super family protein that is involved in sperm-egg fusion. Antibodies blocking Izumo inhibit sperm-egg fusion in vitro [200]. Male *Izumo* null mice develop normally but are sterile [68]. Izumo deficient sperm penetrate the egg zona pellucida but fail to fuse with egg membranes. Overexpression of Izumo in a male *Izumo* null mouse rescues fertility. Thus Izumo is truly essential for mammalian sperm-egg fusion [68].

Despite the absolute requirement of Izumo in gamete fusion, neither the binding partners nor molecular actions of Izumo are known. After the acrosome reaction, Izumo relocates from the anterior head to a new region opposite to the anterior acrosome, including part of the postacrosomal region [70, 201]. Izumo relocates via actin cytoskeleton and lipid rafts which are enriched with cholesterol and gangliosides [70, 71]. Tssk6, a member of the testis-specific serine kinase family of proteins expressed in postmeiotic male germ cells, is required for actin polymerization and translocation of Izumo in sperm [71].

4.4.1.4 CD47/SIRP-α Interaction

CD47 and signal regulatory protein α (SIRP-α) are members of the Ig super family, and their interactions are required for formation of multinucleated giant cells and osteoclasts [202, 203]. CD47 has a single V-type Ig-like extracellular domain. It is a ligand for SIRP-α (also known as macrophage fusion receptor or SHPS-1), which contains three Ig-like domains and is transiently expressed at a high level as macrophage fusion begins [204]. Anti-CD47 antibodies block CD47-SIRP-α interaction, and macrophage-macrophage fusion as well [203]. The number of osteoclasts in *CD47* null mice is significantly reduced. Macrophages isolated from $CD47^{-/-}$ mice form fewer osteoclasts under the stimulation of M-CSF and RANKL [202]. Interaction between SIRP-α and CD47 recruits and activates SHP-1 and SHP-2, which are src homology-2 (SH2)-domain-containing tyrosine phosphatases [205]. Activated SHP-1 and SHP-2 inhibit macrophage phagocytosis [206] and migration [207], promote macrophage adhesion, fusion and formation of giant cells and osteoclasts [202, 203]. Interestingly, CD47 also interacts with integrins to stimulate chemotaxis and migration [208]. Therefore CD47 may promote either macrophage migration or fusion depending on the protein to which it binds.

4.4.2 Cadherins

Cadherins mediate cell adhesion through calcium-dependent, homophilic binding of extracellular cadherin repeats [209]. Intracellular domains of cadherins bind catenins which tether to the actin cytoskeleton. Several cadherins have been implicated in cellular fusion. E-cadherin facilitates fusion of macrophages. N-cadherin facilitates fusion of gametes and fusion between myoblasts and myotubes. M-cadherin also participates in myoblast fusion. Cadehrin-11 facilitates fusion of cytotrophoblasts.

4.4.2.1 N-Cadherin

N-cadherin is expressed in many tissues including the embryonic developing skeletal muscle and adult regenerative myofibers [210] and spermatozoa [211]. Anti N-cadherin antibodies inhibit sperm-oocyte fusion [212]. N-cadherin-deficient mice do not survive long enough for analysis of skeletal muscle development, but primary myoblasts isolated from these mice differentiate and fuse normally [213, 214]. N-cadherin-deficient myoblasts express M-cadherin and cadherin 11, which may compensate for the absence of N-cadherin. N-cadherin binds the Ig super family protein Cdo in *cis*, activating Cdc42 and p38 MAP kinase which in turn regulate the expression of specific genes. N-cadherin also inhibits myoblast migration by activating FAK or $\alpha5\beta1$ integrins [215].

4.4.2.2 M-Cadherin

M-cadherin is implicated in fusion of myoblasts, perhaps by facilitating adhesion and activating Rac1, a key regulator of the actin cytoskeleton. M-cadherin co-immunoprecipitates with Trio, a guanine nucleotide exchange factor for Rac1. In C2C12 myoblasts, M-cadherin-mediated adhesion activates Rac1 through binding with Trio [216]. M-cadherin is enriched in lipid rafts in myoblasts and translocated at sites of cell contact prior to fusion [217]. Expression of M-cadherin is regulated by MyoD and other myogenic factors [218]. Inhibition of M-cadherin functions by blocking peptides, neutralizing antibodies or RNAi reduces myotube growth in vitro [219, 220]. In contrast, M-cadherin deficient mice develop skeletal muscle and regenerate damaged muscle normally, and primary myoblasts derived from null mice can form myotubes normally in vitro [221].

4.4.2.3 E-Cadherin and Cadherin-11

E-cadherin enables macrophages to form osteoclasts or multinucleated giant cells [20, 21, 222]. E-cadherins and catenins form functional complexes at the sites of cell contact [21]. Through homotypic interactions, these complexes join neighboring macrophages. Transcription factor STAT6 induces expression of E-cadherin in response to IL-4 stimulation [20]. STAT6 deficient macrophages do not fuse to form giant cells [20]. However, deficiency of E-cadherin has much milder defects in multinucleated giant cell formation than deficiency of STAT6 [21], which is consistent with the observations that STAT6 has much broader functions than E-cadherin, including turning on fusion-necessary proteins DC-STAMP and MMP-9 [22, 23].

The expression of cadherin-11 in cytotrophoblasts is opposite from that of E-cadherin. E-cadherin facilitates clustering of cytotrophoblasts, but expression of E-cadherin decreases during subsequent fusion. On the other hand, cadherin-11 increases during cytotrophoblast fusion [223, 224]. Inhibition of cadherin-11 in primary cytotrophoblasts allows continuous expression of E-cadherin and prevents formation of syncytiocytotrophoblast [225]. Overexpression of cadherin-11 in a mononucleated trophoblastic cell line JEG-3 leads to formation of multinucleated syncytium, suggesting that cadherin-11 enhances trophoblast fusion [225].

4.4.3 Tetraspanins

Tetraspanins are a family of membrane proteins with four transmembrane domains, intracellular N- and C-termini and two extracellular loops, one short and one longer. Tetraspanins can interact with each other or with other cell surface molecules such as Ig super family proteins, integrins and membrane-anchored growth factors [226]. The tetraspanins CD9 and CD81 are important for the fusion between sperm and egg, myoblast and myotube, and among macrophages. Despite their importance in cell–cell fusion, the precise mechanism by which CD9 and CD81 mediate membrane fusion is not known.

CD9 and CD81 are required for fusion of egg and sperm [227]. The fertility of CD9 deficient female mice is severely decreased [228–230], and mice deficient both of CD9 and CD81 are sterile [231]. Eggs from CD9 deficient mice can be rescued by injection of wild-type *cd9* mRNA. Micro-injection of *cd81* mRNA, partially restores the ability of CD9 null eggs to fuse with sperm, indicating that CD9 and CD81 are partially redundant [232]. However, if the SFD (173–175) sequence in the large extracellular loop of CD9 is mutated to AAA, the mutated rescue does not occur [233]. CD9 localizes to, and may control, the curvature of the microvilli, since CD9 deficiency increases microvillus diameter [234]. The curvature of lipid bilayers facilitates formation of a fusion-stalk [85].

Opposite to their roles in sperm-egg fusion, CD9 and CD81 inhibit formation of multinucleated giant cells [235, 236]. During inflammation, CD9 deficient mice have more multinucleated foreign body giant cells than wild type mice and mice deficient of both CD9 and CD81 have even more [235]. In contrast, another tetraspanin, CD63, promotes macrophage fusion, because anti-CD63 antibodies strongly inhibit multinucleated giant cell formation [236]. Different from its role in multinucleated giant cell formation, CD9 appears to promote the RANKL-stimulated osteoclastogenesis [237]. RANKL enhances CD9 expression in RAW264.7 macrophages. The cell surface CD9 proteins are then enriched in lipid rafts. Neutralization of CD9 or disruption of lipid rafts strongly inhibits the formation of osteoclasts [237].

CD9 and CD81 may also help myoblasts fuse with nascent myotubes [238]. Inhibition of CD9 and CD81 suppresses myotube growth and overexpression of CD9 causes a four- to eight-fold increase in syncitia formed by human myoblast-derived sarcoma cells. CD9 is associated with β1 integrin in the plasma membrane and the expression of CD9 depends on β1 integrin in C2C12 myoblasts [156].

4.4.4 Integrins

Besides facilitating cell migration, integrins mediate cell adhesion and fusion. In mice with conditional deficiency of β1 integrin, muscle fibers are shorter than normal and unfused cells accumulate [156]. β1 integrin associates with tetraspanin CD9 and controls its expression in myoblasts [156]. Since CD9 mediates fusion between myoblast and myotube [238], β1 integrin probably contributes to cell membrane fusion by organizing proteins, such as CD9, to encourage fusion. The heterodimer of β1 integrin appears to be α3 integrin, because the integrin α3 subunit colocalizes with actin, the integrin β1 subunit and ADAM12 [239]. Manipulating the levels of α3 integrin significantly changes the frequency of myoblast fusion [239].

The integrin β1 subunit may also facilitate fusion of macrophages [240]. Anti β1 integrin antibodies block macrophage fusion induced by mycobacteria lipids but do not affect cell aggregation. However, in IL-4 induced multinucleated giant cell formation, β1 and β2 integrins do not mediate cell–cell adhesion, but rather facilitate adhesion to culture dishes [241].

4.4.5 Glycosyl-Phosphatidylinositol (GPI)-Anchored proteins

GPI-anchored proteins on the egg plasma membrane have been implicated in fusion with sperm and are essential for fertilization. GPI-anchored proteins are often associated with lipid rafts which are critical for fusion during myogenesis [217]. Removing GPI-anchored proteins with phosphatidylinositol-specific phospholipase C reduces sperm-egg fusion by 90% [242]. Female mice with an oocyte-specific knockout of PIG-A, which participates in first steps of GPI synthesis, are infertile because *Pig-a* null eggs are unable to fuse with sperm [243]. However, the specific egg GPI-anchored protein(s) involved and their role(s) during sperm-egg fusion are not known. The loss of GPI-anchored proteins may disturb the organization and function of lipid rafts on egg plasma membrane, impairing interactions with sperm.

4.4.6 A Disintegrin And Metalloproteinase (ADAM)

Members of the ADAM family of proteins contribute to cell fusion. ADAM members are defined by a common modular ectodomain with cell adhesion and cell fusion motifs (disintegrin and cysteine-rich domains), and a Zn-protease domain capped by a large prodomain [244]. The protease domain in most ADAM proteins is not catalytically functional, but rather facilitates adhesion and/or fusion proteins. Early works using specific monoclonal antibodies suggested many ADAMs participate in cell fusion: ADAMs 1, 2 and 3 in sperm-oocyte fusion, ADAM 9 in macrophage giant cell formation, ADAM 12 in myoblast, macrophage and trophoblast fusion, and ADAM 8 in osteoclast formation [244]. However, mice deficient of various members of the ADAM family exhibit only mild phenotypes and none is indispensible for cellular fusion [3, 245]. These mild phenotypes may be the result of functional redundancy among the ADAMs, since many ADAMs have similar adhesion-mediating disintegrin and cysteine-rich domains.

4.4.7 Dendritic Cell-Specific Transmembrane Protein (DC-STAMP)

DC-STAMP is a putative seven-transmembrane protein essential for the formation of osteoclasts and multinucleated giant cells. DC-STAMP is highly expressed during the formation of osteoclasts [25, 246]. Expression can be stimulated by RANKL, IL-4, DAP12, or MCP1 [24, 37, 144]. Inhibition of DC-STAMP by RNAi or by antibodies prevents the formation of osteoclasts, while overexpression of DC-STAMP enhances RANKL-induced osteoclastogenesis [246]. Both osteoclast and foreign body giant cell formation are completely absent in DC-STAMP deficient mice [25]. Wild-type and DC-STAMP deficient osteoclast precursors can undergo heterotypic fusion in the presence of M-CSF and RANKL, suggesting that DC-STAMP might interact with a yet unknown molecule in neighbor cells [25].

The distribution of DC-STAMP is critical for osteoclast formation [247]. Macrophage RAW 264.7 cells express DC-STAMP even without RANKL treatment [247, 248]. However, RANKL induces the internalization of membrane DC-STAMP in some of the RAW 264.7 cells, and thus generates two cell populations: a population with low levels of membrane DC-STAMP due to the internalization of DC-STAMP, and a population with high levels of membrane DC-STAMP without experiencing internalization [247]. Cells with low membrane DC-STAMP levels express high levels of *cd9, cd47, Trap, Oc-stamp* and *Dc-stamp* itself, facilitating cell fusion. These macrophages can fuse with each other or with cells carrying high levels of DC-STAMP on their membranes. In contrast, the macrophages with high membrane DC-STAMP levels do not fuse with each other. Thus, RANKL may stimulate the binding between DC-STAMP and an unknown ligand, causing internalization of surface DC-STAMP, in turn inducing expression of genes facilitating fusion [247].

A RANKL upregulated gene that encodes a protein with DC-STAMP-like C-terminal was recently identified, and termed as osteoclast stimulatory transmembrane protein (OC-STAMP) [249]. OC-STAMP has six transmembrane helices and inhibition of OC-STAMP prevents multinucleated osteoclast formation. Overexpression of OC-STAMP promotes fusion. Whether OC-STAMP interacts with other molecules during cell fusion is not yet known.

4.5 Fusion Pore Formation and Expansion

How cell membranes fuse is still not clear. Most models are based on the events enabling fusion of viruses with cells. In these models, the curvature of opposing membranes brings the outer leaflets into close contact, and then mixes lipids to form a hemifusion intermediate. Tension in the extending diaphragm promotes fusion of the inner leaflets and the formation of a fusion pore [2, 250, 251]. After formation of the initial pore, expansion of the fusion pore must occur to allow mixing of cellular content [252].

The plasma membranes of neighboring cells usually do not fuse spontaneously because the fusion process requires energy [252]. It is thus assumed that some fusion-specific proteins, i.e. "fusogens", mediate the plasma membrane fusion. Fusogens lower the energy barrier by directly rearranging the lipid bilayer and lead to cell membrane fusion [252, 253]. Several criteria have been suggested in identifying true fusogens, such as they can mediate fusion of cells that normally do not fuse in situ, in heterologous cells or in biochemically constructed liposomes [251]. A few viral fusogens and two *C. elegans* fusogens (EFF-1 and AFF-1) have been identified [250]. As the discovery of mammalian fusogen has been difficult to prove, we will propose an alternative model of cell membrane fusion, in which lipids, adhesion proteins and actin filaments are coordinated to accomplish cell membrane fusion.

4.5.1 Actin Cytoskeleton

While refashioning of the actin cytoskeleton is a critical event in cell migration and cell fusion, how it does so remains unclear. Some potential functions of the actin cytoskeleton in fusion have been suggested by obsrevations in *Drosophila*, mammalian myogenesis, and virus mediated cell fusion. In contrast, *C. elegans* does not seem to require fully functional actin cytoskeleton for cell fusion, as mutants with defects in actin network develop normal cell fusion [254–256].

Recent observations on actin dynamic in rat L6 myoblasts suggest that a transient cortical actin "wall" structure may provide a rigid "supportive platform" for alignment between myoblasts [257]. In differentiating myoblasts, non-muscle myosin motor activity organizes cortical actin filaments into thick bundles parallel to the plasma membrane. This "wall-like" structure establishes the bipolar shape of myoblasts [257, 258]. Knockdown of non-muscle myosin not only impairs formation of the actin wall, but also distorts the shape of the plasma membrane and impairs fusion [257]. Interestingly, the actin wall is assembled only on one side of a myoblast. The plasma membrane with an actin wall usually aligns with a membrane without an actin wall in a neighboring myoblast (Fig. 4.1a).

As cell fusion proceeds, the actin wall disassembles to form actin-free gaps populated by membrane bound vesicles of unknown origin. In some cases, fusion pore structures are observed at the actin-free gaps. Similarly, during the fusion of virus and cell, the cortical actin filaments are also disassembled before fusion pore formation and pore expansion [259–261].

Actin polymerization may generate the force needed for forming and expanding fusion pores as revealed in *Drosophila* [262–264]. During myogenesis in *Drosophila*, the interactions of the Ig super family proteins in funder cells and in fusion competent myoblasts activate two actin-nucleating complexes, SCAR and WASP [5]. Both complexes regulate the Arp2/3 complex which initiates new actin filament branches from an existing filament. Polymerizing actin filaments generate forces to push against plasma membranes [265, 266]. Fusion pores do not form in the cells deficient in Arp3 [262]. While both SCAR and WASP regulate the Arp2/3 complex, SCAR is required for the formation of fusion pore and WASP for expansion of the pore [263, 264]. Thus the same actin polymerization-based force drives pore formation or expansion, depending on spatial and temporal factors. SCAR-regulated polymerization might push the plasma membrane vertically curving the opposing membranes and facilitating fusion pore formation. In contrast, WASP may orient horizontally the force generated by actin polymerization, pushing the fused membranes laterally and expanding fusion pores (Fig. 4.1c and d).

Actin polymerization may also promote the expansion of fusion pore in mammalian cells, for example, HIV-1 mediated cell–cell fusion requires Arp2/3-dependent actin polymerization [267]. Inhibition of Arp2/3 regulators, such as Rac and Abl, arrests HIV-1 mediated cell fusion at the hemifusion step, suggesting that Abl-regulated actin polymerization is important for fusion pore expansion during viral fusogen mediated cell–cell fusion [267].

The regulation of actin cytoskeleton for cell fusion is best demonstrated during myogenesis in *Drosophila* [5]. The signals transduced by small GTPase Rac1 play a key role in rearranging actin networks during myoblast fusion. Cell–cell contacts activate guanine nucleotide exchange factors (GEFs) that activate Rac1. Rac1 in turn stimulates the SCAR actin-nucleating complex which brings together Arp2/3 complex with an actin monomer on the side of a filament to nucleate a branch. In addition, a Rac1-independent actin-nucleating factor WASP also regulates F-actin polymerization through Arp2/3 [5].

Rac1 is also critical for myoblast fusion during primary myogenesis in mammals [173]. Mice lacking Rac1 have short, thin muscle fibers and myoblast fusion is blocked. Absence of fusion may reflect a deficit in the recruitment of actin fibers, vinculin and Arp2/3 to myoblast contact sites. However, migration of myogenic precursor cells appears normal as Rac1 deficient mice have normal numbers of these cells at the sites where myogenesis should occur [173]. In mammals, there are two GEFs that activate Rac1, Trio and Dock1 (also called Dock180). Mice lacking Dock1 exhibit a dramatic decrease in skeletal muscle tissue due to the deficiency in myoblast fusion [268]. The Dock1 initiated actin filament rearrangements also function in macrophages, as mice lacking Dock1 do not form multinucleated giant cells [269]. Trio is another Rac1GEF needed for cell fusion. *Trio*-deficient mice appear normal until E14.5, but then myogenesis fails because myoblasts do not fuse with nascent myofibers during secondary myogenesis [270]. Trio resides in a complex with M-cadherin and Rac1 in myoblasts undergoing fusion, suggesting that M-cadherin may also activate Rac1 through Trio during myoblast fusion [216]. As in *Drosophila*, the SCAR complex downstream of the Rac1 signal is essential for mouse myogenesis. Knock-down Nap1, a conserved member of the SCAR complex, mimics the phenotypes of Rac1 or Dock1 deficiency [171].

Brag2 is a GEF which activates the GTPase ARF6 in mammals. Activation of ARF6 translocates Rac1 to sites of cell fusion in *Drosophila* [271]. Mice deficient in Brag2 exhibit impaired myoblast fusion [269]. A few cells deficient in Brag2 do fuse, forming "stubby" syncytia in which elongation fails and nuclei are centrally clustered. This elongation failure is not seen in Dock1 deficiency [269]. The stubby morphology may be caused by the lack of physical association between paxillin and β1 integrin. This association helps the translocation of paxillin to focal adhesion sites [269, 272].

Besides Rac1, the small GTPase Cdc42 is another important regulator of actin cytoskeleton. Interestingly, Cdc42 is required for myoblast fusion in mice, but not in flies [173]. Myoblasts in Cdc42 deficient mice do not fuse and do not accumulate F-actin, vinvulin and Ena-Vasp at sites of cell contact. Cdc42 and Rac1 appear to function in a non-redundant manner during the fusion process, as Cdc42 mutant cells have normal recruitment of Arp2/3 complex [173].

The regulation of Cdc42 activity during myoblast fusion is not fully understood. The Ig super family protein Cdo binds N-cadherin *in cis*. Together they recruit and activate Cdc42 as well as its downstream p38α/β MAP kinase through scaffold proteins [187, 188]. The activated p38α/β in turn induces MyoD-dependent, muscle-specific gene expression [189]. However, the formation of filopodia in C2C12 myoblasts plated on N-cadherin substrate depends on Cdc42 and Cdo, but not p38α/β activity [186]. Filopodia has been suggested to be important for cell–cell contact and subsequent membrane fusion [273].

4.5.2 Lipid Rafts

In plasma membranes, cholesterol and sphingolipids form dynamic nanoscale assemblies called lipid rafts [274]. Lipid rafts act as membrane organization centers for protein sorting, membrane trafficking, signal transduction, actin network rearrangements and changing membrane fluidity. Disruption of lipid rafts by removing cholesterol from plasma membrane prevents the formation of multinucleated myotubes and osteoclasts, even though cell–cell adhesion still remains [217, 237].

The requirement for lipid rafts in membrane fusion would seem to contradict early reports that myoblast fusion occurs within cholesterol-free sites where membrane fluidity is increased [275, 276]. However, careful observations on the dynamics of lipid rafts at the sites of myoblast fusion have resolved this apparent discrepancy [217]. Lipid rafts carrying adhesion proteins, such as M-cadherin, cluster at the leading edge of lamellipodia and facilitate the adhesion of adjacent myoblasts. After adhesion, lipid rafts disperse rapidly, allowing cholesterol- and adhesion molecule-free domains to fill the site where the apposing membranes fuse (Fig. 4.1a and b). Thus lipid raft clustering and rapid dispersion are the key events in myoblast fusion [217].

Fusion of other cell types may also require clustering and dispersion of the lipid rafts. Scrambling of phosphatidylserine facilitates the redistribution and extraction of cholesterol from plasma membrane, as demonstrated during sperm capacitation [51, 52]. GPI-anchored proteins, which mainly associate with lipid rafts, have been implicated in sperm-egg fusion [242, 243]. Tetraspanin CD9, localized in lipid rafts, is implicated in RANKL-stimulated osteoclastogenesis [237]. Inhibiting CD9 or disrupting lipid rafts significantly reduces the formation of osteoclasts.

4.5.3 A Novel Model of Plasma Membrane Fusion

We here describe our own model to explain fusion of cell membranes based on observations of adhesion proteins, actin cytoskeleton and lipid rafts during cell fusion, especially in mammalian myoblast fusion. The model is portrayed as a series of events summarized in Fig. 4.1. We postulate that initially, adhesion molecules are enriched in lipid rafts on cell surfaces forming rigid structures in fusion competent cells. The highly organized, myosin-dependent cortical actin walls assist the alignment and adhesion of apposing membranes. Many adhesion proteins associate with and regulate the actin network. Next, the interaction of adhesion molecules initiates signaling pathways regulating local dissembling of actin walls and actin cytoskeleton rearrangement. Next, the lipid rafts disperse rapidly, allowing highly fluidic components of the phospholipid bilayer to flow into the site. Arp2/3-dependent actin polymerization generates force pushing plasma membrane outward to generate curvature that eventually leads to direct contact between opposing membranes. Next, the lipids on outer leaflets of opposing membranes mix, and the mixture of inner leaflet lipids causes formation of a fusion pore. Actin polymerization then pushes the fused membrane laterally, expanding the fusion pore.

This lipid raft and actin coordinated model of plasma membrane fusion does not rely on fusogens. The model does depend on adhesion molecules to bring opposing membranes into close proximity. However, the nature of these adhesion molecules is not restricted. Hence many combinations of adhesion proteins can promote fusion. Because many adhesion proteins can function redundantly, deficiency or inhibition of one protein may not impair fusion. The model further predicts that fusion depends on positioning of adhesion proteins in lipid rafts. These events are complex and a single protein may not be sufficient to "coordinate" the process, hence few if any proteins will meet the criteria of being a true fusogen, that is being able to mediate fusion of cells that normally do not fuse in situ, and fusion between heterologous cells or between biochemically constructed liposomes [251].

The fusogens identified so far are mostly viral glycoproteins. Viral fusogens appear to be able to coordinate cell adhesion, actin polymerization and the clustering/dispersion cycling of lipid rafts. For example, the mature form of HIV Env contains an attachment subunit, gp120, and a fusion subunit,

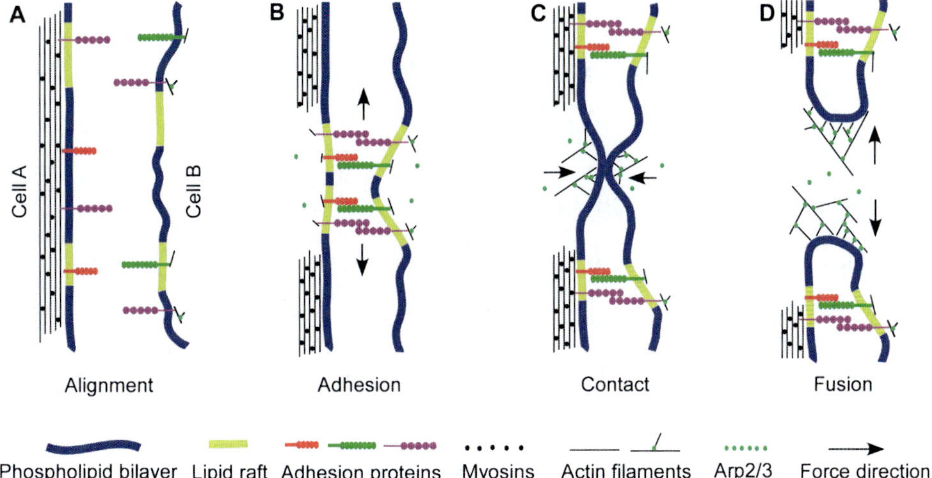

Fig. 4.1 A model of events causing fusion of cell membranes. (**a**) Lipid rafts recruit adhesion proteins such as IgSF proteins and cadherins at the sites of cell–cell alignment. Many of the adhesion proteins are associated with actin network. The highly organized and myosin-dependent cortical actin wall in one cell facilitates membrane alignment and adhesion. (**b**) Interaction of adhesion molecules brings lipid rafts to the sites of cell contact and causes the local dissembling of actin walls and actin cytoskeleton rearrangement. (**c**) The lipid rafts quickly disperse laterally, allowing the highly fluidic standard phospholipid bilayer to fill in the site. Arp2/3-dependent actin polymerization creates force, pushing the lipid bilayer outward to form a membrane curvature that eventually leads to direct contact between the opposing membranes. (**d**) Fusion pore formed after the mix of lipid bilayers. Actin filaments push membrane laterally expanding the fusion pore

gp41. Binding of gp120 to its receptor CD4 and a coreceptor located in lipid rafts of the host membrane [277], stimulates a dramatic conformational change of gp41. This conformational change leads to the formation of rigid coiled-coil structures and exposes amphipathic fusion peptides that later insert into the host membrane functioning as adhesion molecules. Folding back of the helical coil into a hairpin structure and dispersion of lipid rafts brings opposing membranes together and allows fusion to proceed [277, 278]. Binding of Env to receptors in target cells also activates Rac, which initiates Arp2/3-dependent actin polymerization [267]. Inhibition of Rac, or Abl, a downstream target of Rac, arrests fusion at the hemifusion step, suggesting that the Env-promoted actin polymerization generates the force needed for fusion pore expansion [267, 279].

Two *bona fide* fusogens, AFF-1 and EFF-1, are found in *C. elegans* [280, 281]. In contrast with the single-membrane localization of viral fusogens, AFF-1 and EFF-1 must be expressed in both fusing membranes. How AFF-1 and EFF-1 promote fusion is still poorly understood. Since cell fusion in *C. elegans* does not require rearrangement of actin cytoskeleton [254–256], these fusogens may function in a simpler way than viral fusogens.

4.5.4 Syncytin-1 and -2

Syncytin-1 and -2 are encoded by human endogenous retrovirus elements HER-W and HER-FRD respectively. They mediate placental fusion and are considered *bona fide* fusogens [282, 283]. Syncytins are structurally similar to HIV Env fusogen. Cytotrophoblasts cultured with antisense *syscytin-1* exhibit little fusion [284]. Expression of syncytin-1 in other cell types can also generate multinucleated cells, indicating sycytin-1 is sufficient to mediate cell fusion [282, 285]. Like viral fusogens, syncytin-1 binds to receptors, in this case to ASCT1 or ASCT2, before inducing membrane fusion [285, 286]. Expression of syncytin-2 can also induce cell fusion [283]. The receptor for syncytin-2 is not yet identified.

4.6 Post-fusion Resetting and Cell Survival

After fusion, cells modify their cellular components either to prevent further fusion events or to be ready for another round of fusion [5]. As an example of the former, sperm-egg fusion triggers Ca^{2+} oscillations that induce exocytosis of the cortical granule contents of the egg, causing the zona pellucida to become refractory to binding and penetration of sperm [287]. As an example of the latter, formation of multinucleated cells and syncytiotrophoblast depend on repeated fusion events. In the absence of a continuous fusion with cytotrophoblasts, the syncytiotrophoblast dies necrotically within a few days [288].

4.6.1 Myoferlin

One of the post-fusion resetting challenges is to sequester surplus plasma membrane to maintain optimum surface tension without losing cell surface proteins important for another round of fusion. Myoferlin may accomplish this task. Myoferlin promotes endocytic recycling and thus revives the fusion competence of plasma membrane [289, 290]. Myoferlin is a member of the ferlin family of proteins. It has six C2 domains and is homologous to dysferlin. Myoferlin is highly expressed in myoblasts undergoing fusion [291] and is upregulated in response to myofiber damage in a NFAT-dependent manner [292]. Myoferlin null mice have small myofibers and reduced muscle mass, and myoferlin deficient myoblasts form myotubes inefficiently in vitro [291]. Myoferlin directly interacts with the eps15 homology domain protein EHD2 which is implicated in endocytic recycling of membrane proteins [290]. Myoferlin-null myoblasts do not recycle insulin-like growth factor 1 (IGF1) receptors to the cell surface after its internalization and these receptors accumulate into endosomes/lysosomes [289]. Therefore, myoferlin-null myofibers do not respond well to IGF1-treatment. Whether other fusion-promoting proteins on cell surfaces are also recycled in a myoferlin/EHD2-dependent manner is an important question still to be accessed.

4.6.2 Bcl-2 and c-Flip

Initiator caspases 8 and 9 have been implicated to regulate the differentiation of cytotrophoblasts and myoblasts, respectively [48, 88]. It is thus important to diminish the activities of these initiator caspases to avoid the risk of going through the whole apoptotic cascade post cell fusion. Anti-apoptotic proteins Bcl-2 (B-cell lymphoma 2) and C-Flip have been suggested to down-regulate caspase activities in humans [48]. Bcl-2 prevents the oligomerisation and activation of caspases and thus antagonizes apoptosis [293]. Bcl-2 is expressed in the syncytiotrophoblast throughout gestation, and in differentiating Jeg-3 choriocarcinoma cells [294]. C-Flip directly regulates the activity of caspase 8 [295]. Fixed tissue sections taken from early and late stages of pregnancy demonstrated that C-Flip are present in caspase-8 positive cells, suggesting that C-Flip may protect trophoblast cells from cell death [296].

4.7 Conclusions

Fusion of one cell with another generates a "multi-nucleated" cell. Although multi-nucleated cells might be easily identified by light microscopy, microscopic techniques inevitably underestimate the contribution of cell fusion to the development and structure of tissues. Failure of microscopy reflects in part the possibility that many multi-nucleated cells, such as multi-nucleated giant cells, hepatocytes and muscle cells, die or are shed after recovery from injury. Failure of microscopy might also reflect the possibility that in some cases nuclei might also fuse or be shed [2]. And, the frequency of nuclear fusion and shedding is quite difficult to estimate by static morphologic or genetic analysis [297].

The contribution of cell fusion to the architecture and physiology of tissues is better identified by depletion or inhibition of molecules thought to be involved in the cellular processes that generate fusion. Targeting molecules by inhibitors, antibodies, RNA interference, mutation or gene deletions, etc., the sole function of which is fusion, might provide an incisive picture of how often and in what ways fusion contributes to the structure and function of tissues. In this communication we provide a list and cursory description of some of the molecules thought to contribute to cell fusion.

This list and the putative function of these molecules in cell fusion must be taken as provisionary. Most of the molecules we discuss have functions besides the promoting of cell fusion. Some of these "other" functions may be necessary for life, some may be necessary for all together different processes (e.g. prevention of apoptosis) that incidentally support cell fusion. Some of these molecules support multiple steps in cell fusion and some provide functions that can be replaced. For example, M-cadherin supports many processes of morphogenesis and is associated with lipid rafts and accumulates at the sites of cell contact [217]. In addition to mediating cell–cell adhesion, M-cadherin also activates Rac1, the key regulator of actin polymerization [216]. Inhibition of M-cadherin functions by blocking peptides, neutralizing antibodies or RNAi reduces myotube growth in vitro [219, 220]. However, mice without *Cdh15* gene, which encodes M-cadherin, develop normal skeletal muscle [221]. Other cadherins, such as N-cadherin and cadherin11 may compensate the loss of M-cadherin [221].

The complexity of cell fusion makes it difficult to identify molecules promoting fusion through gain-of-function experiments. If cell fusion is essential for normal development, natural fusogens, akin to viral fusogens, have thus far eluded detection [250]. Of course the difficulty of finding natural fusogens could mean that "master fusogenic proteins" do not exist in physiology and that the viral proteins are unique.

Among the molecules implicated in cellular fusion, many are specific for one cell type, especially those engaged in early steps of fusion. Many of these cell type-specific molecules are also involved in cell differentiation that is prior to cell fusion. In contrast, pore formation and expansion mainly reflect the functions of molecules such as lipids, actin, adhesion proteins and their regulators, that are common among the plasma membrane fusion events in many cell types. Understanding what regulates the later steps in cell fusion is thus particularly challenging.

We propose a new model for the later events in cell fusion. Our model is based on recent advances in the field concerning involvement of lipid rafts, adhesion proteins and actin rearrangement in cell fusion (Section 4.5.3 and Fig. 4.1). This model does not rely on fusogens. The cell membrane fusion model includes four steps: (i) recruitment of adhesion molecules onto lipid rafts and alignment of opposing membranes; (ii) membrane adhesion and actin rearrangement; (iii) quick lateral dispersion of lipid rafts and formation of membrane curvatures; and (iv) fusion pore formation and expansion driven by actin polymerization.

Our model may help to address key questions of cell fusion. Do all kinds of fusion require the same function and dispersion of lipid rafts? Experiments such as depletion of lipid raft component cholesterol from cell membrane in various cell types may provide a straightforward answer to this question. Must adhesion proteins involved in cell fusion reside in lipid rafts? Several adhesion proteins, such as CD9, M-cadherin, N-cadherin and NCAM, accumulate at membrane fusion sites in a lipid raft-dependent manner [217, 237]. However, other adhesion proteins thought to contribute to cell fusion have not yet been tested. Proteome analysis of lipid raft-associated proteins might identify novel fusion-promoting proteins. What controls the rapid lateral dispersion of lipid rafts? Since many adhesion proteins associate with and regulate the actin network, membrane adhesion proteins might regulate the actin-myosin network and thus control the movement of lipid rafts. How does actin rearrangement facilitate fusion pore formation and expansion? In *Drosophila*, actin polymerization provides the force needed for the formation of fusion pore and the expansion of the pore [262–264]. The mechanical procedure of doing so is still poorly understood.

Fusion between progenitor cells and somatic cells has been implied in regeneration of skeletal muscle, liver, intestine, cardiomyocytes and brain [134–136, 298, 299]. How cell fusion contributes to regeneration is still unclear because the frequency of fusion might be low [8]. Nevertheless, inflammation and injury dramatically increases this frequency [134–136], but the impact of this increase is not clear. The chemokine SDF-1 is present at high levels at the site of damaged tissues [126–131]. Perhaps cells expressing SDF-1 receptors, CXCR4, may be recruited and there may fuse with endangered cells [132, 133]. Further understanding the molecular mechanisms of recruiting and fusing stem cells and progenitor cells with somatic cells will eventually shed light on how fusion participates in tissue responses and whether it is to be encouraged or suppressed for therapeutic applications.

Acknowledgments We apologize to many authors whose work has not been cited directly but through other review articles and whose work could not be cited owing to space limitations. Supported by a grant from the National Institutes of Health (HL52297).

References

1. Schwann T (1839) Mikroskopische Untersuchungen uber die Uebereunstimmung in der Struktur und dem Wachsten der Thiere und Pflanzen. Sander'schen Buchhanlung, Berlin
2. Ogle BM, Cascalho M, Platt JL (2005) Biological implications of cell fusion. Nat Rev Mol Cell Biol 6:567–575
3. Vjugina U, Evans JP (2008) New insights into the molecular basis of mammalian sperm-egg membrane interactions. Front Biosci 13:462–476
4. Huppertz B, Bartz C, Kokozidou M (2006) Trophoblast fusion: fusogenic proteins, syncytins and ADAMs, and other prerequisites for syncytial fusion. Micron 37:509–517
5. Rochlin K, Yu S, Roy S et al (2010) Myoblast fusion: when it takes more to make one. Dev Biol 341:66–83
6. Helming L, Gordon S (2007) The molecular basis of macrophage fusion. Immunobiology 212:785–793
7. Alvarez-Dolado M (2007) Cell fusion: biological perspectives and potential for regenerative medicine. Front Biosci 12:1–12
8. Lluis F, Cosma MP (2010) Cell-fusion-mediated somatic-cell reprogramming: a mechanism for tissue regeneration. J Cell Physiol 223:6–13
9. Helming L, Gordon S (2009) Molecular mediators of macrophage fusion. Trends Cell Biol 19:514–522
10. DeFife KM, Jenney CR, McNally AK et al (1997) Interleukin-13 induces human monocyte/macrophage fusion and macrophage mannose receptor expression. J Immunol 158:3385–3390
11. Enelow RI, Sullivan GW, Carper HT et al (1992) Induction of multinucleated giant cell formation from in vitro culture of human monocytes with interleukin-3 and interferon-gamma: comparison with other stimulating factors. Am J Respir Cell Mol Biol 6:57–62
12. McNally AK, Anderson JM (1995) Interleukin-4 induces foreign body giant cells from human monocytes/macrophages. Differential lymphokine regulation of macrophage fusion leads to morphological variants of multinucleated giant cells. Am J Pathol 147:1487–1499
13. Coury F, Annels N, Rivollier A et al (2008) Langerhans cell histiocytosis reveals a new IL-17A-dependent pathway of dendritic cell fusion. Nat Med 14:81–87
14. McInnes A, Rennick DM (1988) Interleukin 4 induces cultured monocytes/macrophages to form giant multinucleated cells. J Exp Med 167:598–611
15. Horsley V, Jansen KM, Mills ST et al (2003) IL-4 acts as a myoblast recruitment factor during mammalian muscle growth. Cell 113:483–494
16. Weinberg JB, Hobbs MM, Misukonis MA (1984) Recombinant human gamma-interferon induces human monocyte polykaryon formation. Proc Natl Acad Sci USA 81:4554–4557
17. Belosevic M, Finbloom DS, Van Der Meide PH et al (1989) Administration of monoclonal anti-IFN-gamma antibodies in vivo abrogates natural resistance of C3H/HeN mice to infection with Leishmania major. J Immunol 143:266–274

18. Chensue SW, Terebuh PD, Warmington KS et al (1992) Role of IL-4 and IFN-gamma in Schistosoma mansoni egg-induced hypersensitivity granuloma formation. Orchestration, relative contribution, and relationship to macrophage function. J Immunol 148:900–906
19. Helming L, Gordon S (2007) Macrophage fusion induced by IL-4 alternative activation is a multistage process involving multiple target molecules. Eur J Immunol 37:33–42
20. Moreno JL, Mikhailenko I, Tondravi MM et al (2007) IL-4 promotes the formation of multinucleated giant cells from macrophage precursors by a STAT6-dependent, homotypic mechanism: contribution of E-cadherin. J Leukoc Biol 82:1542–1553
21. Van den Bossche J, Bogaert P, van Hengel J et al (2009) Alternatively activated macrophages engage in homotypic and heterotypic interactions through IL-4 and polyamine-induced E-cadherin/catenin complexes. Blood 114:4664–4674
22. Jones JA, McNally AK, Chang DT et al (2008) Matrix metalloproteinases and their inhibitors in the foreign body reaction on biomaterials. J Biomed Mater Res A 84:158–166
23. MacLauchlan S, Skokos EA, Meznarich N et al (2009) Macrophage fusion, giant cell formation, and the foreign body response require matrix metalloproteinase 9. J Leukoc Biol 85:617–626
24. Yagi M, Ninomiya K, Fujita N et al (2007) Induction of DC-STAMP by alternative activation and downstream signaling mechanisms. J Bone Miner Res 22:992–1001
25. Yagi M, Miyamoto T, Sawatani Y et al (2005) DC-STAMP is essential for cell-cell fusion in osteoclasts and foreign body giant cells. J Exp Med 202:345–351
26. Stein M, Keshav S, Harris N et al (1992) Interleukin 4 potently enhances murine macrophage mannose receptor activity: a marker of alternative immunologic macrophage activation. J Exp Med 176:287–292
27. McNally AK, DeFife KM, Anderson JM (1996) Interleukin-4-induced macrophage fusion is prevented by inhibitors of mannose receptor activity. Am J Pathol 149:975–985
28. Jansen KM, Pavlath GK (2006) Mannose receptor regulates myoblast motility and muscle growth. J Cell Biol 174:403–413
29. Lacey DL, Timms E, Tan HL et al (1998) Osteoprotegerin ligand is a cytokine that regulates osteoclast differentiation and activation. Cell 93:165–176
30. Yasuda H, Shima N, Nakagawa N et al (1998) Osteoclast differentiation factor is a ligand for osteoprotegerin/osteoclastogenesis-inhibitory factor and is identical to TRANCE/RANKL. Proc Natl Acad Sci USA 95:3597–3602
31. Takayanagi H, Kim S, Koga T et al (2002) Induction and activation of the transcription factor NFATc1 (NFAT2) integrate RANKL signaling in terminal differentiation of osteoclasts. Dev Cell 3:889–901
32. Kim K, Lee SH, Ha Kim J et al (2008) NFATc1 induces osteoclast fusion via up-regulation of Atp6v0d2 and the dendritic cell-specific transmembrane protein (DC-STAMP). Mol Endocrinol 22:176–185
33. Feng H, Cheng T, Steer JH et al (2009) Myocyte enhancer factor 2 and microphthalmia-associated transcription factor cooperate with NFATc1 to transactivate the V-ATPase d2 promoter during RANKL-induced osteoclastogenesis. J Biol Chem 284:14667–14676
34. Lee SH, Rho J, Jeong D et al (2006) v-ATPase V0 subunit d2-deficient mice exhibit impaired osteoclast fusion and increased bone formation. Nat Med 12:1403–1409
35. Simonet WS, Lacey DL, Dunstan CR et al (1997) Osteoprotegerin: a novel secreted protein involved in the regulation of bone density. Cell 89:309–319
36. Yasuda H, Shima N, Nakagawa N et al (1998) Identity of osteoclastogenesis inhibitory factor (OCIF) and osteoprotegerin (OPG): a mechanism by which OPG/OCIF inhibits osteoclastogenesis in vitro. Endocrinology 139:1329–1337
37. Helming L, Tomasello E, Kyriakides TR et al (2008) Essential role of DAP12 signaling in macrophage programming into a fusion-competent state. Sci Signal 1:ra11
38. Paloneva J, Mandelin J, Kiialainen A et al (2003) DAP12/TREM2 deficiency results in impaired osteoclast differentiation and osteoporotic features. J Exp Med 198:669–675
39. Kaifu T, Nakahara J, Inui M et al (2003) Osteopetrosis and thalamic hypomyelinosis with synaptic degeneration in DAP12-deficient mice. J Clin Invest 111:323–332
40. Peng Q, Malhotra S, Torchia JA et al (2010) TREM2- and DAP12-dependent activation of PI3K requires DAP10 and is inhibited by SHIP1. Sci Signal 3:ra38
41. Turnbull IR, Colonna M (2007) Activating and inhibitory functions of DAP12. Nat Rev Immunol 7:155–161
42. Zou W, Reeve JL, Liu Y et al (2008) DAP12 couples c-Fms activation to the osteoclast cytoskeleton by recruitment of Syk. Mol Cell 31:422–431
43. Sessions A, Horwitz AF (1983) Differentiation-related differences in the plasma membrane phospholipid asymmetry of myogenic and fibrogenic cells. Biochim Biophys Acta 728:103–111
44. Adler RR, Ng AK, Rote NS (1995) Monoclonal antiphosphatidylserine antibody inhibits intercellular fusion of the choriocarcinoma line, JAR. Biol Reprod 53:905–910

45. Gadella BM, Harrison RA (2000) The capacitating agent bicarbonate induces protein kinase A-dependent changes in phospholipid transbilayer behavior in the sperm plasma membrane. Development 127:2407–2420
46. Helming L, Winter J, Gordon S (2009) The scavenger receptor CD36 plays a role in cytokine-induced macrophage fusion. J Cell Sci 122:453–459
47. Leventis PA, Grinstein S (2010) The distribution and function of phosphatidylserine in cellular membranes. Annu Rev Biophys 39:407–427
48. Gauster M, Huppertz B (2010) The paradox of caspase 8 in human villous trophoblast fusion. Placenta 31:82–88
49. Rote NS, Wei BR, Xu C et al (2010) Caspase 8 and human villous cytotrophoblast differentiation. Placenta 31:89–96
50. Harrison RA, Gadella BM (2005) Bicarbonate-induced membrane processing in sperm capacitation. Theriogenology 63:342–351
51. Gadella BM, Harrison RA (2002) Capacitation induces cyclic adenosine 3′,5′-monophosphate-dependent, but apoptosis-unrelated, exposure of aminophospholipids at the apical head plasma membrane of boar sperm cells. Biol Reprod 67:340–350
52. Flesch FM, Brouwers JF, Nievelstein PF et al (2001) Bicarbonate stimulated phospholipid scrambling induces cholesterol redistribution and enables cholesterol depletion in the sperm plasma membrane. J Cell Sci 114: 3543–3555
53. Harrison RA, Ashworth PJ, Miller NG (1996) Bicarbonate/CO2, an effector of capacitation, induces a rapid and reversible change in the lipid architecture of boar sperm plasma membranes. Mol Reprod Dev 45:378–391
54. Litvin TN, Kamenetsky M, Zarifyan A et al (2003) Kinetic properties of "soluble" adenylyl cyclase. Synergism between calcium and bicarbonate. J Biol Chem 278:15922–15926
55. Harrison RA, Miller NG (2000) cAMP-dependent protein kinase control of plasma membrane lipid architecture in boar sperm. Mol Reprod Dev 55:220–228
56. Harrison RA (1996) Capacitation mechanisms, and the role of capacitation as seen in eutherian mammals. Reprod Fertil Dev 8:581–594
57. de Vries KJ, Wiedmer T, Sims PJ et al (2003) Caspase-independent exposure of aminophospholipids and tyrosine phosphorylation in bicarbonate responsive human sperm cells. Biol Reprod 68:2122–2134
58. Callahan MK, Williamson P, Schlegel RA (2000) Surface expression of phosphatidylserine on macrophages is required for phagocytosis of apoptotic thymocytes. Cell Death Differ 7:645–653
59. MacKenzie A, Wilson HL, Kiss-Toth E et al (2001) Rapid secretion of interleukin-1beta by microvesicle shedding. Immunity 15:825–835
60. Van den Eijnde SM, Boshart L, Reutelingsperger CP et al (1997) Phosphatidylserine plasma membrane asymmetry in vivo: a pancellular phenomenon which alters during apoptosis. Cell Death Differ 4:311–316
61. van den Eijnde SM, van den Hoff MJ, Reutelingsperger CP et al (2001) Transient expression of phosphatidylserine at cell-cell contact areas is required for myotube formation. J Cell Sci 114:3631–3642
62. Lyden TW, Vogt E, Ng AK et al (1992) Monoclonal antiphospholipid antibody reactivity against human placental trophoblast. J Reprod Immunol 22:1–14
63. Lyden TW, Ng AK, Rote NS (1993) Modulation of phosphatidylserine epitope expression by BeWo cells during forskolin treatment. Placenta 14:177–186
64. Vogt E, Ng AK, Rote NS (1996) A model for the antiphospholipid antibody syndrome: monoclonal antiphosphatidylserine antibody induces intrauterine growth restriction in mice. Am J Obstet Gynecol 174:700–707
65. Daleke DL (2007) Phospholipid flippases. J Biol Chem 282:821–825
66. Florman HM, Jungnickel MK, Sutton KA (2008) Regulating the acrosome reaction. Int J Dev Biol 52:503–510
67. Wassarman PM, Litscher ES (2008) Mammalian fertilization is dependent on multiple membrane fusion events. Methods Mol Biol 475:99–113
68. Inoue N, Ikawa M, Isotani A et al (2005) The immunoglobulin superfamily protein Izumo is required for sperm to fuse with eggs. Nature 434:234–238
69. Myles DG, Koppel DE, Cowan AE et al (1987) Rearrangement of sperm surface antigens prior to fertilization. Ann N Y Acad Sci 513:262–273
70. Miranda PV, Allaire A, Sosnik J et al (2009) Localization of low-density detergent-resistant membrane proteins in intact and acrosome-reacted mouse sperm. Biol Reprod 80:897–904
71. Sosnik J, Miranda PV, Spiridonov NA et al (2009) Tssk6 is required for Izumo relocalization and gamete fusion in the mouse. J Cell Sci 122:2741–2749
72. Bleil JD, Wassarman PM (1983) Sperm–egg interactions in the mouse: sequence of events and induction of the acrosome reaction by a zona pellucida glycoprotein. Dev Biol 95:317–324
73. Arnoult C, Zeng Y, Florman HM (1996) ZP3-dependent activation of sperm cation channels regulates acrosomal secretion during mammalian fertilization. J Cell Biol 134:637–645
74. Florman HM, Tombes RM, First NL et al (1989) An adhesion-associated agonist from the zona pellucida activates G protein-promoted elevations of internal Ca^{2+} and pH that mediate mammalian sperm acrosomal exocytosis. Dev Biol 135:133–146

75. Arnoult C, Cardullo RA, Lemos JR et al (1996) Activation of mouse sperm T-type Ca^{2+} channels by adhesion to the egg zona pellucida. Proc Natl Acad Sci USA 93:13004–13009
76. Fukami K, Nakao K, Inoue T et al (2001) Requirement of phospholipase Cdelta4 for the zona pellucida-induced acrosome reaction. Science 292:920–923
77. Fukami K, Yoshida M, Inoue T et al (2003) Phospholipase Cdelta4 is required for Ca^{2+} mobilization essential for acrosome reaction in sperm. J Cell Biol 161:79–88
78. Jungnickel MK, Sutton KA, Wang Y et al (2007) Phosphoinositide-dependent pathways in mouse sperm are regulated by egg ZP3 and drive the acrosome reaction. Dev Biol 304:116–126
79. Achuthan A, Masendycz P, Lopez JA et al (2008) Regulation of the endosomal SNARE protein syntaxin 7 by colony-stimulating factor 1 in macrophages. Mol Cell Biol 28:6149–6159
80. Tomes CN, De Blas GA, Michaut MA et al (2005) alpha-SNAP and NSF are required in a priming step during the human sperm acrosome reaction. Mol Hum Reprod 11:43–51
81. Tomes CN, Michaut M, De Blas G et al (2002) SNARE complex assembly is required for human sperm acrosome reaction. Dev Biol 243:326–338
82. Lamkanfi M, Festjens N, Declercq W et al (2007) Caspases in cell survival, proliferation and differentiation. Cell Death Differ 14:44–55
83. Black S, Kadyrov M, Kaufmann P et al (2004) Syncytial fusion of human trophoblast depends on caspase 8. Cell Death Differ 11:90–98
84. Gauster M, Siwetz M, Orendi K et al (2010) Caspases rather than calpains mediate remodelling of the fodrin skeleton during human placental trophoblast fusion. Cell Death Differ 17:336–345
85. Martens S, McMahon HT (2008) Mechanisms of membrane fusion: disparate players and common principles. Nat Rev Mol Cell Biol 9:543–556
86. Mandal D, Mazumder A, Das P et al (2005) Fas-, caspase 8-, and caspase 3-dependent signaling regulates the activity of the aminophospholipid translocase and phosphatidylserine externalization in human erythrocytes. J Biol Chem 280:39460–39467
87. Gauster M, Siwetz M, Huppertz B (2009) Fusion of villous trophoblast can be visualized by localizing active caspase 8. Placenta 30:547–550
88. Murray TV, McMahon JM, Howley BA et al (2008) A non-apoptotic role for caspase-9 in muscle differentiation. J Cell Sci 121:3786–3793
89. Fernando P, Kelly JF, Balazsi K et al (2002) Caspase 3 activity is required for skeletal muscle differentiation. Proc Natl Acad Sci USA 99:11025–11030
90. Goll DE, Thompson VF, Li H et al (2003) The calpain system. Physiol Rev 83:731–801
91. Balcerzak D, Poussard S, Brustis JJ et al (1995) An antisense oligodeoxyribonucleotide to m-calpain mRNA inhibits myoblast fusion. J Cell Sci 108 (Pt 5):2077–2082
92. Kuchay SM, Kim N, Grunz EA et al (2007) Double knockouts reveal that protein tyrosine phosphatase 1B is a physiological target of calpain-1 in platelets. Mol Cell Biol 27:6038–6052
93. Honda M, Masui F, Kanzawa N et al (2008) Specific knockdown of m-calpain blocks myogenesis with cDNA deduced from the corresponding RNAi. Am J Physiol Cell Physiol 294:C957–965
94. Kramerova I, Kudryashova E, Tidball JG et al (2004) Null mutation of calpain 3 (p94) in mice causes abnormal sarcomere formation in vivo and in vitro. Hum Mol Genet 13:1373–1388
95. Kramerova I, Kudryashova E, Wu B et al (2006) Regulation of the M-cadherin-beta-catenin complex by calpain 3 during terminal stages of myogenic differentiation. Mol Cell Biol 26:8437–8447
96. Charrasse S, Comunale F, Grumbach Y et al (2006) RhoA GTPase regulates M-cadherin activity and myoblast fusion. Mol Biol Cell 17:749–759
97. Kim YS, Nakanishi G, Lewandoski M et al (2003) GLIS3, a novel member of the GLIS subfamily of Kruppel-like zinc finger proteins with repressor and activation functions. Nucleic Acids Res 31:5513–5525
98. Duan H, Nguyen HT (2006) Distinct posttranscriptional mechanisms regulate the activity of the Zn finger transcription factor lame duck during Drosophila myogenesis. Mol Cell Biol 26:1414–1423
99. Duan H, Skeath JB, Nguyen HT (2001) Drosophila Lame duck, a novel member of the Gli superfamily, acts as a key regulator of myogenesis by controlling fusion-competent myoblast development. Development 128: 4489–4500
100. Furlong EE, Andersen EC, Null B et al (2001) Patterns of gene expression during Drosophila mesoderm development. Science 293:1629–1633
101. Ruiz-Gomez M, Coutts N, Suster ML et al (2002) myoblasts incompetent encodes a zinc finger transcription factor required to specify fusion-competent myoblasts in Drosophila. Development 129:133–141
102. Watanabe N, Hiramatsu K, Miyamoto R et al (2009) A murine model of neonatal diabetes mellitus in Glis3-deficient mice. FEBS Lett 583:2108–2113
103. Kang HS, Kim YS, ZeRuth G et al (2009) Transcription factor Glis3, a novel critical player in the regulation of pancreatic beta-cell development and insulin gene expression. Mol Cell Biol 29:6366–6379

104. Brand-Saberi B, Muller TS, Wilting J et al (1996) Scatter factor/hepatocyte growth factor (SF/HGF) induces emigration of myogenic cells at interlimb level in vivo. Dev Biol 179:303–308
105. Corti S, Salani S, Del Bo R et al (2001) Chemotactic factors enhance myogenic cell migration across an endothelial monolayer. Exp Cell Res 268:36–44
106. Lafreniere JF, Mills P, Bouchentouf M et al (2006) Interleukin-4 improves the migration of human myogenic precursor cells in vitro and in vivo. Exp Cell Res 312:1127–1141
107. Lafreniere JF, Mills P, Tremblay JP et al (2004) Growth factors improve the in vivo migration of human skeletal myoblasts by modulating their endogenous proteolytic activity. Transplantation 77:1741–1747
108. Webb SE, Lee KK, Tang MK et al (1997) Fibroblast growth factors 2 and 4 stimulate migration of mouse embryonic limb myogenic cells. Dev Dyn 209:206–216
109. Robertson TA, Maley MA, Grounds MD et al (1993) The role of macrophages in skeletal muscle regeneration with particular reference to chemotaxis. Exp Cell Res 207:321–331
110. Minniti CP, Luan D, O'Grady C et al (1995) Insulin-like growth factor II overexpression in myoblasts induces phenotypic changes typical of the malignant phenotype. Cell Growth Differ 6:263–269
111. Germani A, Di Carlo A, Mangoni A et al (2003) Vascular endothelial growth factor modulates skeletal myoblast function. Am J Pathol 163:1417–1428
112. Ratajczak MZ, Majka M, Kucia M et al (2003) Expression of functional CXCR4 by muscle satellite cells and secretion of SDF-1 by muscle-derived fibroblasts is associated with the presence of both muscle progenitors in bone marrow and hematopoietic stem/progenitor cells in muscles. Stem Cells 21:363–371
113. Allen DL, Teitelbaum DH, Kurachi K (2003) Growth factor stimulation of matrix metalloproteinase expression and myoblast migration and invasion in vitro. Am J Physiol Cell Physiol 284:C805–815
114. Torrente Y, El Fahime E, Caron NJ et al (2003) Tumor necrosis factor-alpha (TNF-alpha) stimulates chemotactic response in mouse myogenic cells. Cell Transplant 12:91–100
115. Chowdhury SR, Muneyuki Y, Takezawa Y et al (2009) Synergic stimulation of laminin and epidermal growth factor facilitates the myoblast growth through promoting migration. J Biosci Bioeng 108:174–177
116. Adams JC, Schwartz MA (2000) Stimulation of fascin spikes by thrombospondin-1 is mediated by the GTPases Rac and Cdc42. J Cell Biol 150:807–822
117. Turner DC, Lawton J, Dollenmeier P et al (1983) Guidance of myogenic cell migration by oriented deposits of fibronectin. Dev Biol 95:497–504
118. Yao CC, Ziober BL, Sutherland AE et al (1996) Laminins promote the locomotion of skeletal myoblasts via the alpha 7 integrin receptor. J Cell Sci 109 (Pt 13):3139–3150
119. Bischoff R (1997) Chemotaxis of skeletal muscle satellite cells. Dev Dyn 208:505–515
120. Rosario M, Birchmeier W (2003) How to make tubes: signaling by the Met receptor tyrosine kinase. Trends Cell Biol 13:328–335
121. Odemis V, Lamp E, Pezeshki G et al (2005) Mice deficient in the chemokine receptor CXCR4 exhibit impaired limb innervation and myogenesis. Mol Cell Neurosci 30:494–505
122. Forde S, Tye BJ, Newey SE et al (2007) Endolyn (CD164) modulates the CXCL12-mediated migration of umbilical cord blood CD133+ cells. Blood 109:1825–1833
123. Bae GU, Gaio U, Yang YJ et al (2008) Regulation of myoblast motility and fusion by the CXCR4-associated sialomucin, CD164. J Biol Chem 283:8301–8309
124. Miller RJ, Banisadr G, Bhattacharyya BJ (2008) CXCR4 signaling in the regulation of stem cell migration and development. J Neuroimmunol 198:31–38
125. Nagasawa T (2007) The chemokine CXCL12 and regulation of HSC and B lymphocyte development in the bone marrow niche. Adv Exp Med Biol 602:69–75
126. Lazarini F, Tham TN, Casanova P et al (2003) Role of the alpha-chemokine stromal cell-derived factor (SDF-1) in the developing and mature central nervous system. Glia 42:139–148
127. Stumm RK, Rummel J, Junker V et al (2002) A dual role for the SDF-1/CXCR4 chemokine receptor system in adult brain: isoform-selective regulation of SDF-1 expression modulates CXCR4-dependent neuronal plasticity and cerebral leukocyte recruitment after focal ischemia. J Neurosci 22:5865–5878
128. Yamaguchi J, Kusano KF, Masuo O et al (2003) Stromal cell-derived factor-1 effects on *ex vivo* expanded endothelial progenitor cell recruitment for ischemic neovascularization. Circulation 107:1322–1328
129. Fox JM, Chamberlain G, Ashton BA et al (2007) Recent advances into the understanding of mesenchymal stem cell trafficking. Br J Haematol 137:491–502
130. Hill WD, Hess DC, Martin-Studdard A et al (2004) SDF-1 (CXCL12) is upregulated in the ischemic penumbra following stroke: association with bone marrow cell homing to injury. J Neuropathol Exp Neurol 63:84–96
131. Togel F, Isaac J, Hu Z et al (2005) Renal SDF-1 signals mobilization and homing of CXCR4-positive cells to the kidney after ischemic injury. Kidney Int 67:1772–1784
132. Blanchet MR, McNagny KM (2009) Stem cells, inflammation and allergy. Allergy Asthma Clin Immunol 5:13

133. Kucia M, Ratajczak J, Reca R et al (2004) Tissue-specific muscle, neural and liver stem/progenitor cells reside in the bone marrow, respond to an SDF-1 gradient and are mobilized into peripheral blood during stress and tissue injury. Blood Cells Mol Dis 32:52–57
134. Davies PS, Powell AE, Swain JR et al (2009) Inflammation and proliferation act together to mediate intestinal cell fusion. PLoS ONE 4:e6530
135. Nygren JM, Liuba K, Breitbach M et al (2008) Myeloid and lymphoid contribution to non-haematopoietic lineages through irradiation-induced heterotypic cell fusion. Nat Cell Biol 10:584–592
136. Johansson CB, Youssef S, Koleckar K et al (2008) Extensive fusion of haematopoietic cells with Purkinje neurons in response to chronic inflammation. Nat Cell Biol 10:575–583
137. Horsley V, Friday BB, Matteson S et al (2001) Regulation of the growth of multinucleated muscle cells by an NFATC2-dependent pathway. J Cell Biol 153:329–338
138. Sotiropoulos A, Ohanna M, Kedzia C et al (2006) Growth hormone promotes skeletal muscle cell fusion independent of insulin-like growth factor 1 up-regulation. Proc Natl Acad Sci USA 103:7315–7320
139. Griffin CA, Kafadar KA, Pavlath GK (2009) MOR23 promotes muscle regeneration and regulates cell adhesion and migration. Dev Cell 17:649–661
140. Fukuda N, Yomogida K, Okabe M et al (2004) Functional characterization of a mouse testicular olfactory receptor and its role in chemosensing and in regulation of sperm motility. J Cell Sci 117:5835–5845
141. Tangirala RK, Murao K, Quehenberger O (1997) Regulation of expression of the human monocyte chemotactic protein-1 receptor (hCCR2) by cytokines. J Biol Chem 272:8050–8056
142. Kyriakides TR, Foster MJ, Keeney GE et al (2004) The CC chemokine ligand, CCL2/MCP1, participates in macrophage fusion and foreign body giant cell formation. Am J Pathol 165:2157–2166
143. Jay SM, Skokos E, Laiwalla F et al (2007) Foreign body giant cell formation is preceded by lamellipodia formation and can be attenuated by inhibition of Rac1 activation. Am J Pathol 171:632–640
144. Miyamoto K, Ninomiya K, Sonoda KH et al (2009) MCP-1 expressed by osteoclasts stimulates osteoclastogenesis in an autocrine/paracrine manner. Biochem Biophys Res Commun 383:373–377
145. Harper CV, Barratt CL, Publicover SJ (2004) Stimulation of human spermatozoa with progesterone gradients to simulate approach to the oocyte. Induction of [Ca(2+)](i) oscillations and cyclical transitions in flagellar beating. J Biol Chem 279:46315–46325
146. Teves ME, Barbano F, Guidobaldi HA et al (2006) Progesterone at the picomolar range is a chemoattractant for mammalian spermatozoa. Fertil Steril 86:745–749
147. Teves ME, Guidobaldi HA, Unates DR et al (2010) Progesterone sperm chemoattraction may be modulated by its corticosteroid-binding globulin carrier protein. Fertil Steril 93:2450–2452
148. Teves ME, Guidobaldi HA, Unates DR et al (2009) Molecular mechanism for human sperm chemotaxis mediated by progesterone. PLoS ONE 4:e8211
149. Osman RA, Andria ML, Jones AD et al (1989) Steroid induced exocytosis: the human sperm acrosome reaction. Biochem Biophys Res Commun 160:828–833
150. Hood JD, Cheresh DA (2002) Role of integrins in cell invasion and migration. Nat Rev Cancer 2:91–100
151. Ocalan M, Goodman SL, Kuhl U et al (1988) Laminin alters cell shape and stimulates motility and proliferation of murine skeletal myoblasts. Dev Biol 125:158–167
152. Sonnenberg A, Modderman PW, Hogervorst F (1988) Laminin receptor on platelets is the integrin VLA-6. Nature 336:487–489
153. Kramer RH, McDonald KA, Vu MP (1989) Human melanoma cells express a novel integrin receptor for laminin. J Biol Chem 264:15642–15649
154. Siegel AL, Atchison K, Fisher KE et al (2009) 3D timelapse analysis of muscle satellite cell motility. Stem Cells 27:2527–2538
155. Mayer U, Saher G, Fassler R et al (1997) Absence of integrin alpha 7 causes a novel form of muscular dystrophy. Nat Genet 17:318–323
156. Schwander M, Leu M, Stumm M et al (2003) Beta1 integrins regulate myoblast fusion and sarcomere assembly. Dev Cell 4:673–685
157. Echtermeyer F, Schober S, Poschl E et al (1996) Specific induction of cell motility on laminin by alpha 7 integrin. J Biol Chem 271:2071–2075
158. Glenn HL, Wang Z, Schwartz LM (2010) Acheron, a Lupus antigen family member, regulates integrin expression, adhesion, and motility in differentiating myoblasts. Am J Physiol Cell Physiol 298:C46–55
159. Wang Z, Glenn H, Brown C et al (2009) Regulation of muscle differentiation and survival by Acheron. Mech Dev 126:700–709
160. Calderwood DA, Shattil SJ, Ginsberg MH (2000) Integrins and actin filaments: reciprocal regulation of cell adhesion and signaling. J Biol Chem 275:22607–22610
161. Alvarez B, Stroeken PJ, Edel MJ et al (2008) Integrin Cytoplasmic domain-Associated Protein-1 (ICAP-1) promotes migration of myoblasts and affects focal adhesions. J Cell Physiol 214:474–482

162. Boissy P, Machuca I, Pfaff M et al (1998) Aggregation of mononucleated precursors triggers cell surface expression of alphavbeta3 integrin, essential to formation of osteoclast-like multinucleated cells. J Cell Sci 111 (Pt 17):2563–2574
163. McHugh KP, Hodivala-Dilke K, Zheng MH et al (2000) Mice lacking beta3 integrins are osteosclerotic because of dysfunctional osteoclasts. J Clin Invest 105:433–440
164. Lakkakorpi PT, Bett AJ, Lipfert L et al (2003) PYK2 autophosphorylation, but not kinase activity, is necessary for adhesion-induced association with c-Src, osteoclast spreading, and bone resorption. J Biol Chem 278: 11502–11512
165. Miyazaki T, Sanjay A, Neff L et al (2004) Src kinase activity is essential for osteoclast function. J Biol Chem 279:17660–17666
166. Brazier H, Pawlak G, Vives V et al (2009) The Rho GTPase Wrch1 regulates osteoclast precursor adhesion and migration. Int J Biochem Cell Biol 41:1391–1401
167. Brazier H, Stephens S, Ory S et al (2006) Expression profile of RhoGTPases and RhoGEFs during RANKL-stimulated osteoclastogenesis: identification of essential genes in osteoclasts. J Bone Miner Res 21:1387–1398
168. Kim T, Ha HI, Kim N et al (2009) Adrm1 interacts with Atp6v0d2 and regulates osteoclast differentiation. Biochem Biophys Res Commun 390:585–590
169. Husnjak K, Elsasser S, Zhang N et al (2008) Proteasome subunit Rpn13 is a novel ubiquitin receptor. Nature 453:481–488
170. Stricker J, Falzone T, Gardel ML (2010) Mechanics of the F-actin cytoskeleton. J Biomech 43:9–14
171. Nowak SJ, Nahirney PC, Hadjantonakis AK et al (2009) Nap1-mediated actin remodeling is essential for mammalian myoblast fusion. J Cell Sci 122:3282–3293
172. Kawamura K, Takano K, Suetsugu S et al (2004) N-WASP and WAVE2 acting downstream of phosphatidylinositol 3-kinase are required for myogenic cell migration induced by hepatocyte growth factor. J Biol Chem 279:54862–54871
173. Vasyutina E, Martarelli B, Brakebusch C et al (2009) The small G-proteins Rac1 and Cdc42 are essential for myoblast fusion in the mouse. Proc Natl Acad Sci USA 106:8935–8940
174. Wells CM, Walmsley M, Ooi S et al (2004) Rac1-deficient macrophages exhibit defects in cell spreading and membrane ruffling but not migration. J Cell Sci 117:1259–1268
175. Wheeler AP, Wells CM, Smith SD et al (2006) Rac1 and Rac2 regulate macrophage morphology but are not essential for migration. J Cell Sci 119:2749–2757
176. Rohatgi R, Ho HY, Kirschner MW (2000) Mechanism of N-WASP activation by CDC42 and phosphatidylinositol 4, 5-bisphosphate. J Cell Biol 150:1299–1310
177. Morishima S, Morita I, Tokushima T et al (2003) Expression and role of mannose receptor/terminal high-mannose type oligosaccharide on osteoclast precursors during osteoclast formation. J Endocrinol 176:285–292
178. Taylor ME, Bezouska K, Drickamer K (1992) Contribution to ligand binding by multiple carbohydrate-recognition domains in the macrophage mannose receptor. J Biol Chem 267:1719–1726
179. Chabot V, Reverdiau P, Iochmann S et al (2006) CCL5-enhanced human immature dendritic cell migration through the basement membrane in vitro depends on matrix metalloproteinase-9. J Leukoc Biol 79:767–778
180. Rahat MA, Marom B, Bitterman H et al (2006) Hypoxia reduces the output of matrix metalloproteinase-9 (MMP-9) in monocytes by inhibiting its secretion and elevating membranal association. J Leukoc Biol 79: 706–718
181. Barclay AN (2003) Membrane proteins with immunoglobulin-like domains–a master superfamily of interaction molecules. Semin Immunol 15:215–223
182. Xu Z, Jin B (2010) A novel interface consisting of homologous immunoglobulin superfamily members with multiple functions. Cell Mol Immunol 7:11–19
183. Sohn RL, Huang P, Kawahara G et al (2009) A role for nephrin, a renal protein, in vertebrate skeletal muscle cell fusion. Proc Natl Acad Sci USA 106:9274–9279
184. Donoviel DB, Freed DD, Vogel H et al (2001) Proteinuria and perinatal lethality in mice lacking NEPH1, a novel protein with homology to NEPHRIN. Mol Cell Biol 21:4829–4836
185. Garg P, Verma R, Nihalani D et al (2007) Neph1 cooperates with nephrin to transduce a signal that induces actin polymerization. Mol Cell Biol 27:8698–8712
186. Krauss RS (2010) Regulation of promyogenic signal transduction by cell-cell contact and adhesion. Exp Cell Res 316:3042–3049
187. Lu M, Krauss RS (2010) N-cadherin ligation, but not Sonic hedgehog binding, initiates Cdo-dependent p38alpha/beta MAPK signaling in skeletal myoblasts. Proc Natl Acad Sci USA 107:4212–4217
188. Kang JS, Bae GU, Yi MJ et al (2008) A Cdo-Bnip-2-Cdc42 signaling pathway regulates p38alpha/beta MAPK activity and myogenic differentiation. J Cell Biol 182:497–507
189. Guasconi V, Puri PL (2009) Chromatin: the interface between extrinsic cues and the epigenetic regulation of muscle regeneration. Trends Cell Biol 19:286–294

190. Takaesu G, Kang JS, Bae GU et al (2006) Activation of p38alpha/beta MAPK in myogenesis via binding of the scaffold protein JLP to the cell surface protein Cdo. J Cell Biol 175:383–388
191. Cole F, Zhang W, Geyra A et al (2004) Positive regulation of myogenic bHLH factors and skeletal muscle development by the cell surface receptor CDO. Dev Cell 7:843–854
192. Bae GU, Yang YJ, Jiang G et al (2009) Neogenin regulates skeletal myofiber size and focal adhesion kinase and extracellular signal-regulated kinase activities in vivo and in vitro. Mol Biol Cell 20:4920–4931
193. De Vries M, Cooper HM (2008) Emerging roles for neogenin and its ligands in CNS development. J Neurochem 106:1483–1492
194. Quach NL, Biressi S, Reichardt LF et al (2009) Focal adhesion kinase signaling regulates the expression of caveolin 3 and beta1 integrin, genes essential for normal myoblast fusion. Mol Biol Cell 20:3422–3435
195. Dickson G, Gower HJ, Barton CH et al (1987) Human muscle neural cell adhesion molecule (N-CAM): identification of a muscle-specific sequence in the extracellular domain. Cell 50:1119–1130
196. Povlsen GK, Ditlevsen DK (2010) The neural cell adhesion molecule NCAM and lipid rafts. Adv Exp Med Biol 663:183–198
197. Dickson G, Peck D, Moore SE et al (1990) Enhanced myogenesis in NCAM-transfected mouse myoblasts. Nature 344:348–351
198. Baldwin TJ, Fazeli MS, Doherty P et al (1996) Elucidation of the molecular actions of NCAM and structurally related cell adhesion molecules. J Cell Biochem 61:502–513
199. Charlton CA, Mohler WA, Blau HM (2000) Neural cell adhesion molecule (NCAM) and myoblast fusion. Developmental Biology 221:112–119
200. Okabe M, Adachi T, Takada K et al (1987) Capacitation-related changes in antigen distribution on mouse sperm heads and its relation to fertilization rate in vitro. J Reprod Immunol 11:91–100
201. Ellerman DA, Pei J, Gupta S et al (2009) Izumo is part of a multiprotein family whose members form large complexes on mammalian sperm. Mol Reprod Dev 76:1188–1199
202. Lundberg P, Koskinen C, Baldock PA et al (2007) Osteoclast formation is strongly reduced both in vivo and in vitro in the absence of CD47/SIRPalpha-interaction. Biochem Biophys Res Commun 352:444–448
203. Han X, Sterling H, Chen Y et al (2000) CD47, a ligand for the macrophage fusion receptor, participates in macrophage multinucleation. J Biol Chem 275:37984–37992
204. Saginario C, Qian HY, Vignery A (1995) Identification of an inducible surface molecule specific to fusing macrophages. Proc Natl Acad Sci USA 92:12210–12214
205. Matozaki T, Murata Y, Okazawa H et al (2009) Functions and molecular mechanisms of the CD47-SIRPalpha signalling pathway. Trends Cell Biol 19:72–80
206. Oldenborg PA, Gresham HD, Lindberg FP (2001) CD47-signal regulatory protein alpha (SIRPalpha) regulates Fcgamma and complement receptor-mediated phagocytosis. J Exp Med 193:855–862
207. Motegi S, Okazawa H, Ohnishi H et al (2003) Role of the CD47-SHPS-1 system in regulation of cell migration. EMBO J 22:2634–2644
208. Brown EJ, Frazier WA (2001) Integrin-associated protein (CD47) and its ligands. Trends Cell Biol 11:130–135
209. Borghi N, James Nelson W (2009) Intercellular adhesion in morphogenesis: molecular and biophysical considerations. Curr Top Dev Biol 89:1–32
210. Krauss RS, Cole F, Gaio U et al (2005) Close encounters: regulation of vertebrate skeletal myogenesis by cell-cell contact. J Cell Sci 118:2355–2362
211. Rufas O, Fisch B, Ziv S et al (2000) Expression of cadherin adhesion molecules on human gametes. Mol Hum Reprod 6:163–169
212. Marin-Briggiler CI, Lapyckyj L, Gonzalez Echeverria MF et al (2010) Neural cadherin is expressed in human gametes and participates in sperm-oocyte interaction events. Int J Androl 33:e228–239
213. Radice GL, Rayburn H, Matsunami H et al (1997) Developmental defects in mouse embryos lacking N-cadherin. Dev Biol 181:64–78
214. Charlton CA, Mohler WA, Radice GL et al (1997) Fusion competence of myoblasts rendered genetically null for N-cadherin in culture. J Cell Biol 138:331–336
215. Huttenlocher A, Lakonishok M, Kinder M et al (1998) Integrin and cadherin synergy regulates contact inhibition of migration and motile activity. J Cell Biol 141:515–526
216. Charrasse S, Comunale F, Fortier M et al (2007) M-cadherin activates Rac1 GTPase through the Rho-GEF trio during myoblast fusion. Mol Biol Cell 18:1734–1743
217. Mukai A, Kurisaki T, Sato SB et al (2009) Dynamic clustering and dispersion of lipid rafts contribute to fusion competence of myogenic cells. Exp Cell Res 315:3052–3063
218. Hsiao SP, Chen SL (2010) Myogenic regulatory factors regulate M-cadherin expression by targeting its proximal promoter elements. Biochem J 428:223–233
219. Zeschnigk M, Kozian D, Kuch C et al (1995) Involvement of M-cadherin in terminal differentiation of skeletal muscle cells. J Cell Sci 108 (Pt 9):2973–2981

220. Kuch C, Winnekendonk D, Butz S et al (1997) M-cadherin-mediated cell adhesion and complex formation with the catenins in myogenic mouse cells. Exp Cell Res 232:331–338
221. Hollnagel A, Grund C, Franke WW et al (2002) The cell adhesion molecule M-cadherin is not essential for muscle development and regeneration. Mol Cell Biol 22:4760–4770
222. Mbalaviele G, Chen H, Boyce BF et al (1995) The role of cadherin in the generation of multinucleated osteoclasts from mononuclear precursors in murine marrow. J Clin Invest 95:2757–2765
223. Floridon C, Nielsen O, Holund B et al (2000) Localization of E-cadherin in villous, extravillous and vascular trophoblasts during intrauterine, ectopic and molar pregnancy. Mol Hum Reprod 6:943–950
224. MacCalman CD, Furth EE, Omigbodun A et al (1996) Regulated expression of cadherin-11 in human epithelial cells: a role for cadherin-11 in trophoblast-endometrium interactions? Dev Dyn 206:201–211
225. Getsios S, MacCalman CD (2003) Cadherin-11 modulates the terminal differentiation and fusion of human trophoblastic cells in vitro. Dev Biol 257:41–54
226. Le Naour F, Andre M, Boucheix C et al (2006) Membrane microdomains and proteomics: lessons from tetraspanin microdomains and comparison with lipid rafts. Proteomics 6:6447–6454
227. Ziyyat A, Rubinstein E, Monier-Gavelle F et al (2006) CD9 controls the formation of clusters that contain tetraspanins and the integrin alpha 6 beta 1, which are involved in human and mouse gamete fusion. J Cell Sci 119:416–424
228. Kaji K, Oda S, Shikano T et al (2000) The gamete fusion process is defective in eggs of Cd9-deficient mice. Nat Genet 24:279–282
229. Le Naour F, Rubinstein E, Jasmin C et al (2000) Severely reduced female fertility in CD9-deficient mice. Science 287:319–321
230. Miyado K, Yamada G, Yamada S et al (2000) Requirement of CD9 on the egg plasma membrane for fertilization. Science 287:321–324
231. Rubinstein E, Ziyyat A, Prenant M et al (2006) Reduced fertility of female mice lacking CD81. Dev Biol 290: 351–358
232. Kaji K, Oda S, Miyazaki S et al (2002) Infertility of CD9-deficient mouse eggs is reversed by mouse CD9, human CD9, or mouse CD81; polyadenylated mRNA injection developed for molecular analysis of sperm-egg fusion. Dev Biol 247:327–334
233. Zhu GZ, Miller BJ, Boucheix C et al (2002) Residues SFQ (173-175) in the large extracellular loop of CD9 are required for gamete fusion. Development 129:1995–2002
234. Runge KE, Evans JE, He ZY et al (2007) Oocyte CD9 is enriched on the microvillar membrane and required for normal microvillar shape and distribution. Dev Biol 304:317–325
235. Takeda Y, Tachibana I, Miyado K et al (2003) Tetraspanins CD9 and CD81 function to prevent the fusion of mononuclear phagocytes. J Cell Biol 161:945–956
236. Parthasarathy V, Martin F, Higginbottom A et al (2009) Distinct roles for tetraspanins CD9, CD63 and CD81 in the formation of multinucleated giant cells. Immunology 127:237–248
237. Ishii M, Iwai K, Koike M et al (2006) RANKL-induced expression of tetraspanin CD9 in lipid raft membrane microdomain is essential for cell fusion during osteoclastogenesis. J Bone Miner Res 21:965–976
238. Tachibana I, Hemler ME (1999) Role of transmembrane 4 superfamily (TM4SF) proteins CD9 and CD81 in muscle cell fusion and myotube maintenance. J Cell Biol 146:893–904
239. Brzoska E, Bello V, Darribere T et al (2006) Integrin alpha3 subunit participates in myoblast adhesion and fusion in vitro. Differentiation 74:105–118
240. Puissegur MP, Lay G, Gilleron M et al (2007) Mycobacterial lipomannan induces granuloma macrophage fusion via a TLR2-dependent, ADAM9- and beta1 integrin-mediated pathway. J Immunol 178:3161–3169
241. McNally AK, Anderson JM (2002) Beta1 and beta2 integrins mediate adhesion during macrophage fusion and multinucleated foreign body giant cell formation. Am J Pathol 160:621–630
242. Coonrod SA, Naaby-Hansen S, Shetty J et al (1999) Treatment of mouse oocytes with PI-PLC releases 70-kDa (pI 5) and 35- to 45-kDa (pI 5.5) protein clusters from the egg surface and inhibits sperm-oolemma binding and fusion. Dev Biol 207:334–349
243. Alfieri JA, Martin AD, Takeda J et al (2003) Infertility in female mice with an oocyte-specific knockout of GPI-anchored proteins. J Cell Sci 116:2149–2155
244. Tousseyn T, Jorissen E, Reiss K et al (2006) (Make) stick and cut loose–disintegrin metalloproteases in development and disease. Birth Defects Res C Embryo Today 78:24–46
245. Kurisaki T, Masuda A, Sudo K et al (2003) Phenotypic analysis of Meltrin alpha (ADAM12)-deficient mice: involvement of Meltrin alpha in adipogenesis and myogenesis. Mol Cell Biol 23:55–61
246. Kukita T, Wada N, Kukita A et al (2004) RANKL-induced DC-STAMP is essential for osteoclastogenesis. J Exp Med 200:941–946
247. Mensah KA, Ritchlin CT, Schwarz EM (2010) RANKL induces heterogeneous DC-STAMP(lo) and DC-STAMP(hi) osteoclast precursors of which the DC-STAMP(lo) precursors are the master fusogens. J Cell Physiol 223:76–83

248. Hotokezaka H, Sakai E, Ohara N et al (2007) Molecular analysis of RANKL-independent cell fusion of osteoclast-like cells induced by TNF-alpha, lipopolysaccharide, or peptidoglycan. J Cell Biochem 101:122–134
249. Yang M, Birnbaum MJ, MacKay CA et al (2008) Osteoclast stimulatory transmembrane protein (OC-STAMP), a novel protein induced by RANKL that promotes osteoclast differentiation. J Cell Physiol 215:497–505
250. Sapir A, Avinoam O, Podbilewicz B et al (2008) Viral and developmental cell fusion mechanisms: conservation and divergence. Dev Cell 14:11–21
251. Oren-Suissa M, Podbilewicz B (2007) Cell fusion during development. Trends Cell Biol 17:537–546
252. Chernomordik LV, Zimmerberg J, Kozlov MM (2006) Membranes of the world unite! J Cell Biol 175:201–207
253. Jahn R, Lang T, Sudhof TC (2003) Membrane fusion. Cell 112:519–533
254. Costa M, Raich W, Agbunag C et al (1998) A putative catenin-cadherin system mediates morphogenesis of the Caenorhabditis elegans embryo. J Cell Biol 141:297–308
255. Ding M, Woo WM, Chisholm AD (2004) The cytoskeleton and epidermal morphogenesis in C. elegans. Exp Cell Res 301:84–90
256. McKeown C, Praitis V, Austin J (1998) sma-1 encodes a betaH-spectrin homolog required for Caenorhabditis elegans morphogenesis. Development 125:2087–2098
257. Duan R, Gallagher PJ (2009) Dependence of myoblast fusion on a cortical actin wall and nonmuscle myosin IIA. Dev Biol 325:374–385
258. Swailes NT, Colegrave M, Knight PJ et al (2006) Non-muscle myosins 2A and 2B drive changes in cell morphology that occur as myoblasts align and fuse. J Cell Sci 119:3561–3570
259. Chen A, Leikina E, Melikov K et al (2008) Fusion-pore expansion during syncytium formation is restricted by an actin network. J Cell Sci 121:3619–3628
260. Richard JP, Leikina E, Chernomordik LV (2009) Cytoskeleton reorganization in influenza hemagglutinin-initiated syncytium formation. Biochim Biophys Acta 1788:450–457
261. Wurth MA, Schowalter RM, Smith EC et al (2010) The actin cytoskeleton inhibits pore expansion during PIV5 fusion protein-promoted cell-cell fusion. Virology 404:117–126
262. Berger S, Schafer G, Kesper DA et al (2008) WASP and SCAR have distinct roles in activating the Arp2/3 complex during myoblast fusion. J Cell Sci 121:1303–1313
263. Gildor B, Massarwa R, Shilo BZ et al (2009) The SCAR and WASp nucleation-promoting factors act sequentially to mediate Drosophila myoblast fusion. EMBO Rep 10:1043–1050
264. Massarwa R, Carmon S, Shilo BZ et al (2007) WIP/WASp-based actin-polymerization machinery is essential for myoblast fusion in Drosophila. Dev Cell 12:557–569
265. Pollard TD, Borisy GG (2003) Cellular motility driven by assembly and disassembly of actin filaments. Cell 112:453–465
266. Saarikangas J, Zhao H, Lappalainen P (2010) Regulation of the actin cytoskeleton-plasma membrane interplay by phosphoinositides. Physiol Rev 90:259–289
267. Harmon B, Campbell N, Ratner L (2010) Role of Abl kinase and the Wave2 signaling complex in HIV-1 entry at a post-hemifusion step. PLoS Pathog 6:e1000956
268. Laurin M, Fradet N, Blangy A et al (2008) The atypical Rac activator Dock180 (Dock1) regulates myoblast fusion in vivo. Proc Natl Acad Sci USA 105:15446–15451
269. Pajcini KV, Pomerantz JH, Alkan O et al (2008) Myoblasts and macrophages share molecular components that contribute to cell-cell fusion. J Cell Biol 180:1005–1019
270. O'Brien SP, Seipel K, Medley QG et al (2000) Skeletal muscle deformity and neuronal disorder in Trio exchange factor-deficient mouse embryos. Proc Natl Acad Sci USA 97:12074–12078
271. Chen EH, Pryce BA, Tzeng JA et al (2003) Control of myoblast fusion by a guanine nucleotide exchange factor, loner, and its effector ARF6. Cell 114:751–762
272. Dunphy JL, Moravec R, Ly K et al (2006) The Arf6 GEF GEP100/BRAG2 regulates cell adhesion by controlling endocytosis of beta1 integrins. Curr Biol 16:315–320
273. Faix J, Breitsprecher D, Stradal TE et al (2009) Filopodia: Complex models for simple rods. Int J Biochem Cell Biol 41:1656–1664
274. Lingwood D, Simons K (2010) Lipid rafts as a membrane-organizing principle. Science 327:46–50
275. Prives J, Shinitzky M (1977) Increased membrane fluidity precedes fusion of muscle cells. Nature 268:761–763
276. Sekiya T, Takenawa T, Nozawa Y (1984) Reorganization of membrane cholesterol during membrane fusion in myogenesis in vitro: a study using the filipin-cholesterol complex. Cell Struct Funct 9:143–155
277. Waheed AA, Freed EO (2009) Lipids and membrane microdomains in HIV-1 replication. Virus Res 143:162–176
278. Kielian M, Rey FA (2006) Virus membrane-fusion proteins: more than one way to make a hairpin. Nat Rev Microbiol 4:67–76
279. Pontow SE, Heyden NV, Wei S et al (2004) Actin cytoskeletal reorganizations and coreceptor-mediated activation of rac during human immunodeficiency virus-induced cell fusion. J Virol 78:7138–7147
280. Mohler WA, Shemer G, del Campo JJ et al (2002) The type I membrane protein EFF-1 is essential for developmental cell fusion. Dev Cell 2:355–362

281. Sapir A, Choi J, Leikina E et al (2007) AFF-1, a FOS-1-regulated fusogen, mediates fusion of the anchor cell in C. elegans. Dev Cell 12:683–698
282. Mi S, Lee X, Li X et al (2000) Syncytin is a captive retroviral envelope protein involved in human placental morphogenesis. Nature 403:785–789
283. Blaise S, de Parseval N, Benit L et al (2003) Genomewide screening for fusogenic human endogenous retrovirus envelopes identifies syncytin 2, a gene conserved on primate evolution. Proc Natl Acad Sci USA 100: 13013–13018
284. Frendo JL, Olivier D, Cheynet V et al (2003) Direct involvement of HERV-W Env glycoprotein in human trophoblast cell fusion and differentiation. Mol Cell Biol 23:3566–3574
285. Blond JL, Lavillette D, Cheynet V et al (2000) An envelope glycoprotein of the human endogenous retrovirus HERV-W is expressed in the human placenta and fuses cells expressing the type D mammalian retrovirus receptor. J Virol 74:3321–3329
286. Marin M, Lavillette D, Kelly SM et al (2003) N-linked glycosylation and sequence changes in a critical negative control region of the ASCT1 and ASCT2 neutral amino acid transporters determine their retroviral receptor functions. J Virol 77:2936–2945
287. Sun QY (2003) Cellular and molecular mechanisms leading to cortical reaction and polyspermy block in mammalian eggs. Microsc Res Tech 61:342–348
288. Castellucci M, Kaufmann P, Bischof P (1990) Extracellular matrix influences hormone and protein production by human chorionic villi. Cell Tissue Res 262:135–142
289. Demonbreun AR, Posey AD, Heretis K et al (2010) Myoferlin is required for insulin-like growth factor response and muscle growth. FASEB J 24:1284–1295
290. Doherty KR, Demonbreun AR, Wallace GQ et al (2008) The endocytic recycling protein EHD2 interacts with myoferlin to regulate myoblast fusion. J Biol Chem 283:20252–20260
291. Doherty KR, Cave A, Davis DB et al (2005) Normal myoblast fusion requires myoferlin. Development 132: 5565–5575
292. Demonbreun AR, Lapidos KA, Heretis K et al (2010) Myoferlin regulation by NFAT in muscle injury, regeneration and repair. J Cell Sci 123:2413–2422
293. Adams JM, Cory S (1998) The Bcl-2 protein family: arbiters of cell survival. Science 281:1322–1326
294. Sakuragi N, Matsuo H, Coukos G et al (1994) Differentiation-dependent expression of the BCL-2 proto-oncogene in the human trophoblast lineage. J Soc Gynecol Investig 1:164–172
295. Chang DW, Xing Z, Pan Y et al (2002) c-FLIP(L) is a dual function regulator for caspase-8 activation and CD95-mediated apoptosis. EMBO J 21:3704–3714
296. Ka H, Hunt JS (2006) FLICE-inhibitory protein: expression in early and late gestation human placentas. Placenta 27:626–634
297. Ogle BM, Butters KB, Plummer TB et al (2004) Spontaneous fusion of cells between species yields transdifferentiation and retroviral in vivo. FASEB J 18:548–550
298. Ferrari G, Cusella-De Angelis G, Coletta M et al (1998) Muscle regeneration by bone marrow-derived myogenic progenitors. Science 279:1528–1530
299. Wang X, Willenbring H, Akkari Y et al (2003) Cell fusion is the principal source of bone-marrow-derived hepatocytes. Nature 422:897–900

Chapter 5
Membrane Fusions During Mammalian Fertilization

Bart M. Gadella and Janice P. Evans

Abstract Successful completion of fertilization in mammals requires three different types of membrane fusion events. Firstly, the sperm cell will need to secrete its acrosome contents (acrosome exocytosis; also known as the acrosome reaction); this allows the sperm to penetrate the extracellular matrix of the oocyte (zona pellucida) and to reach the oocyte plasma membrane, the site of fertilization. Next the sperm cell will bind and fuse with the oocyte plasma membrane (also known as the oolemma), which is a different type of fusion in which two different cells fuse together. Finally, the fertilized oocyte needs to prevent polyspermic fertilization, or fertilization by more than one sperm. To this end, the oocyte secretes the contents of cortical granules by exocytotic fusions of these vesicles with the oocyte plasma membrane over the entire oocyte cell surface (also known as the cortical reaction or cortical granule exocytosis). The secreted cortical contents modify the zona pellucida, converting it to a state that is unreceptive to sperm, constituting a block to polyspermy. In addition, there is a block at the level of the oolemma (also known as the membrane block to polyspermy).

5.1 Introduction

Fertilization of the oocyte involves three membrane fusion events [1] namely, (1) a preparative series of secretion membrane fusions at the apical sperm surface known as acrosome exocytosis [2]. The membrane fusions are induced when the sperm cell binds to specific zona binding proteins at the sperm surface [3–7]. The acrosome exocytosis is a multipoint membrane fusion event between the sperm plasma membrane and the outer acrosomal membrane (see Fig. 5.1 [8, 9]) and the exposed acrosomal content is required for sperm to penetrate the zona pellucida [10–12]. This so-called zona drilling effectively takes place because the sperm at this stage also has acquired hyperactivated motility [13]. (2) After zona penetration the sperm enters the perivitelline space where it can bind and fuse with the oocyte plasma membrane [14, 15]. This is the actual fertilization fusion in which the contents of the sperm are delivered into the oocyte cytoplasm. The plasma membrane of the equatorial segment (see Fig. 5.1) is the site where proteins are located that orchestrate sperm-oocyte binding and fusion [16]. (3) In order to prevent polyspermy the oocyte has to activate defense systems to block redundant sperm-oocyte fusion [17]. To this end the first fertilizing sperm delivers activation factors

B.M. Gadella (✉)
Departments of Biochemistry and Cell Biology and of Farm Animal Health, Research Program: Biology of Reproductive Cells, Faculty of Veterinary Medicine, Utrecht University, Utrecht, The Netherlands
e-mail: b.m.gadella@uu.nl

T. Dittmar, K.S. Zänker (eds.), *Cell Fusion in Health and Disease*, Advances in Experimental Medicine and Biology 713, DOI 10.1007/978-94-007-0763-4_5, © Springer Science+Business Media B.V. 2011

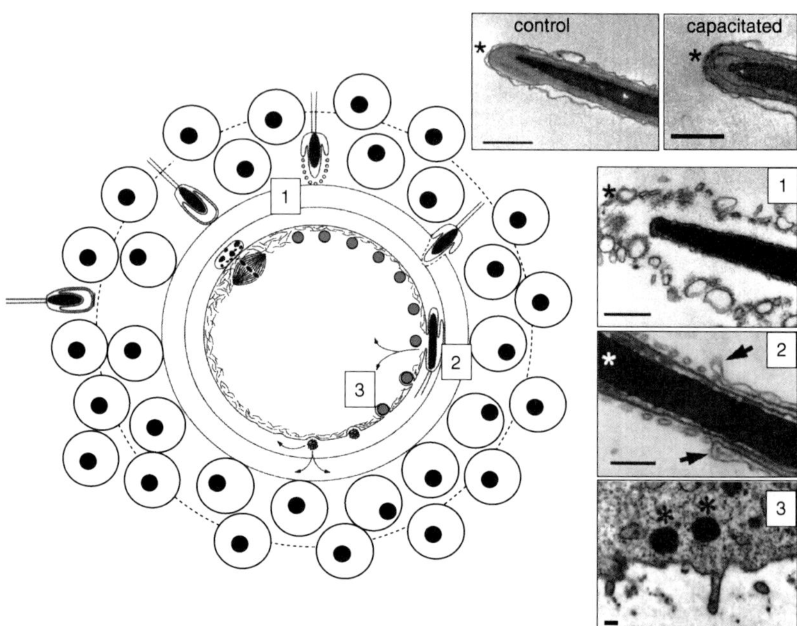

Fig. 5.1 The proposed sequence of events around the three fusions involved in monospermic fertilization. Sperm that have entered the oviduct will shed off decapacitation factors that were adhered peripherally to the sperm surface. During this process the acrosome is docked to the sperm plasma membrane and at the docked area the formation of high affinity zona binding complexes are formed [9, 35]. The control (control) versus capacitated (capacitated) sperm show the very close apposition of the sperm plasma membrane with the outer acrosomal membrane a feature emerging at the apical tip of the sperm head (*astrix*). It is not clear whether the resulting sperm which can associate with the extracellular matrix of the expanded cumulus mass surrounding the unfertilized oocyte induces some early steps of acrosomal fusion (proposed by 27) or that acrosome intact sperm are penetrating through the cumulus by the use of hyperactivated motility in combination with surface proteins [116]. Oviductal secreted proteins are also reported to be important for cumulus and zona pellucida properties [29]. *1*. The recognition of the zona pellucida (primary zona binding to ZP1/ZP3/ZP4) and subsequent initiation of the acrosome reaction (or of the acute secretory phase of it) which is induced by the zona pellucida. The unique multipointfusion of one organelle with the sperm surface generates mixed vesicles at the apical side of the sperm head. *2*. The acrosome reaction causes local modifications of zona proteins and the hyperactivated sperm can penetrate this structure due to secondary zona binding (to ZP2 and ZP3). The surface of the penetrating sperm will be further remodeled and this probably serves to enable the fertilization fusion [30]. Note that the equatorial segment of the sperm head (indicated with *arrows*) has remained resistant to exocytotic fusion. The plasma membrane and the outer acrosomal membrane have become continuous at this point. This needs to be protected from the acrosome reaction as it specifically contains the machinery to fertilize the oocyte. *3*. After the fertilization fusion the cortical reaction (induced by soluble sperm factors now diffusing into the oocyte cortex) causes an overall coating of the oolemma as well as the hardening of the zona pellucida by chemically altering zona proteins. The cleavage of ZP2 and ZP3 appears to be particularly instrumental for the release of sperm from the zona pullicida and to elicit an efficient block to polyspermy [23]. At the time of ovulation the MII phase oocytes have their cortical granules stored just under the oocytes plasma membrane. The fertilization fusion is followed up by a massive series of single point exocytotic fusions of the cortical granules (indicated with *asterix*). The distance bar indicates 50 nm. Panels with transmission electron microscopy micrographs are modified from [9, 113], the line drawing is modified from [7]

into the oocyte cytoplasm. The signaling cascade activated by these factors induce fusions of secretory granules that tightly located under the oocyte plasma membrane (known as cortical granules) [18, 19]. After the secretion of the cortical content into the perivitelline space, this will lead to modifications of the zona pellucida structure [20–23]. In some species, this has been characterized as "zona hardening" defined as resistence to proteolytic digestion resulting in a zona pellucida that cannot be penetrated by acrosome-reacted sperm, and also cannot be recognized by acrosome-intact sperm [24–26].

Therefore, successful fertilization of an oocyte depends on three independent and quite differently organized membrane fusion events. The current understanding of membrane fusion and fertilization will be overviewed in this chapter.

5.2 Surface Remodeling of Gametes Prior to Zona Binding

5.2.1 The Cumulus-Oocyte Complex in the Oviduct

In mammals fertilization takes place in the oviduct near the ampulla region. The oocyte enters this site after ovulation and is surrounded by a thick (7 μm) extracellular matrix called the zona pellucida [27] and by a multi-cellular layer of cumulus mass (cells and extracellular matrix material, see Fig. 5.1). The oviduct probably modifies these extracellular structures to some extent [25, 28, 29].

5.2.2 Sperm Cell Surface Remodeling

Before the sperm cells enter the oviduct they have already passed a lengthy trajectory of transport and coinciding post-testicular modifications [30]. Sperm cells that are released in the testis from the Sertoli cells into the lumen of seminiferous tubules have shut down transcription and translation processes as well as membrane recycling (no endocytosis or exocytosis [31]). With respect to sperm-zona binding it is of special interest the identified transmembrane proteins with affinity for the zona pellucida originate from sperm forming precursor cells in the testis [32–35]. However, more recent approaches revealed that additional proteins are attached to the sperm surface (especially the epididymis, where sperm cells further mature and acquire motility) that serve to bind the zona pellucida [35–40]. After the ejaculation of sperm a specific coat of proteins containing decapacitation factors serve to stabilize sperm [41–45]. This is required to allow maximum sperm survival during their lengthy transport through the female genital tract (cervix, uterus) and to reach the oviduct intact. At this site sperm release their protective coat and become capacitated (i.e. capable to fertilize because they can recognize the zona pellucida). Therefore, the sperm regain fertilization capacity they originally had in the cauda epididymis (prior to ejaculation) and can induce the acrosome reaction after eventual zona pellucida binding. This capacitation process is guided in the oviduct by sperm-oviduct epithelial interactions (for review see [13]). After a certain period the sperm is released from the oviduct epithelial cell and has hyperactivated motility characteristics and demonstrates efficient zona binding behaviour.

During in vitro fertilization the capacitation of ejaculated sperm is mimicked by washing sperm through discontinuous density gradients (to remove decapacitation factors) and to incubate sperm for a couple of hours in an in vitro capacitation medium which mimics the ionic and metabolic composition of oviductal fluid (also known as synthetic oviductal fluid (SOF)) [7, 20, 46]. Mammalian sperm becomes activated by three principle capacitation factors namely (1) bicarbonate which activates adenylate cyclase/protein kinase A and tyrosine kinase signaling pathways, (2) albumin which specifically extracts sterols from the sperm plasma membrane and (3) extracellular calcium allowing Ca^{2+} mediated signaling cascades [46, 47]. For some species additional glycosaminoglycans are required to remove persistent decapacitation factors from the sperm surface [48]. Taken together sperm capacitation results in the induction of glycolysis in the sperm tail required for the hyperactivated motility (more instant and local production of ATP in the lengthy sperm tail which does not contain mitochondria [49–51]). In the sperm head it causes the redistribution of surface molecules. Most notably this results in the aggregation of lipid rafts and therein the formation of a functional zona pellucida binding protein complex [30, 46, 52–54]. This zona binding complex not only functionally allows sperm zona binding but also mediates the acrosome exocytosis after this binding (see Section 5.3).

5.3 Zona Binding and Initiation of the Acrosome Reaction

5.3.1 Zona Pellucida Contains Acrosome Exocytosis Inducing Binding Sites

Traditionally it was thought that sperm-zona binding is a simple ligand receptor like interaction in which one zona receptor (namely ZP3) binds to one sperm ligand [55]. This concept appears to be oversimplified. The sperm surface has recently been shown to bind to at least three of the four human zona proteins (namely ZP1, ZP3 and ZP4) and most likely the species-specific zona protein matrix quartenary native state is important for sperm recognition [7]. The sperm cell also binds to this zona protein matrix with multiple proteins, most likely organized into zona binding protein complexes. Some of the identified proteins may be required for the induction of the acrosome reaction. For instance the presence of a potassium channel [35] may indicate that zona binding could induce a K^+ dependent sperm membrane hyperpolarization which in turn allows the opening or a voltage dependent Ca^{2+} channel and by doing so cause elevated cytosolic Ca^{2+} levels required for initiation of acrosome exocytosis [56]. Beyond this, the presence of a phosphatase [35] may indicate that binding may activate specific signaling events that are required for the induction of the acrosome reaction.

5.3.2 Acrosome Exocytosis

Acrosome exocytosis itself is the result of SNARE interactions between the outer acrosomal membrane and the plasma membrane of the so-called pre-equatorial region of the sperm head [57, 58]. Remarkably the two membranes fuse with each other at this entire surface domain which encompasses more than half of the sperm head surface [8, 9]. The multipoint fusion secretion event results in the generation of mixed vesicles that contain acrosomal outer membrane and plasma membrane material. The remaining unfused acrosomal membranes (at the equatorial area of the acrosome and the acrosomal inner membrane covering the apical part of the sperm nucleus) now take over the surface function of the plasma membrane [15, 16]. The vesiculated part of the apical acrosome membrane and plasma membrane are removed from the sperm. The group of Gadella has studied how SNARE proteins are orchestrating this multiple membrane fusion event. In freshly ejaculated sperm SNARE interactions between the apical sperm plasma membrane and the outer acrosomal membrane are not yet established [9]. However, during sperm capacitation these two membranes become stably docked by the formation of a *trans* ternary SNARE complex of proteins from the sperm plasma membrane as well as from the outer acrosmal membrane. The complex consisted of syntaxin1, VAMP1 and SNAP23 in a 1:1 stoichiometry [9]. The docked membranes could even be isolated as bilamellar structures. Related to this stability the capacitated acrosome becomes docked but does not fuse with the plasma membrane. For the execution of the acrosome fusions additional Ca^{2+} entry (in vitro by use of Ca^{2+} ionophores, in vivo after zona binding) is required [9]. Diverse groups have shown that SNARE complex interacting proteins such as complexins [9, 59, 60], dynamins [61], Rab 3A [62], synaptotagmins [63], multi-PDZ domain protein MUPP1, Calmodulin and CaMKIIalpha [64, 65], Rab-2a, syntaxin binding proteins and Munc-18 (Tsai et al., unpublished results) have been discovered in sperm (see Fig. 5.2). When and how they interact with SNARE proteins and whether they are involved in stabilizing the *trans* SNARE complex or are involved in the Ca^{2+} conversion to *cis* complexes (thus eliciting the acrosome plasma membrane fusions) is matter of future research (see also Fig. 5.2).

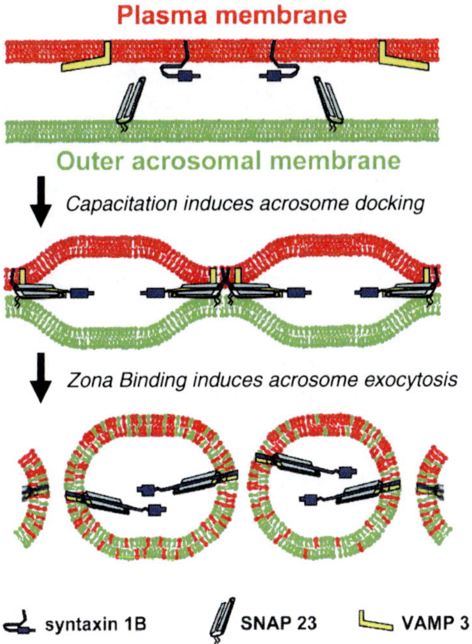

Fig. 5.2 Two step model for SNARE mediated acrosome exocytosis of the sperm. *Sperm capacitation* induces the stable docking of the sperm plasma membrane with the outer acrosomal membrane. The multiple docking of these two membranes does not lead to premature exocytosis. The identified interaction partners are for porcine sperm [9] but may differ between mammalian species. Two mechanisms have been described to stabilize the *trans* ternary SNARE protein complexes. (i) During sperm capacitation an aggregation of lipid rafts at the apical ridge area of the sperm head. This is the site where the sperm binds to the zona pellucida and where the acrosome exocytosis as a response of that binding is initiated [46]. In that area within the aggregating lipid rafts MUPP1/CaMKIIα have been reported to interact with the *trans* ternary SNARE protein complex and this association functions as a fusion clamp [64, 65]. (ii) The important factor in mouse sperm is the phosphorylated form of synaptotagmin, which appears to be important for preventing the acrosome exocytosis [63]. Beyond these factors also complexin and dynamin are interacting with the trans SNARE complex [59–61] but are not able at this stage to induce the *trans* to *cis* conformational shift of the complex. Munc18b is also associated to the *trans* ternary SNARE complex (unpublished observation). It is not clear whether or not Rab3A [62] is already associated to the *trans* SNARE complex at this stage. The current concept is that complexin, Munc18b (which can bind to syntaxin) are stabilizing the *trans* SNARE protein complex and prevent spontaneous acrosome exocytosis. *Zona binding* evokes Ca^{2+} entry (see Section 5.3.1) and this causes both the dissociation of the MUPP1/CaMKII fusion clamp [64] and a calcineurin-mediated dephosphorylation of synaptotagmin VI [63]. The dephosphorylation of synaptotagmins also appear to be essential for the acrosome reaction [59, 62, 117, 63] (unpublished observation). The role of Rab3A [62] and Rab2A (unpublished observation) in the formation of the *cis* ternary SNARE complex conformation is not yet clear. It is possible that these GTPase forming proteins were already recruited during sperm capacitation to the *trans* complex and by the zona-induced changes help to create the *cis* configuration either by dissociation (of Munc 18b) and the coinciding intrinsic Ca^{2+} sensing properties of the dephosphorylated synaptotagmin with the aid of Rab3A [62] or Rab2A. Figure is modified from [9]

5.4 Zona Penetration After the Acrosome Reaction

A result of the above described acrosome exocytosis (Section 5.3) is exposure of the acrosomal content at the front surface of the sperm head where primary zona binding initiated the acrosome reaction. The now exposed intra-acrosomal layer of proteins consists of an array of proteins that interact with the zona pellucida (for review see [7]). This so-called secondary zona binding was thought to be specific to ZP2 [66–68] but recently it has been shown that the intra-acrosomal protein sp56 binds to ZP3 [4, 12].

Beyond the more massive secondary zona binding (compared to the primary zona recognition binding at the sperm surface), the exposed intra-acrosomal proteins also cause a local enzymatic cleavage of this network of 3–4 heavily glycosylated zona proteins [66, 69–75]. For a review on secondary zona binding proteins from the acrosome see [7]. Note that acrosome exocytosis and consequent zona drilling coincides with the generation of hyperactivated motility of sperm [13]. Together they form the pre-requisites for a recycling modus: secondary zona binding followed by local digestion of the ZP network, zona penetration and subsequent rebinding to the ZP. Thus acrosome exocytosis enables the sperm to reach the perivitelline space and exposes the inner acrosomal membrane and the equatorial segment of the sperm head (i.e., that area where the acrosome outer membrane was inert to fuse with the sperm plasma membrane), which is required for the sperm to be capable of fusing with the oocyte membrane (see Section 5.5).

5.5 Gamete Membrane Fusion and the Oocyte-to-Embryo Transition

Sperm-oocyte fusion is one of the best-known extracellular membrane fusion events, and yet it is one of the most poorly understood. Especially in comparison to other types of extracellular fusion events and to SNARE-mediated vesicle fusion events, relatively little is known about the mechanisms underlying sperm-oocyte fusion in any species, particularly mammalian. The identification of fusion-mediating factors in fertilization also has been difficult, but there are multiple possible explanations for this. From the standpoint of genetic/knockout studies, perhaps there are multiple factors with substantial functional overlap, and this redundancy has made it difficult to identify gamete fusion-defective phenotypes. Alternatively, the fusion-mediating factors may play critical roles in other cell types, making it impossible to assess gamete fusion (e.g., embryonic or neonatal lethality) without use of a conditional knockout. From the standpoint of biochemistry or developing function-blocking antibodies as a means to identify these fusion-mediating factors, it is possible that these factors are few in number, unstable, and/or only transiently exposed. These considerations are also valid for proteins involved in sperm-zona interactions described in Section 4.3.1 and 4.4.

Only two proteins, the tetraspanin CD9 on the mouse oocyte and the immunoglobulin superfamily member IZUMO1 (previously known as Izumo) on mouse sperm, have been shown by gene knockout studies as being essential specifically for sperm-egg interaction (Fig. 5.3). Note: It is unclear if CD9 and/or different tetraspanins function in other species' oocytes in gamete membrane fusion [76]. Other mouse knockouts have less severe defects in gamete interactions or fertilization, or have multiple gamete function defects (e.g., [77]). The discovery of CD9's role in murine fertilization occurred rather serendipitously, when the knockout mouse lacking this member of the protein family was found to have greatly reduced female fertility. This is rather remarkable, since CD9 is expressed in numerous cell types in the body, but there is only an obvious phenotype with oocytes showing a significantly reduced ability to fuse with sperm [78–80]. The discovery of the role of IZUMO1 came as a result of persistence and hard work, with 17 years between the report of the function-blocking activity of the monoclonal antibody OBF13 on sperm-oocyte fusion [81] and the report of

Fig. 5.3 (continued) 14 (SAMP14; also known as sperm acrosome associated 4, SPACA4), SAMP32 (also known as SPACA1), and Sperm Lysosomal-Like Protein 1 (SLLP1; also known as SPACA3) [70, 135–137], all of which are novel proteins. Finally, zinc metalloprotease (MP) activity has been implicated by the finding that mouse sperm-egg fusion is reduced in the presence of various metalloprotease inhibitors [138]. Reagents that disrupt the action of enzymes that mediate thiol-disulfide exchanged in proteins (protein disulfide isomerases, PDIs) also reduce the incidence of sperm-egg fusion [139, 140]. The results with N-ethyl-maleimide (NEM) and 5,5′-dithiobis (2-nitro-benzoic acid) (DTNB) [139, 140] suggest that sulfhydryl groups may be a common element involved in fusion systems, as they are in vesicle fusion [141], and certain viral fusion events [142, 143]

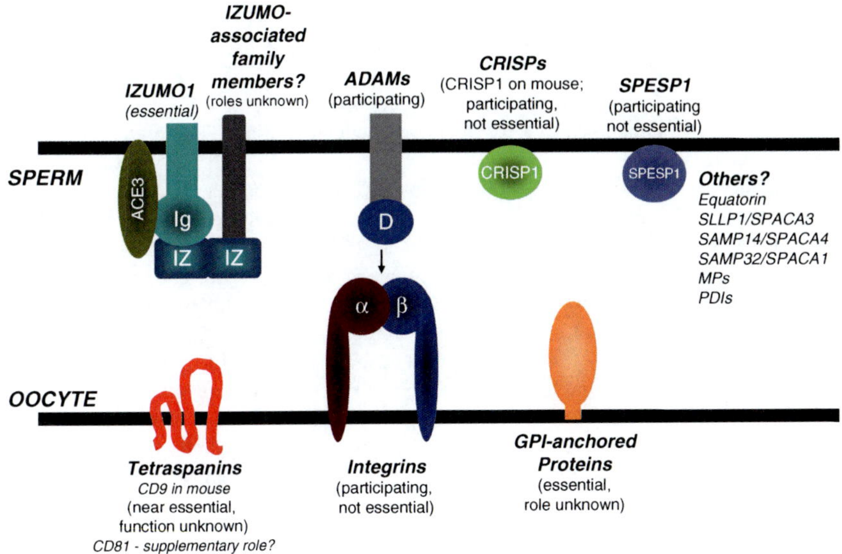

Fig. 5.3 Schematic diagram illustrating molecules implicated in gamete membrane binding and fusion. This diagram illustrates the molecules proposed to participate in sperm-oocyte membrane interactions (i.e., binding and/or fusion). CD9 is the major player identified thus far on the oocyte. $CD9^{-/-}$ females are severely subfertile (only 60% of $CD9^{-/-}$ females become pregnant, litters sizes are 75% smaller, and achieving these pregnancies takes nearly four times longer than it does for control mice [84]). IVF assays show that sperm binding to *CD9*$^{-/-}$ oocytes appears to be unaffected, but sperm rarely fuse [78, 79, 118]. Mouse egg CD9 is likely to function in conjunction with another tetraspanin, CD81; CD9/CD81 double knockout female mice are completely infertile, suggesting that CD9 and CD81 play complementary roles in fertilization [84]. GPI-anchored proteins on the oocyte are also implicated. Mice with oocytes deficient in GPI-anchored proteins are infertile, and these oocytes poorly support sperm-oocyte binding and fusion [86], but it remains unknown which GPI-anchored protein(s) are crucial and what role(s) oocyte GPI-anchored proteins could play. The last main candidates on oocytes are members of the integrin family. Integrins are heterodimeric membrane proteins, made up of an α and a β subunit, with 18 α subunits and eight β subunits combining to make at least 24 different combinations. α6, α3, and β1 in oocytes are not essential for fertility [119, 120], but in vitro studies of certain subunits have revealed defects in sperm-oocyte binding [121–123]. Oocytes with reduced amounts of α9 support sperm binding and fusion less well than do control eggs [123], in agreement with the finding that several ADAMs can interact with α9β1 [124]. Oocytes deficient in β1-deficient show defects in sperm-oocyte binding [121]. On the sperm, IZUMO1 has been shown to be essential for sperm-oocyte fusion [125]. IZUMO1 is member of the immunoglobulin superfamily (IgSF) proteins, and contains an immunoglobulin-like domain (Ig). IZUMO1 also has a ~150 amino acid domain that has been called the Izumo domain (IZ in the diagram), and this domain has been found in three other proteins [87]. Interestingly, IZUMO1 is associated with other Izumo domain proteins, although the functions of these in gamete fusion are not known. ACE3 (Angiotensin Converting Enzyme 3) is another IZUMO1-associated protein [82]; the *Ace3* knockout did not show any defects in male fertility or sperm function in vitro, although there is a slight abnormality in the localization of IZUMO1 [82]. Several sperm ADAMs have been implicated in sperm-oocyte interaction; while no single ADAM is essential, there appears a correlation between the ability of sperm to bind and fuse with the egg membrane and the levels of certain ADAM proteins (see [15] for more information). This suggests that ADAMs could function in redundant roles, consistent with the fact that ADAMs have similar adhesion-mediating motifs to interact with integrins via their disintegrin domain (D in the diagram) [124, 126–128]. Cysteine-rich Secretory Protein 1 was implicated in sperm-oocyte fusion by antibody studies in the 1980s [129, 130], and the *Crisp1* knockout was recently reported. Sperm from *Crisp1*$^{-/-}$ males show a modest decrease in sperm-oocyte fusion in in vitro fertilization assays, although male fertility appears normal [131]. Finally, as noted in the text, acrosome exocytosis exposes the inner acrosomal membrane and the equatorial segment of the sperm head, rendering the sperm capable of interacting with the oolemma. Thus, proteins in the equatorial segment of the sperm head have been of interest. SPESP1 (sperm equatorial segment protein 1; [132]) is a candidate, based on the finding that *Spesp1*$^{-/-}$ males produce slightly smaller litters than wild type controls (22%), and have sperm with reduced (although not completely deficient) ability to undergo sperm-oocyte fusion [77]. Other proteins associated with sperm-oocyte fusion and exposure or rearrangement after acrosome exocytosis include equatorin [133, 134], Sperm Acrosomal Membrane-Associated

the phenotype of the knockout [82]. Numerous other molecules have also been suggested to participate in mammalian gamete membrane interaction (sperm binding and/or sperm-oocyte membrane fusion; Fig. 5.3). Mouse knockouts have been made of several of these, and many have less dramatic phenotypes that the *Cd9*$^{-/-}$ and *Izumo1*$^{-/-}$ mice, namely often only partial loss of function in sperm-oocyte interaction (and sometimes little or no impairment of fertility). These are addressed in the figure legend for Fig. 5.3.

One theme that seems to be emerging in mammalian gamete membrane fusion that is conserved with other examples of membrane fusion is that membrane order, multimeric protein complexes, and the gamete surface proteome may prove to play critical roles. CD9 and other tetraspanins are known to function as organizers of membrane domains, known as tetraspanin-enriched microdomains [83]. Mouse egg CD9 is likely to function in conjunction with another tetraspanin, CD81. CD9/CD81 double knockout female mice are completely infertile, suggesting that CD9 and CD81 play complementary roles in fertilization [84]. Interestingly, glycosylphosphatidyl inositol (GPI)-anchored proteins in oocytes have been implicated by an oocyte-specific knockout of *Piga*, a subunit of an N-acetyl glucosaminyl transferase that participates in first steps of the synthesis of GPI-anchored proteins [85]; female mice with this oocyte-specific *Piga* knockout are infertile [86]. It is possible that the *Piga* deficiency and the resulting lack of GPI-anchored proteins in the oocyte membrane alters membrane composition and/or organization so that sperm interactions are not favored. GPI-anchored proteins are enriched in lipid microdomains, raising the possibility that the microdomain structure of the egg plasma membrane could be perturbed in the absence of GPI-anchored proteins. The importance of membrane order may also extend to sperm. IZUMO1 has recently been described to associate with other membrane proteins [87, 88]. Likewise, members of the ADAM (A Disintegrin and A Metalloprotease domain) family are other sperm proteins implicated in gamete membrane interactions, and the genetic deletion of one *Adam* can affect the expression of multiple ADAM proteins on the sperm surface [15, 89], and protein trafficking during spermatogenesis [90], suggestive of a role of ADAMs in sperm membrane order. Finally, another knockout, *Tssk6*, is defective in sperm-oocyte fusion and has an abnormality in IZUMO1 localization [91], also possibly indicative of aberrant membrane order.

5.6 The Membrane (Oolemma) Block to Polyspermy

5.6.1 Redundant Sperm Around the Fertilized Oocyte

Mammalian oocytes regulate their ability to interact with sperm, namely the membrane block to polyspermy, by altering the receptivity of the oolemma to sperm after fertilization. This was demonstrated by classic studies in which fertilized oocytes recovered from natural matings were found to have extra sperm in the perivitelline space, apparently unable to penetrate the oolemma [92–94]. The numbers of supernumerary perivitelline sperm vary by species, suggestive that there are differences in the reliance on the various polyspermy prevention mechanisms between different species. The oocytes of some species such as rabbit, pika, pocket gopher, and mole have tens to hundreds of sperm in the perivitelline space, suggestive of a highly effective membrane block and a relatively ineffective ZP block. Species in which perivitelline sperm are rare (dog, sheep, field vole) likely have a highly effective ZP block. Numerous species (including mouse, human, rat, guinea pig, cat, pig, cattle) appear to use both blocks to polyspermy; in these oocytes, one or two or up to $\sim$10 sperm are found in the perivitelline space of early zygotes [92–97].

5.6.2 Prevention of Polyspermy at the Oolemma

The basis of the membrane block to polyspermy – i.e., what is different about the zygote membrane that prevents additional sperm fusions – is not known. In mouse oocytes, this membrane transition occurs gradually; the membrane block is not yet established by 0.75 h post-insemination, but is established by 1.5 h post-insemination [17]. Experiments using fluorescent tags in mouse oocytes to track membrane lipids or protein diffusion suggest that fertilization-induced changes do occur, although such changes have not been well characterized [98, 99]. It has recently been shown that cortical tension is higher in zygotes than in unfertilized oocytes [100], although the exact role that this may play in the membrane block to polyspermy remains to be determined. The mechanism by which the membrane block is triggered also is an active area of investigation. One key finding is that this membrane block appears to be largely independent of cortical granule exocytosis, although it is possible the contents of the cortical granules may augment the membrane block, even if cortical granule exocytosis is not an essential component. Oocytes that are activated in ways that induce increased cytosolic Ca^{2+} concentration and the cortical reaction (calcium ionophore, strontium chloride, injection of a soluble sperm extract, or by fertilization by intracytoplasmic sperm injection) maintain membranes that are receptive to sperm [17, 101–104]. The failure of ICSI-generated embryos to establish a membrane block to polyspermy has been interpreted to indicate that sperm membrane incorporation into the oolemma is linked with membrane block establishment [103], or that membrane block establishment occurs as a result of changes in the oocyte occurring with the process of gamete fusion [104]. In the mouse, the sperm head surface area is only ~0.14% of the oocyte surface area, and thus a membrane block mechanism involving dilution of the oolemma with sperm membrane seems unlikely. Instead, establishment of the membrane block may involve signaling occurring with gamete fusion, although the injection of a soluble sperm extract fails to trigger membrane block establishment, indicating that membrane block establishment is not solely controlled by the sperm-induced increase in cytosolic Ca^{2+} [104].

5.7 Cortical Reaction and the Zona Pellucida Block to Polyspermic Fertilization

5.7.1 Cortical Granules Content Can Modify the Zona Pellucida Structure

The concern for a just fertilized oocyte is to prevent additional sperm to bind to and fuse with the oolemma (polyspemy). The just-fused first sperm introduces soluble cytosolic factors like phospholipase C zeta into the oocyte [105]. These factors induce intracellular Ca^{2+} events and the oocyte plasma membrane depolarization and both are triggers for the cortical granule exocytosis [18]. The secretory granules that reside in the cortex (the area just under the oocyte plasma membrane) fuse with the oocyte plasma membrane and the content of the granules is released into the perivitelline space [106, 107]. Although the contents of the cortical granules are very poorly characterized [106, 108–111], it is known that the release of these materials results in the cleavage of ZP2 and ZP3 into the truncated $ZP2_f$ and $ZP3_f$ forms [20–22]. These alterations are associated with zona hardening (defined as resistance to proteolysis in certain in vitro assays). As a result sperm stop penetrating the hardened zona and do not show affinity for the zona pellucida.

5.7.2 Maturation Dependent Exocytotic Fusion Machinery of the Cortical Reaction

Cortical granule exocytosis resembles to some extent the exocytosis of the acrosome (Section 5.3.2) in that in both cases secretory granule exocytosis takes place at the surface of a gamete. The main difference between the two exocytosis events is that cortical granule exocytosis is a series of single point fusion events of many cortical granules with the oolemma, while the acrosome exocytosis is a multiple point fusion event of one acrosome with the sperm plasma membrane. The majority of secretory granules migrate towards the oocyte plasma membrane during pre-ovulatory maturation of oocytes somewhere between the germinal vesicle stage (GV, oocytes that are arrested at the prophase of meiosis I) and the arrested stage at metaphase of meiosis II (MII) [107, 112]. Nevertheless, some of the secretory granules already reside in the cortex of GV oocytes but fail to be competent to fuse with the oocyte surface at that stage which appears to depent on too low activity of calcium/calmodulin dependent kinase II activity which becomes activated in MII oocytes [113] and related to this MAPK activity seems to be inviolved in activating the cortical granule exocytosis as well [114]. In addition we have shown recently that cortical granules become docked at the oocytes plasma membrane during these two meiotic maturation stages [107]. The SNARE proteins SNAP23, VAMP-1 and syntaxin 2 are involved and probably form a similar trimeric *trans* complex prior to fertilization (which therefore is analogous to acrosome docking during sperm capacitation). The docked cortical granules are also decorated with complexins (probably stabilizing the SNARE complex) and with clathrin (Fig. 5.4). After fertilization the Ca^{2+} mobilization and related signaling and cytoskeletal rearrangements [18, 115] cause the *trans* to *cis* ternary configuration of the trimeric SNARE complex is turned on and explains the cortical reaction. Remarkably complexin and clathrin dissociate from the cortex and relocate to intracellular structures and this is probably required to re-establish endocytosis and membrane recycling [107]. In case of polyspermic fertilization of pig oocytes, we observed a normal cortical reaction but this is not followed by a release of clathrin which may indicate that at the level of the oolemma fusion the inhibition of endocytosis may have a relationship with the fusion properties of the oolemma [107]. This observation confirms the finding that the polyspermy block is at least not immediately dependent on cortical exocytosis (see Section 5.6.2). As noted above, little to nothing is known about the content of the cortical granules although in general its content should resemble that of other secretory vesicles. Like the acrosomal enzymes also the cortical granule enzymes are capable to alter the zona pellucida structure. But the cortical granule enzymes differ from acrosomal enzymes in that acrosomal enzymes digest the zona pellucida matrix locally (allowing sperm penetration) whereas the cortical granule enzymes make the zona pellucida impermeable for acrosome-reacted sperm (the so called slow polyspermy block).

5.8 Conclusion

This chapter provides an overview about the three fusion events involved in mammalian fertilization. Acrosome exocytosis is first hurdle, allowing the sperm to fertilize the oocyte by resulting in localized digestion of the zona pellucida and thus permitting the sperm to gain access to the oocyte plasma membrane. The first sperm to interact with the oocyte plasma membrane and to execute actual fertilization fusion delivers its male haploid genome to the oocyte. This sperm also activates the oocyte, leading to the oocyte-to-embryo transition, including the establishment of blocks to polyspermy, with the zona pellucida block being mediated by the third membrane fusion event of fertilization (i.e., cortical granule exocytosis). Both the secretion of the acrosome and the cortical granules can be considered as classical exocytotic events in which trimeric SNARE complex formation cause vesicle docking to the gametes plasma membrane and Ca^{2+}-dependent configuration to a *cis* trimeric SNARE complex causes exocytosis. However, the acrosome exocytosis is unique in showing multiple fusions of only one large secretory vesicle with the sperm plasma membrane, whereas the cortical granule exocytosis likely initiaties with single point fusions of an array of cortical vesicles over the entire the oocyte plasma membrane. The regulation machinery for vesicle docking and fusion with the plasma membrane for both gametes needs to be studied into greater detail. In between the two exocytotic membrane fusions lays the actual sperm-oocyte membrane fusion, which remains to be poorly understood. The fusion between two gamete plasma membranes may share similarities to other extracellular fusion events (e.g., myofibril formation, syncytia-forming transformed cancer cells in culture), or viral fusion.

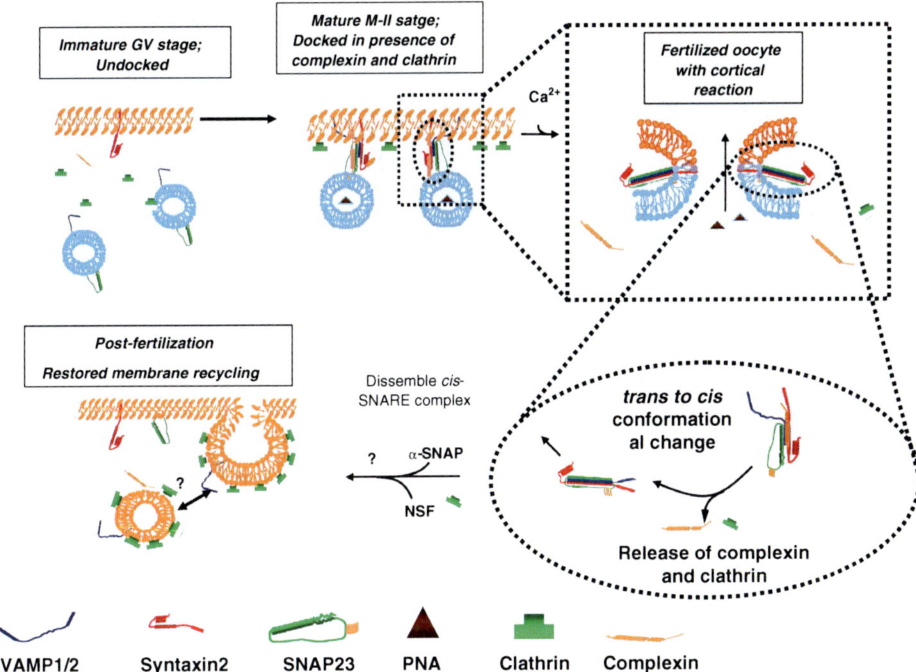

Fig. 5.4 Model for SNARE mediated cortical exocytosis of the fertilized oocyte. At the germinal vesicle stage (GV) the majority of the secretory granules (*blue* vesicles) are not residing in the cortex of the oocyte and are therefore do not interact with the oocyte plasma membrane (*yellow* membrane). During later meiotic maturation at the metaphase II stage (MII) about all granules have migrated towards the cortex region and strong co-localization of oocyte plasma membrane and cortical granule content has been demonstrated. Therefore, it is possible that the cortical granules are docked to the oocyte plasma membrane and that this interaction is stabilized in an analogous way to that of the acrosome. This would explain why the premature cortical exocytosis is not observed. Complexin and clathrin are at this stage exclusively present in the area where the cortical granules and the plasma membrane are interacting. It is noteworthy to mention that membrane recycling is silenced in MII oocytes and no exocytosis as well as endocytosis can be observed. The concentration of complexin and clathrin at the cortex may well have to do with this. Once the sperm has fertilized the oocyte, it will introduce oocyte activation factors into the oocyte that allow Ca^{2+} mobilization (see Section 5.7.1) and this results in at least the dissociation of clathrin and complexin. This dissociation has not been found in polyspermic fertilized oocytes that still retain membrane recycling blocked at the oocyte surface. Perhaps this explains why redundant sperm can cause polyspermic fertilization in those oocytes. In the monospermic oocytes the release of complexin and clathrin coincides with the onset of membrane recycling by means of endocytosis and exocytosis and thus to further embryo development. This figure summarizes the studies of Tsai et al. [107]

Taken together three membrane fusion events serve to achieve optimal monospermic fertilization of mammalian oocytes. However, it is surprising that, despite of decades of research, the actual molecular understanding of the fertilization of the mammalian oocyte is still relatively limited.

References

1. Wassarman PM, Litscher ES (2008) Mammalian fertilization is dependent on multiple membrane fusion events. Methods Mol Biol 475:99–113
2. Florman HM, Jungnickel MK, Sutton KA (2008) Regulating the acrosome reaction. Int J Dev Biol 52:503–510
3. Chiu PC, Wong BS, Chung MK et al (2008) Effects of native human zona pellucida glycoproteins 3 and 4 on acrosome reaction and zona pellucida binding of human spermatozoa. Biol Reprod 79:869–877

4. Buffone MG, Rodriguez-Miranda E, Storey BT et al (2009) Acrosomal exocytosis of mouse sperm progresses in a consistent direction in response to zona pellucida. J Cell Physiol 220:611–620
5. Gupta SK, Bansal P, Ganguly A et al (2009) Human zona pellucida glycoproteins: functional relevance during fertilization. J Reprod Immunol 83:50–55
6. Ganguly A, Bukovsky A, Sharma RK et al (2010) In humans, zona pellucida glycoprotein-1 binds to spermatozoa and induces acrosomal exocytosis. Hum Reprod 25:1643–1656
7. Gadella BM (2010) Interaction of sperm with the zona pellucida. Soc Reprod Fertil 67:267–287
8. Zanetti N, Mayorga LS (2009) Acrosomal swelling and membrane docking are required for hybrid vesicle formation during the human sperm acrosome reaction. Biol Reprod 81:396–405
9. Tsai PS, Garcia-Gil N, van Haeften T et al (2010) How pig sperm prepares to fertilize: stable acrosome docking to the plasma membrane. PLoS One 18:e11204
10. Kim KS, Gerton GL (2003) Differential release of soluble and matrix components: evidence for intermediate states of secretion during spontaneous acrosomal exocytosis in mouse sperm. Dev Biol 264:141–152
11. Kim E, Yamashita M, Kimura M et al (2008) Sperm penetration through cumulus mass and zona pellucida. Int J Dev Biol 52:677–682
12. Buffone MG, Foster JA, Gerton GL (2008) The role of the acrosomal matrix in fertilization. Int J Dev Biol 52:511–522
13. Suarez SS (2008) Regulation of sperm storage and movement in the mammalian oviduct. Int J Dev Biol 52: 455–462
14. Primakoff P, Myles DG (2007) Cell-cell membrane fusion during mammalian fertilization. FEBS Lett 581: 2174–2180
15. Vjugina U, Evans JP (2008) New insights into the molecular basis of mammalian sperm-egg membrane interactions. Front Biosci 13:462–476
16. Vigil P (1989) Gamete membrane fusion in hamster spermatozoa with reacted equatorial segment. Gamete Res 23:203–213
17. Gardner AJ, Williams CJ, Evans JP (2007) Establishment of the mammalian membrane block to polyspermy: Evidence for calcium-dependent and -independent regulation. Reproduction 133:383–393
18. McAvey BA, Wortzman GB, Williams CJ et al (2002) Involvement of calcium signaling and the actin cytoskeleton in the membrane block to polyspermy in mouse eggs. Biol Reprod 67:1342–1352
19. Gardner AJ, Evans JP (2006) Mammalian membrane block to polyspermy: new insights into how mammalian eggs prevent fertilisation by multiple sperm. Reprod Fertil Dev 18:53–61
20. Bleil JD, Wassarman PM (1983) Sperm-egg interactions in the mouse: sequence of events and induction of the acrosome reaction by a zona pellucida glycoprotein. Dev Biol 95:317–324
21. Schroeder AC, Schultz RM, Kopf GS et al (1990) Fetuin inhibits zona pellucida hardening and conversion of ZP2 to ZP2f during spontaneous mouse oocyte maturation in vitro in the absence of serum. Biol Reprod 43: 891–897
22. Kalab P, Kopf GS, Schultz RM (1991) Modifications of the mouse zona pellucida during oocyte maturation and egg activation: effects of newborn calf serum and fetuin. Biol Reprod 45:783–787
23. Gahlay G, Gauthier L, Baibakov B et al (2010) Gamete recognition in mice depends on the cleavage status of an egg's zona pellucida protein. Science 329:216–219
24. Coy P, Gadea J, Romar R et al (2002) Effect of in vitrofertilization medium on the acrosome reaction, cortical reaction, zona pellucida hardening and in vitro development in pigs. Reproduction 124:279–288
25. Coy P, Avilés M (2010) What controls polyspermy in mammals, the oviduct or the oocyte? Biol Rev Camb Philos Soc 85:593–605
26. Papi M, Brunelli R, Sylla L et al (2010) Mechanical properties of zona pellucida hardening. Eur Biophys J 39:987–92
27. Wassarman PM, Litscher ES (2009) The multifunctional zona pellucida and mammalian fertilization. J Reprod Immunol 83:45–49
28. Canovas S, Romar R, Grullon LA et al (2009) Pre-fertilization zona pellucida hardening by different cross-linkers affects IVF in pigs and cattle and improves embryo production in pigs. Reproduction 137:803–812
29. Avilés M, Gutiérrez-Adán A, Coy P (2010) Oviductal secretions: Will they be key factors for the future ARTs? Mol Hum Reprod 16:896–906
30. Tsai PS, Gadella BM (2009) Molecular kinetics of proteins at the surface of porcine sperm before and during fertilization. Soc Reprod Fertil Suppl 66:23–36
31. Boerke A, Dieleman SJ, Gadella BM (2007) A possible role for sperm RNA in early embryo development. Theriogenology 68:S147–S155
32. Cho C, Ge H, Branciforte D et al (2000) Analysis of mouse fertilin in wild-type and fertilin beta(-/-) sperm: evidence for C-terminal modification, alpha/beta dimerization, and lack of essential role of fertilin alpha in sperm-egg fusion. Dev Biol 222:289–295

33. Ghosh I, Datta K (2003)Sperm surface hyaluronan binding protein (HABP1) interacts with zona pellucida of water buffalo (Bubalus bubalis) through its clustered mannose residues. Mol Reprod Dev 64:235–244
34. Rodeheffer C, Shur BD (2004) Sperm from beta1,4-galactosyltransferase I-null mice exhibit precocious capacitation. Development 131:491–501
35. van Gestel RA, Brewis IA, Ashton PR et al (2007) Multiple proteins present in purified porcine sperm apical plasma membranes interact with the zona pellucida of the oocyte. Mol Hum Reprod 13:445–454
36. Montfort L, Frenette G, Sullivan R (2002) Sperm-zona pellucida interaction involves a carbonyl reductase activity in the hamster. Mol Reprod Dev 61:113–119
37. Shur BD, Rodeheffer C, Ensslin MA et al (2006) Identification of novel gamete receptors that mediate sperm adhesion to the egg coat. Mol Cell Endocrinol 250:137–148
38. Busso D, Cohen DJ, Maldera JA et al (2007) A novel function for CRISP1 in rodent fertilization: involvement in sperm-zona pellucida interaction. Biol Reprod 77:848–854
39. Copland SD, Murphy AA, Shur BD (2009) The mouse gamete adhesin, SED1, is expressed on the surface of acrosome-intact human sperm. Fertil Steril 92:2014–2019
40. Marín-Briggiler CI, González-Echeverría MF, Munuce MJ et al (2010) Glucose-regulated protein 78 (Grp78/BiP) is secreted by human oviduct epithelial cells and the recombinant protein modulates sperm-zona pellucida binding. Fertil Steril 93:1574–1584
41. Ensslin M, Calvete JJ, Thole HH et al (1995) Identification by affinity chromatography of boar sperm membrane-associated proteins bound to immobilized porcine zona pellucida. Mapping of the phosphorylethanolamine-binding region of spermadhesin AWN. Biol Chem Hoppe Seyler 376:733–738
42. Caballero I, Vázquez JM, Rodríguez-Martínez H et al (2005) Influence of seminal plasma PSP-I/PSP-II spermadhesin on pig gamete interaction. Zygote 13:11–16
43. Nixon B, MacIntyre DA, Mitchell LA et al (2006) The identification of mouse sperm-surface-associated proteins and characterization of their ability to act as decapacitation factors. Biol Reprod 74:275–287
44. Manjunath P, Bergeron A, Lefebvre J et al (2007) Seminal plasma proteins: functions and interaction with protective agents during semen preservation. Soc Reprod Fertil Suppl 65:217–228
45. Vadnais ML, Roberts KP 2010 Seminal plasma proteins inhibit in vitro- and cooling-induced capacitation in boar spermatozoa. Reprod Fertil Dev 22:893–900
46. Gadella BM, Tsai PS, Boerke A et al (2008) Sperm head membrane reorganisation during capacitation. Int J Dev Biol 52:473–480
47. Gadella BM, Visconti PE (2006) Regulation of capacitation. In: de Jonge C, Barrett C (eds) The sperm cell, production maturation fertilization regeneration. Cambridge University Press, Cambridge
48. Parrish JJ, Susko-Parrish J, Winer MA et al (1988) Capacitation of bovine sperm by heparin. Biol Reprod 38:1171–1180
49. Hung PH, Miller MG, Meyers SA et al (2008) Sperm mitochondrial integrity is not required for hyperactivated motility, zona binding, or acrosome reaction in the rhesus macaque. Biol Reprod 79:367–375
50. Kumar V, Kota V, Shivaji S (2008) Hamster sperm capacitation: role of pyruvate dehydrogenase A and dihydrolipoamide dehydrogenase. Biol Reprod 79:190–199
51. Eddy EM (2006) The spermatozoon. In: Knobil E, Neill JD (eds) Physiology of reproduction, (3rd edn). Elsevier, Amsterdam, The Netherlands
52. van Gestel RA, Brewis IA, Ashton PR et al (2005) Capacitation-dependent concentration of lipid rafts in the apical ridge head area of porcine sperm cells. Mol Hum Reprod 11:583–590
53. Thaler CD, Thomas M, Ramalie JR (2006) Reorganization of mouse sperm lipid rafts by capacitation. Mol Reprod Dev 73:1541–1549
54. Selvaraj V, Buttke DE, Asano A et al (2007) GM1 dynamics as a marker for membrane changes associated with the process of capacitation in murine and bovine spermatozoa. J Androl 28:588–599
55. Wassarman PM (1988) Zona pellucida glycoproteins. Ann Rev Biochem 57:415–442
56. Benoff S, Chu CC, Marmar JL et al (2007) Voltage-dependent calcium channels in mammalian spermatozoa revisited. Front Biosci 12:1420–1449
57. Ramalho-Santos J, Moreno RD, Sutovsky P et al (2000) SNAREs in mammalian sperm: possible implications for fertilization. Dev Biol 223:54–69
58. De Blas GA, Roggero CM, Tomes CN et al (2005) Dynamics of SNARE assembly and disassembly during sperm acrosomal exocytosis. PLoS Biol 3:e323
59. Roggero CM, De Blas GA, Dai H et al (2007) Complexin/synaptotagmin interplay controls acrosomal exocytosis. J Biol Chem 282:26335–26343
60. Zhao L, Burkin HR, Shi X et al (2007) Complexin I is required for mammalian sperm acrosomal exocytosis. Dev Biol 309:236–244
61. Zhao L, Shi X, Li L, Miller DJ (2007) Dynamin 2 associates with complexins and is found in the acrosomal region of mammalian sperm. Mol Reprod Dev 74:750–757

62. Lopez CI, Belmonte SA, De Blas GA et al (2007) Membrane-permeant Rab3A triggers acrosomal exocytosis in living human sperm. FASEB J 21:4121–4130
63. Castillo Bennett J, Roggero CM, Mancifesta FE et al (2010) Calcineurin-mediated dephosphorylation of synaptotagmin VI is necessary for acrosomal exocytosis. J Biol Chem 285:26269–26278
64. Ackermann F, Zitranski N, Heydecke D et al (2008) The Multi-PDZ domain protein MUPP1 as a lipid raft-associated scaffolding protein controlling the acrosome reaction in mammalian spermatozoa. J Cell Physiol 214:757–768
65. Ackermann F, Zitranski N, Borth H et al (2009) CaMKIIalpha interacts with multi-PDZ domain protein MUPP1 in spermatozoa and prevents spontaneous acrosomal exocytosis. J Cell Sci 122:4547–4557
66. Dunbar BS, Dudkiewicz AB, Bundman DS (1985) Proteolysis of specific porcine zona pellucida glycoproteins by boar acrosin. Biol Reprod 32:619–630
67. Bleil JD, Greve JM, Wassarman PM (1988) Identification of a secondary sperm receptor in the mouse egg zona pellucida: role in maintenance of binding of acrosome-reacted sperm to eggs. Dev Biol 128:376–385
68. Howes E, Pascall JC, Engel W, Jones R (2001) Interactions between mouse ZP2 glycoprotein and proacrosin; a mechanism for secondary binding of sperm to the zona pellucida during fertilization. J Cell Sci 114:4127–4136
69. Miller DJ, Gong X, Shur BD (1993) Sperm require beta-N-acetylglucosaminidase to penetrate through the egg zona pellucida. Development 118:1279–1289
70. Hao Z, Wolkowicz MJ, Shetty J et al (2002) SAMP32, a testis-specific, isoantigenic sperm acrosomal membrane-associated protein. Biol Reprod 66:735–744
71. Bi M, Hickox JR, Winfrey VP et al (2003) Processing, localization and binding activity of zonadhesin suggest a function in sperm adhesion to the zona pellucida during exocytosis of the acrosome. Biochem J 375:477–488
72. Shetty J, Wolkowicz MJ, Digilio LC et al (2003) SAMP14, a novel, acrosomal membrane-associated, glycosylphosphatidylinositol-anchored member of the Ly-6/urokinase-type plasminogen activator receptor super-family with a role in sperm-egg interaction. J Biol Chem 278:30506–30515
73. Hayashi M, Yonezawa N, Katsumata T et al (2004) Activity of exoglycosidases in ejaculated spermatozoa of boar and bull. Zygote12:105–109
74. Jagadish N, Rana R, Selvi R et al (2005) Characterization of a novel human sperm-associated antigen 9 (SPAG9) having structural homology with c-Jun N-terminal kinase-interacting protein. Biochem J 389:73–82
75. Yu Y, Xu W, Yi YJ et al (2006) The extracellular protein coat of the inner acrosomal membrane is involved in zona pellucida binding and penetration during fertilization: characterization of its most prominent polypeptide (IAM38) Dev Bio 290:32–43
76. Ziyyat A, Rubinstein E, Monier-Gavelle F et al (2006) CD9 controls the formation of clusters that contain tetraspanins and the integrin a_6b_1, which are involved in human and mouse gamete fusion. J Cell Sci 119: 416–424
77. Fujihara Y, Murakami M, Inoue N et al (2010) Sperm equatorial segment protein 1, SPESP1, is required for fully fertile sperm in the mouse. J Cell Si 123:531–1536
78. Kaji K, Oda S, Shikano T et al (2000) The gamete fusion process is defective in eggs of CD9-deficient mice. Nat Genet 24:279–282
79. Le Naour F, Rubinstein E, Jasmin C et al (2000) Severely reduced female fertility in CD9-deficient mice. Science 287:319–321
80. Maleszewski M, Kimura Y, Yanagimachi R (1996) Sperm membrane incorporation into oolemma contributes to the oolemma block to sperm penetration: evidence based on intracytoplasmic sperm injection experiments in the mouse. Mol Reprod Dev 44:256–259
81. Okabe M, Yagasaki M, Oda H et al (1988) Effect of a monoclonal anti-mouse sperm antibody (OBF13) on the interaction of mouse sperm with zona-free mouse and hamster eggs. J Reprod Immunol 13:211–219
82. Inoue N, Kasahara T, Ikawa M et al (2010) Identification and disruption of sperm-specific angiotensin converting enzyme-3 (ACE3) in mouse. PLoS ONE 5: e10301
83. Yáñez-Mó M, Barreiro O, Gordon-Alonso M et al (2009) Tetraspanin-enriched microdomains; a functional unit in cell plasma membranes. Trends Cell Biol 19:434–446
84. Rubinstein E, Ziyyat A, Prenant M et al (2006) Reduced fertility of female mice lacking CD81. Dev Biol 290: 351–358
85. Tiede A, Nischan C, Schubert J et al (2000) Characterisation of the enzymatic complex for the first step in glycosylphosphatidylinositol biosynthesis. Int J Biochem Cell Biol 32:339–350
86. Alfieri JA, Martin AD, Takeda J et al (2003) Infertility in female mice with an oocyte-specific knockout of GPI-anchored proteins. J Cell Sci 116:2149–2155
87. Ellerman DA, Pei J, Gupta S et al (2009) Izumo is part of a multiprotein family whose members form large complexes on mammalian sperm. Mol Reprod Dev 76:1188–1199
88. Inoue N, Ikawa M, Isotani A et al (2005) The immunoglobulin superfamily protein Izumo is required for sperm to fuse with eggs. Nature 434:234–238

89. Han C, Choi E, Park I et al (2009) Comprehensive analysis of reproductive ADAMs: Relationship of ADAM4 and ADAM6 with an ADAM complex required for fertilization in mice. Biol Reprod 80:1001–1008
90. Stein KK, Go JC, Primakoff P et al (2005) Defects in secretory pathway trafficking during sperm development in Adam2 knockout mice. Biol Reprod 73:1032–1038
91. Sosnik J, Miranda PV, Spiridonov NA et al (2009) Tssk6 is required for Izumo relocalization and gamete fusion in the mouse. J Cell Sci 122:2741–2749
92. Lewis WH, Wright ES (1935) On the early development of the mouse egg. Carnegie Inst Contrib Embryol 25:113–143
93. Odor DL, Blandau RJ (1949) The frequency of supernumerary sperm in rat ova. Anat Rec 104:1–11
94. Austin CR (1961) The mammalian egg. Charles C. Thomas, Springfield, IL
95. Hunter RH (1990) Fertilization of pig eggs in vivo and in vitro. J Reprod Fertil Suppl 40:211–226
96. Sengoku K, Tamate K, Horikawa M et al (1995) Plasma membrane block to polyspermy in human oocytes and preimplantation embryos. J Reprod Fertil 105:85–90
97. Hunter RH, Vajta G, Hyttel P (1998) Long-term stability of the bovine block to polyspermy. J Exp Zool 280: 182–188
98. Wolf DE, Edidin M, Handyside AH (1981) Changes in the organization of the mouse egg plasma membrane upon fertilization and first cleavage: Indications from the lateral diffusion rates of fluorescent lipid analogues. Dev Biol 85:195–198
99. Wolf DE, Ziomek CA (1983) Regionalization and lateral diffusion of membrane proteins in unfertilized and fertilized mouse eggs. J Cell Biol 96:1786–1790
100. Larson SM, Lee HJ, Hung PH et al (2010) Cortical mechanics and meiosis II completion in mammalian oocytes are mediated by myosin-II and Ezrin-Radixin-Moesin (ERM) proteins. Mol Biol Cell 21:3182–3192
101. Wolf DP, Nicosia SV, Hamada M (1979) Premature cortical granule loss does not prevent sperm penetration of mouse eggs. Dev Biol 71:22–32
102. Horvath PM, Kellom T, Caulfield J et al (1993) Mechanistic studies of the plasma membrane block to polyspermy in mouse eggs. Mol Reprod Dev 34:65–72
103. Maleszewski M, Kimura Y, Yanagimachi R (1996) Sperm membrane incorporation into oolemma contributes to the oolemma block to sperm penetration: evidence based on intracytoplasmic sperm injection experiments in the mouse. Mol Reprod Dev 44:256–259
104. Wortzman-Show GB, Kurokawa M et al (2007) Calcium and sperm components in the establishment of the membrane block to polyspermy: studies of ICSI and activation with sperm factor. Mol Human Reprod 13:557–565
105. Dale B, Wilding M, Coppola G, Tosti E (2010) How do spermatozoa activate oocytes? Reprod Biomed Online 21:1–3
106. Hoodbhoy T, Talbot P (2001) Characterization, fate, and function of hamster cortical granule components. Mol Reprod Dev 58:223–235
107. Tsai PS, van Haeften T, Gadella BM (2010) Preparation of the cortical reaction: maturation depdent migration of SNARE proteins, clathrin and complexin to the porcine oocyte's surface blocks membrane traffic until fertilization. Biol Reprod doi: 10.1095/biolreprod.110.085647
108. Gwatkin RB, Williams DT, Hartmann JF et al (1973) The zona reaction of hamster and mouse eggs: production in vitro by a trypsin-like protease from cortical granules. J Reprod Fertil 32:259–265
109. Wolf DP, Hamada M (1977) Induction of zonal and egg plasma membrane blocks to sperm penetration in mouse eggs with cortical granule exudate. Biol Reprod 17:350–354
110. Ducibella T (1996) The cortical reaction and development of activation competence in mammalian oocytes. Hum Reprod Update 2:29–42
111. Hoodbhoy T, Talbot P (1994) Mammalian cortical granules: contents, fate, and function. Mol Reprod Dev 39: 439–448
112. Izadyar F, Hage WJ, Colenbrander B et al (1988) The promotory effect of growth hormone on the developmental competence of in vitro matured bovine oocytes is due to improved cytoplasmic maturation. Mol Reprod Dev 49:444–453
113. Abbott AL, Fissore RA, Ducibella T (2001) Identification of a translocation deficiency in cortical granule secretion in preovulatory mouse oocytes. Biol Reprod 65:1640–1647
114. Tae JC, Kim EY, Jeon K et al (2008) A MAPK pathway is involved in the control of cortical granule reaction and mitosis during bovine fertilization. Mol Reprod Dev 75:1300–1306
115. Sun QY (2003) Cellular and molecular mechanisms leading to cortical reaction and polyspermy block in mammalian eggs. Microsc Res Tech 61:342–348
116. Florman HM, Ducibella T (2006) Fertilization in mammals. In: Knobil E, Neill JD (eds) Physiology of reproduction, 3rd edn. Elsevier, Amsterdam, The Netherlands
117. Zarelli VE, Ruete MC, Roggero CM et al (2009) PTP1B dephosphorylates N-ethylmaleimide-sensitive factor and elicits SNARE complex disassembly during human sperm exocytosis. J Biol Chem 284:10491–10503

118. Miyado K, Yamada G, Yamada S et al (2000) Requirement of CD9 on the egg plasma membrane for fertilization. Science 287:321–324
119. He ZY, Brakebusch C, Fassler R et al (2003) None of the integrins known to be present on the mouse egg or to be ADAM receptors are essential for sperm–egg binding and fusion. Dev Biol 254:226–237
120. Miller BJ, Georges-Labouesse E, Primakoff P et al (2000) Normal fertilization occurs with eggs lacking the integrin $alpha_6beta_1$ and is CD9-dependent. J Cell Biol 149:1289–1295
121. Baessler K, Lee Y, Sampson N (2009) Beta1 integrin is an adhesion protein for sperm binding to eggs. ACS Chem Biol 4:357–366
122. Evans JP (2009) Egg integrins: back in the game of mammalian fertilization. ACS Chem Biol 4:321–323
123. Vjugina U, Zhu X, Oh E et al (2009) Reduction of mouse egg surface integrin alpha9 subunit (ITGA9) reduces the egg's ability to support sperm-egg binding and fusion. Biol Reprod 80:833–841
124. Eto K, Huet C, Tarui T et al (2002) Functional classification of ADAMs based on a conserved motif for binding to integrin $alpha_9beta_1$: implications for sperm-egg binding and other cell interactions. J Biol Chem 277: 17804–17810
125. Inoue N, Ikawa M, Isotani A et al (2005) The immunoglobulin superfamily protein Izumo is required for sperm to fuse with eggs. Nature 434:234–238
126. Bigler D, Takahashi Y, Chen MS et al (2000) Sequence-specific interaction between the disintegrin domain of mouse ADAM2 (fertilin b) and murine eggs: Role of the a_6 integrin subunit. J Biol Chem 275:11576–11584
127. Takahashi Y, Bigler D, Ito Y et al (2001) Sequence-specific interaction between the disintegrin domain of mouse ADAM3 and murine eggs: Role of the $beta_1$ integrin-associated proteins CD9, CD81, and CD98. Mol Biol Cell 12:809–820
128. Zhu X, Bansal NP, Evans JP (2000) Identification of key functional amino acids of the mouse fertilin beta (ADAM2) disintegrin loop for cell-cell adhesion during fertilization. J Biol Chem 275:7677–7683
129. Cuasnicu PS, Conesa D, Rochwerger L (1990) Potential contraceptive use of an epididymal protein that participates in fertilization. In: Alexander NJ, Griffin D, Speiler JM, Waites GMH (eds) Gamete Interaction: Prospects for Immunocontraception. Wiley-Liss, New York, NY, pp. 143–153
130. Cuasnicu, PS, Gonzalez-Echeverria MF, Piazza AD et al (1984) Antibody against epididymal glycoprotein blocks fertilizing ability in rats. J Reprod Fert 72:467–471
131. Da Ros VG, Maldera JA, Willis WD et al (2008) Impaired sperm fertilizing ability in mice lacking Cysteine-RIch Secretory Protein 1 (CRISP1). Dev Biol 320:12–8
132. Wolkowicz MJ, Shetty J, Westbrook A (2003) Equatorial segment protein defines a discrete acrosomal subcompartment persisting throughout acrosomeal biogenesis. Biol Reprod 69:735–745
133. Toshimori K, Saxena DK, Tanii I et al (1998) An MN9 antigenic molecule, equatorin, is required for successful sperm-oocyte fusion in mice. Biol Reprod 59:22–29
134. Yamatoya K, Yoshida K, Ito C et al (2009) Equatorin: identification and characterization of the epitope of the MN9 antibody in mouse. Biol Reprod 81:889–897
135. Herrero MB, Mandal A, Digilio LC et al (2005) Mouse SLLP1, a sperm lysozyme-like protein involved in sperm-egg binding and fertilization. Dev Biol 284:126–42
136. Mandal A, Klotz KL, Shetty J et al (2003) SLLP1, A unique, intra-acrosomal, non-bacteriolytic, c lysozyme-like protein of human spermatozoa. Biol Reprod 68:1525–1537
137. Shetty J, Wolkowicz MJ, Digilio LC et al (2003) SAMP14, a novel, acrosomal membrane-associated, glycosylphosphatidylinositol-anchored member of the Ly-6/urokinase-type plasminogen activator receptor superfamily with a role in sperm-egg interaction. J Biol Chem 278:30506–30515
138. Correa LM, Cho C, Myles DG et al (2000) A role for a TIMP-3-sensitive Zn^{2+}-dependent metalloprotease in mammalian gamete membrane fusion. Dev Biol 225:124–134
139. Ellerman DA, Myles DG, Primakoff P (2006) A role for sperm surface protein disulfide isomerase activity in gamete fusion: evidence for the participation of ERp57. Dev Cell 10:831–837
140. Mammoto A, Masumoto N, Tahara M et al (1997) Involvement of a sperm protein sensitive to sulfhydry-depleting reagents in mouse sperm-egg fusion. J Exp Zool 278:178–188
141. Wilson DW, Wilcox CA, Flynn GC (1989) A fusion protein required for vesicle-mediated transport in both mammalian cells and yeast. Nature 339:355–359
142. Fenouillet E, Barbouche R, Courageot J et al (2001) The catalytic activity of protein disulfide isomerase is involved in human immunodeficiency virus envelope-mediated membrane fusion after CD4 cell binding. J Infect Dis 183:744–752
143. Ou W, Silver J (2006) Role of protein disulfide isomerase and other thiol-reactive proteins in HIV-1 envelope protein-mediated fusion. Virology 350:406–417

Chapter 6
Trophoblast Fusion

Berthold Huppertz and Martin Gauster

Abstract The villous trophoblast of the human placenta is the epithelial cover of the fetal chorionic villi floating in maternal blood. This epithelial cover is organized in two distinct layers, the multinucleated syncytiotrophoblast directly facing maternal blood and a second layer of mononucleated cytotrophoblasts. During pregnancy single cytotrophoblasts continuously fuse with the overlying syncytiotrophoblast to preserve this end-differentiated layer until delivery. Syncytial fusion continuously supplies the syncytiotrophoblast with compounds of fusing cytotrophoblasts such as proteins, nucleic acids and lipids as well as organelles. At the same time the input of cytotrophoblastic components is counterbalanced by a continuous release of apoptotic material from the syncytiotrophoblast into maternal blood. Fusion is an essential step in maintaining the syncytiotrophoblast. Trophoblast fusion was shown to be dependant on and regulated by multiple factors such as fusion proteins, proteases and cytoskeletal proteins as well as cytokines, hormones and transcription factors. In this chapter we focus on factors that may be involved in the fusion process of trophoblast directly or that may prepare the cytotrophoblast to fuse.

6.1 Introduction

Syncytial fusion is a general process in animal tissues characterized by dissolution of the separating parts of the plasma membranes of two neighboring cells. Hence, syncytial fusion leads to the assembly of multinucleated structures derived and maintained by continuous fusion of and with mononucleated cells. Such multinucleated structures are no longer termed a cell but rather are referred to as a syncytium. Typical examples in the human are the placental syncytiotrophoblast and myoblast-derived skeletal muscle fibers [1].

During the process of syncytial fusion a large array of different and interdependent intracellular pathways is activated and results in the close interaction of the plasma membranes of two neighboring cells. Finally the interacting parts of the plasma membranes dissolve and allow exchange of cytoplasmic contents (lipids, proteins and RNA) as well as organelles such as mitochondria, the endoplasmic reticulum and the nucleus. Beside this cell–cell fusion process, fusion is a general phenomenon within a cell used for membrane traffic and release or engulfment of vesicles. Fusion between two cells requires a number of prerequisites since plasma membranes do not fuse easily to maintain the individuality of a cell. Hence, syncytial fusion of two cells requires a specific repertoire of players to prepare the cells to fuse and requires the presence and activation of specific fusogenic proteins.

B. Huppertz (✉)
Institute of Cell Biology, Histology and Embryology, Center for Molecular Medicine, Medical University of Graz, 8010 Graz, Austria
e-mail: berthold.huppertz@medunigraz.at

T. Dittmar, K.S. Zänker (eds.), *Cell Fusion in Health and Disease*, Advances in Experimental Medicine and Biology 713, DOI 10.1007/978-94-007-0763-4_6, © Springer Science+Business Media B.V. 2011

6.2 The Trophoblast

In the human as well as other mammals fusion processes are fundamental and compulsory for the start and continuation of pregnancy. At the morula stage the trophoblast is the first cell line to differentiate during embryogenesis. The trophoblast is essential for the development of the placenta and is not a constituent of the embryo proper. Following the morula stage, the trophoblast differentiates into the outer layer of the blastocyst, thus surrounding the early embryo [2].

At the time of implantation cells from the trophectoderm, the trophoblastic cover of the blastocyst, differentiate into the first syncytiotrophoblast (Fig. 6.1). Those trophectoderm cells in contact with the embryoblast as well as the uterine epithelium fuse to generate the very first syncytiotrophoblast. Only this tissue seems to be able to penetrate through the uterine epithelium to enable implantation of the human embryo. At the time of implantation the blastocyst shows a highly specific orientation with the embryonic pole towards the uterine epithelium. If a proper orientation does not occur (Fig. 6.1), spontaneous abortion or fetal growth restriction may be the consequences. After about 2 weeks of development, between d15 and d21 post conception (pc), the early trophoblast further differentiates into separate subtypes, leading to the lineages of the extravillous trophoblast to invade maternal tissues and the villous trophoblast to set up the placental barrier.

6.2.1 The Villous Trophoblast

The villous trophoblast is the epithelial coverage of the placental villous trees. It is the outermost fetal layer of the placenta and comes into direct contact with maternal blood. This tissue consists of two layers, a layer of mononucleated progenitor cells and a second multinucleated outer layer.

The mononucleated villous cytotrophoblasts are in direct contact to their basement membrane which separates this epithelial layer form the connective tissues of the villous stroma. Cytotrophoblasts represent the pool of trophoblast progenitor cells that proliferate, leave the cell cycle, differentiate and finally fuse with the overlying multinucleated layer. This outer layer is the syncytiotrophoblast, a multinucleated layer without any lateral cell borders.

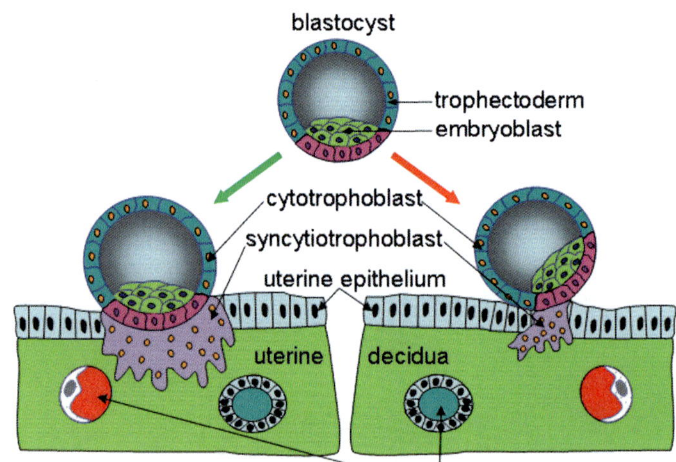

Fig. 6.1 Putative way of blastocyst implantation in the human

In each placenta there is a single syncytiotrophoblast covering all villi of the placenta and representing the placental barrier between fetal tissues and maternal blood [2–4]. The syncytiotrophoblast is essential to guarantee growth of the fetus and thus is responsible for functions such as transport of oxygen, nutrients and waste products, hormone production and immune tolerance [2]. The syncytiotrophoblast is highly differentiated and has lost its proliferative capacity. Moreover, similar to most highly differentiated cells, the nuclei within the syncytiotrophoblast only show moderate rates of RNA synthesis [5]. Hence, to maintain this huge layer throughout pregnancy, continuous syncytial fusion of cytotrophoblasts with the syncytiotrophoblast is needed. Nuclei and other organelles as well as proteins and RNA originating form the cytotrophoblasts are incorporated into the syncytiotrophoblast by syncytial fusion and hence, maintenance of the multinucleated layer is guaranteed.

In recent years a number of studies has shown that preservation of a healthy pregnancy requires a strict coordination and appropriate control of syncytial fusion within the villous trophoblast layer. Alterations and especially dysregulation of syncytial fusion in the trophoblast may directly link to pathological conditions such as intra uterine growth restriction and preeclampsia [6, 7].

6.3 Initiation of Syncytial Fusion in Villous Trophoblast

6.3.1 Fusogenic Proteins

Five criteria have been defined to decide whether or not a molecule belongs to the family of genuine fusogens [8].

1. The molecule needs to be indispensable for fusion of two membranes.
2. The molecule needs to be expressed at the right time and be located at the right place.
3. If expressed in an originally non fusogenic cell, the molecule should be able to induce fusion in such cells.
4. If expressed in heterologous cells, the molecule should be able to induce fusion in such cells.
5. The molecule has to initiate fusion of liposomes as well.

Such criteria clearly limit the number of candidate fusogens known to be involved in syncytial fusion of villous trophoblasts. To date, the only fusogens described to be expressed in human villous trophoblast are retroviral proteins of the syncytin family. DNA regions of retroviral origin were incorporated during evolution and comprise about 8% of the human genome [9]. Generally, such regions are not translated into mRNA or proteins, while few regions are translated into functional proteins. In the human placenta retroviral elements are expressed which are related to trophoblast fusion: the envelope genes (env regions) of ERV-3, HERV-W, and HERV-FRD [10].

Syncytin-1, an envelope protein encoded by the HERV-W gene [11], and syncytin-2, an envelope protein encoded by the HERV-FRD gene [12], are members of the syncytin protein family and have been proposed to be important players in trophoblast fusion. However, their importance for syncytial fusion of trophoblasts is still in doubt [10]. Syncytin-1 meets at least some of the above mentioned criteria and thus could be a candidate fusogen in villous trophoblast [11–15]:

Criterion 1: Intercellular fusion of BeWo cells and primary cytotrophoblasts is impaired in the presence of an anti-syncytin-1 antiserum [11] or syncytin targeting antisense oligonucleotides [14].

Criterion 2: So far, available data on the localization of syncytin-1 in villous trophoblast are rather heterogeneous. The protein was detected in the cytotrophoblast layer and/or the syncytiotrophoblast [11–13, 15].

Criterion 3: Expression of syncytin-1 in non fusogenic COS cells leads to fusion of such cells [11].
Criterion 4: Expression of syncytin-1 results in syncytial fusion between heterologous BeWo and COS cells [11].
Criterion 5: Liposomes (containing GFP expression plasmids) fused with syncytin 1 over-expressing COS cells [11].

Also syncytin-2 was suggested to be involved in trophoblast fusion [16, 17]. However, even though the two syncytins fulfil some of the criteria of genuine fusogens, so far their role in trophoblast fusion is still unclear.

As detailed above the localization of syncytin-1 is still ambiguous, while syncytin-2 has been localized to the cytotrophoblast layer [18–20]. Beside conflicting data on their localization, the number of syncytin expressing cytotrophoblasts is much higher than the actual number of fusing trophoblasts. This makes it hard to draw any conclusions on the physiological relevance of syncytins in the trophoblastic fusion process. The syncytins may well be involved in trophoblast fusion; however, the genuine fusogens involved in villous trophoblast fusion may still wait for their detection.

6.3.2 Preparation of Syncytial Fusion in Villous Trophoblast

Fusogenic molecules are only one aspect of the whole process of syncytial fusion. Additional factors and players need to be in place to get the cell ready for fusion and to transform it to enable a specific fusion event. A villous cytotrophoblast should only fuse with the overlying syncytiotrophoblast rather than with a neighboring cytotrophoblast or any other cell in close vicinity to it. Respective factors and players to control this fusion event are mostly molecules that carry out their functions prior or even after the formation of the fusion pore by the fusogen.

These include regulatory proteins that trigger trophoblast fusion such as protein kinases [21–23], transcription factors like glial cell missing-1 (GCM1) [15, 24] as well as intracellular proteases [25, 26] (Fig. 6.2).

6.3.2.1 Growth Factors and Cytokines

Growth factors and cytokines derived from the maternal as well as the fetal environment obviously influence trophoblast fusion (Table 6.1). Epidermal growth factor (EGF) [27], colony stimulating factor (CSF)-1 [28], granulocyte-macrophage colony-stimulating factor (GM-CSF) [28] as well as leukemia-inhibitory factor (LIF) [29] and transforming growth factor (TGF)-α [29] have been described to induce syncytial fusion of primary cytotrophoblasts in vitro resulting in augmented secretion of human chorionic gonadotropin (hCG) and human placental lactogen (hPL). Vascular endothelial growth factor (VEGF) increased number and size of syncytia in primary first trimester cytotrophoblasts in vitro [30]. Transfection of antisense oligonucleotides against macrophage inhibitory cytokine 1 (MIC-1) into term cytotrophoblasts led to inhibition of syncytium

Fig. 6.2 (continued) syncytin-1 results in the formation of a fusion pore between the cytotrophoblast and the overlying syncytiotrophoblast (ST). Active caspase 8 is transferred into the syncytiotrophoblast during the fusion process. (**c**) After fusion all cellular components of the cytotrophoblast, including proteins, nucleic acids and lipids as well as organelles are incorporated into the syncytiotrophoblast. Also the fragments of the fodrin network are transferred. Caspase 8 is still active at the time of transfer and needs to be inhibited by the action of anti-apoptotic proteins such as Bcl-2 and c-Flip

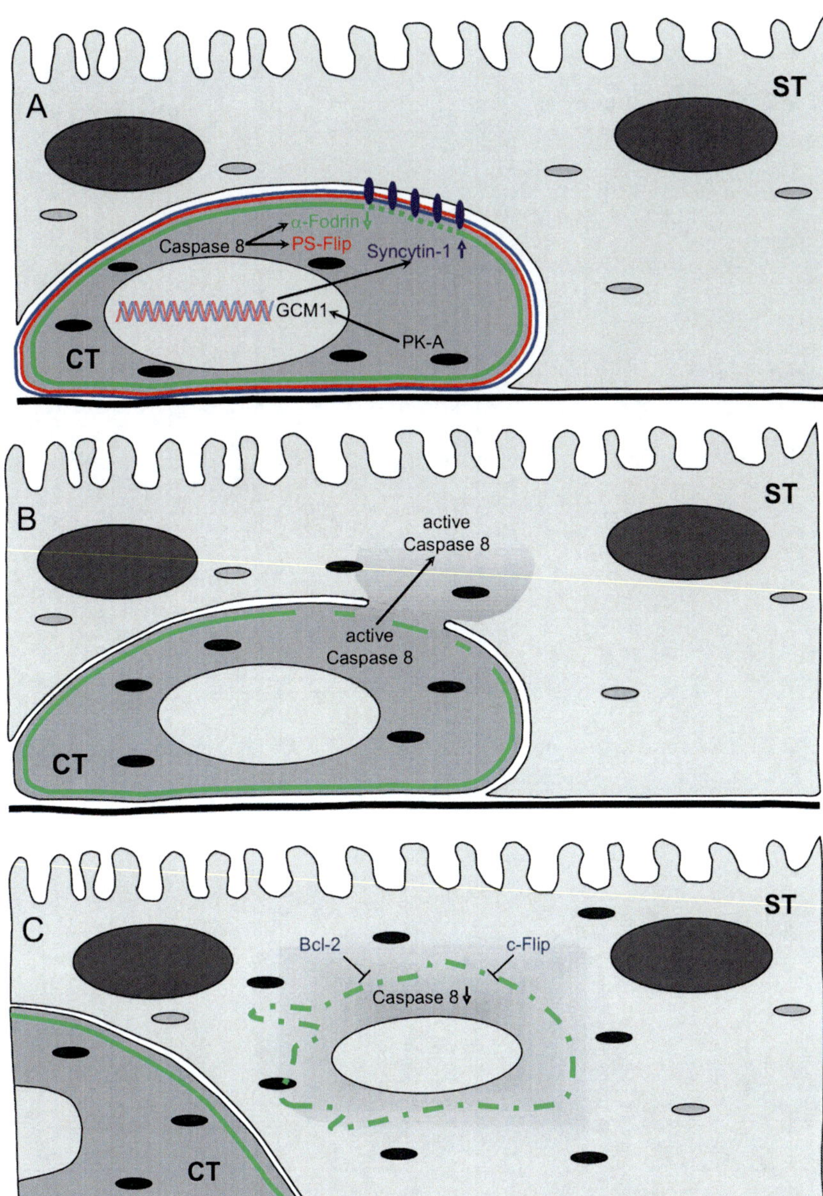

Fig. 6.2 Preparation of fusion and fate of caspase 8 after villous trophoblast fusion. (**a**) Prior to fusion several mechanisms are in place to prepare the cytotrophoblast (CT) for fusion. Pro-caspase 8 is conversed into its active form that subsequently leads to cleavage of the sub-membranous cytoskeletal protein alpha-fodrin (*green*). Moreover, caspase 8 may directly or indirectly mediate the flip of phosphatidylserine from the inner (*red*) to the outer (*blue*) leaflet of the plasma membrane. Activation of protein kinase A (PK-A) results in activation of the transcription factor GCM1 that in turn upregulates expression of the fusogenic protein syncytin-1 (*violet*). (**b**) During fusion fragmentation and remodeling of the fodrin cytoskeleton in combination with the PS-flip and fusogenic proteins such as

Table 6.1 Proteins promoting (+) or hindering (–) syncytial fusion of trophoblasts and trophoblast-derived cells

Cytokines, growth factors and hormones

Type	Factor, full name	Abbreviation	Effect on fusion	Trimester	References
Cytokine	Leukemia inhibitory factor	LIF	+	3T	[29]
Cytokine	Macrophage inhibitory cytokine 1	MIC-1	+	3T	[31]
Cytokine	Tumor necrosis factor alpha	TNF-α	–	3T	[33]
Growth factor	Colony stimulating factor	CSF	+	3T	[28]
Growth factor	Epidermal growth factor	EGF	+	3T	[27]
Growth factor	Granulocyte macrophage colony stimulating factor	GM-CSF	+	3T	[28]
Growth factor	Transforming growth factor alpha	TGF-α	+	3T	[29]
Growth factor	Transforming growth factor beta	TGF-β	–	3T	[75]
Growth factor	Vascular endothelial growth factor	VEGF	+	1T, 3T	[30]
Peptide hormone	Human chorionic gonadotropin	hCG	+	3T	[29]

Transcription factors and protein kinases

Type	Factor/kinase, full name	Abbreviation	Effect on fusion	Trimester	References
Transcription factor	Glial cell missing homolog 1	GCM1	+	BeWo	[34]
Transcription factor	Mammalian achaete/scute homolog 2	Mash-2	–	2T	[76]
Protein kinase	Extracellular signal-regulated kinases 1 and 2	ERK1/2	+	3T	[22]
Protein kinase	p38 isoforms of MAPK	p38	+	3T	[22]
Protein kinase	Protein kinase A	PKA	+	BeWo	[23]

Other proteins

Type	Protein, full name	Abbreviation	Effect on fusion	Trimester	References
Amino acid transporter	Alanine/serine/cysteine/threonine transporter	ASCT1	?		[77, 78]
Amino acid transporter	Alanine/serine/cysteine/threonine transporter	ASCT2	+	BeWo	[77, 78]
Amino acid transporter	Large neutral amino acid transporter	CD98/LAT1	+	BeWo	[79, 80]
Gap junction protein	Connexin 43	Connexin 43	+	3T	[81]
Lectin	Galectin 3	Galectin 3	+	BeWo	[82]
HERV protein	Syncytin 1, 2	Syncytin 1, 2	+	3T, BeWo	[11, 14]
Protease	A disintegrin and metalloprotease 12	ADAM 12	?		[83]
Protease	Initiator caspase 8	Caspase 8	+	VE, BeWo	[26, 49]
Protease	Initiator caspase 10	Caspase 10	?		[53]
Protease	Caspase 14	Caspase 14	?		[84]

Data are derived from in vitro experiments using first trimester trophoblasts (1T), second trimester trophoblasts (2T), third trimester trophoblasts (3T), BeWo cells (BW) or villous explants (VE). Some factors were suggested to play a role in trophoblast fusion without any experimental evidence so far (indicated by a "?").

formation [31]. Also syncytiotrophoblast derived hCG has been shown to have an autocrine effect on the villous trophoblast and to increase syncytium formation [29, 32].

On the contrary, so far only few factors and players have been described to negatively act on trophoblast fusion and to reduce formation of trophoblast syncytia. Tumor necrosis factor (TNF)-α as well as transforming growth factor (TGF)-β have been shown to impair syncytium formation and reduce secretion of hCG and hPL [33].

6.3.2.2 Protein Kinases

Growth factors, cytokines and other environmentally derived factors may bind to their respective receptors on the villous trophoblast. In turn this may turn on downstream signaling pathways activating kinases and transcription factors (Table 6.1). Two members of the family of mitogen-activated protein kinases (MAPKs), p38 and the extracellular signal-regulated kinase1/2 (ERK1/2), were shown to be involved in the regulation of trophoblast differentiation and fusion [22, 23]. In vitro, blockage of ERK1/2 and/or p38 activities by respective inhibitors in primary trophoblasts down regulates differentiation and results in the formation of less syncytia [22]. Transient overexpression of the catalytic subunit of protein kinase A (PKA) led to enhanced fusion of BeWo cells, a trophoblast-derived choriocarcinoma cell line [23].

6.3.2.3 Transcription Factor GCM1

Forskolin is a reagent known to induce fusion in the trophoblast derived cell line BeWo (Fig. 6.3). Administration of forskolin raises intracellular cAMP concentrations in these cells. This in turn leads to the upregulation of glial cell missing homolog 1 (GCM1) [23], a transcription factor belonging to the GCM family of zinc-containing transcription factors [34].

GCM1 has been shown to be directly linked to the initiation of trophoblast fusion. As a transcription factor, GCM1 regulates transcription of genes. In the human placenta, GCM1 has been shown to upregulate the expression of syncytin-1 [15, 19]. At the same time, activity of GCM1 only starts when a cytotrophoblast has left the cell cycle. Thus, GCM1 activity is negatively correlated with trophoblast proliferation [24]. GCM1 is only expressed in the subset of highly differentiated villous cytotrophoblasts [35], which are destined to fuse with the syncytiotrophoblast. This is in agreement with the observation that villous cytotrophoblasts leave the cell cycle before starting their differentiation program resulting in syncytial fusion.

6.3.2.4 Externalization of Phosphatidylserine

The asymmetrical distribution of phospholipids in the two leaflets of mammalian plasma membranes is actively maintained [36]. The inner leaflet contains most of the negatively charged phospholipids such as phosphatidylserine (PS) and phosphatidylethanolamine. However, under certain conditions a cell may redistribute the negatively charged phospholipids, especially PS, from the inner to the outer leaflet of the plasma membrane. This "PS-flip" has been shown to be an early event during the apoptosis cascade which is used as an "eat-me" signal to eliminate apoptotic cells [37]. However, this flip has also been shown to be required for syncytial fusion [38].

The active maintenance of the phospholipid asymmetry is due to at least three different lipid transporters:

1. ATP-independent bidirectional transporters (*scramblases*),
2. ATP-dependent inward flipping aminophospholipid translocases (*flippases, inward translocation*), and
3. ATP-dependent outward flipping transporters (*floppases, outward translocation*).

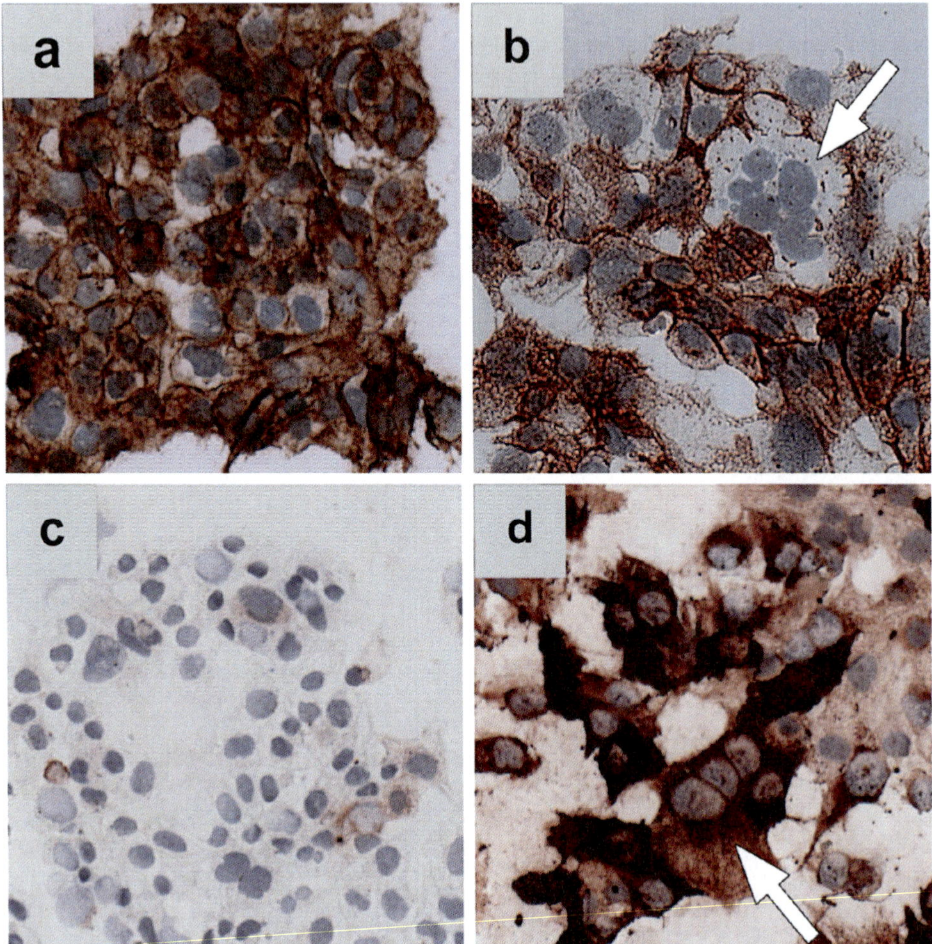

Fig. 6.3 Choriocarcinoma cell line BeWo as an in vitro model to study trophoblast fusion. Forskolin induces morphological and endocrine differentiation of trophoblast derived BeWo cells. While solvent control (DMSO, 0.2%, 24 h) treated cells remain in a mononucleated state (**a**), in the presence of forskolin (20 μM, 24 h) BeWo cells differentiate into multinucleated syncytia (**b**, *arrow*), as visualized by E-cadherin staining. While ßhCG expression is only marginally detected in the presence of the solvent control (**c**), forskolin induces not only syncytialization but also abundant expression of βhCG (**d**, *arrow*)

A *scramblase* transfers lipids from the inner to the outer as well as from the outer to the inner leaflet of a plasma membrane, while the activity of a *flippase* is restricted to the transfer of lipids from the outer to the inner leaflet, and the activity of a *floppase* is restricted to the transport of lipids from the inner to the outer leaflet [39, 40].

The externalization of phosphatidylserine is a prerequisite for syncytial fusion in skeletal muscle and villous trophoblast, the two main multinucleated fusion systems in the human [41, 42]. A short and transient exposition of phosphatidylserine to the outer leaflet of the plasma membrane has been detected in myoblasts at specific cell–cell contact sites just prior to and during fusion into myotubes [42]. In villous trophoblast the PS-flip was detected in a subset of highly differentiated cytotrophoblasts without any other signs of apoptosis [41]. In the trophoblast-derived BeWo choriocarcinoma cell line that is used as an in vitro model for primary trophoblast, the PS-flip is essential for syncytial fusion [38, 43]. Using an antibody directed against PS blocked syncytial fusion of BeWo cells [38, 43].

The reason for the flip of phosphatidylserine to the outer leaflet of the plasma membrane prior to fusion may be as follows: Fusogenic proteins are actively involved in forming the fusion pore. Fusogenic proteins have an alpha-helix structure with an orientation almost parallel to the lipid acyl chains of the plasma membrane. To enable reorientation of the fusion protein to insert it into the lipid bilayer of the neighboring cell, redistribution of phosphatidylserine is required [44]. Only then the highly specific orientation of the fusion protein can induce membrane fusion [45].

6.4 Caspases and Their Roles in Syncytial Fusion

In skeletal muscle as well as placental trophoblast syncytial fusion seems to depend on the activity of intracellular proteases. Hence, such proteases play essential roles in priming mononucleated cells for fusion. In myoblasts, the progenitor cells of skeletal muscle fibers, the activity of calpains, Ca^{2+}-regulated cysteine proteases, is upregulated prior to fusion [46]. Calpains play a critical role in a variety of processes including syncytial fusion [46, 47]. Activity of calpastatin, the endogenous inhibitor of calpains, blocks fusion of myoblasts [48] demonstrating the important role of these proteases in preparing myoblasts for fusion.

Gauster et al. [26] have shown that in villous trophoblast calpains do not seem to play a role in preparing such cells for syncytial fusion [26]. Rather, villous trophoblast seems to make use of a similar but different system of proteases. In the trophoblast activity of initiator caspase 8 has been shown to be essential for the fusion process [26, 49].

Caspases are aspartate-specific cysteine proteases that are known for their crucial roles in the apoptosis cascade. Here they act as initiators or executioners by activating cascade-like events during programmed cell death [50].

However, the belief that caspase activity is limited to apoptosis has become outdated, as several studies supported non-apoptotic functions of caspases. Such functions comprise regulation of inflammatory processes, proliferation, migration as well as differentiation (see [51] for an overview). A good example is the study of Fernando et al. [52] who described a unique role for a caspase 3-mediated signaling cascade in skeletal muscle differentiation and fusion [52].

6.4.1 Localization of Caspase 8

Caspase 8 was initially shown to be expressed in villous trophoblast using immunohistochemistry with antibodies against the active forms and proforms of the protease [53]. Subsequent studies refined the localization of pro-caspase 8 to villous cytotrophoblast in first trimester placental specimens [54, 55]. In an in vitro model making use of first trimester villous explants, an integral role of caspase 8 in trophoblast fusion was suggested. Antisense oligonucleotides to block caspase 8 protein expression or peptide inhibitors (z-IETD-fmk) to block activity of caspase 8 inhibited fusion and led to a multilayered cytotrophoblast layer within 48 h [49, 56].

A recent study from our laboratory revealed that active caspase 8 is present in some post-proliferative, terminally differentiated villous cytotrophoblasts [25]. It was also detected in rare cases of trophoblasts that were located between the two villous trophoblast layers as well as in few sites of the overlying syncytiotrophoblast [25]. Hence, initial activation of caspase 8 takes place in highly differentiated cytotrophoblasts just prior to fusion. The active protein accompanies the content of the fusing cell into the syncytiotrophoblast. Here it stays active for a short while and then becomes inactivated. So far, it is unclear how the spreading of the active protease as well as the inactivation of caspase 8 are regulated.

6.4.2 Contributions of Caspase 8 to Trophoblast Fusion

From its roles and actions in other cell types it becomes clear that caspase 8 is not a genuine fusogen. Rather, it has not yet determined adequately which mechanistic role caspase 8 plays in trophoblast differentiation and fusion. As a protease it will most probably not be able to act in directly fusing two membranes. On the other hand, similar to other factors acting in trophoblast differentiation, caspase 8 may be involved upstream of syncytial fusion. The protease may prime specific parts of the cell for the upcoming fusion event. Such priming processes may comprise the remodeling of membrane architecture initiating the PS-flip, remodeling of the sub-membranous cytoskeleton, as well as initiating complex cell signaling processes.

6.4.2.1 Caspase 8 and Externalization of Phosphatidylserine

As described above (Section 6.3.2.4) loss of phospholipid asymmetry in the plasma membrane, with externalization of PS to the outer leaflet of the membrane, is a hallmark of apoptosis, but was also described as prerequisite of syncytial fusion [38, 57, 58]. Caspase 8 has already been shown to play a role in externalization of phosphatidylserine in erythrocytes [59] and a squamous cell carcinoma cell line [60].

A direct action of caspase 8 in PS externalization during trophoblast fusion has not yet been examined in detail yet. In the trophoblast-derived choriocarcinoma cell line BeWo, caspase 3 was excluded to be involved in PS externalization since administration of a multi-caspase inhibitor (Z-VAD-fmk) to these cells did not change the capacity of the cells to bind FITC-annexin V [36]. Moreover, caspase 3 activity and cytochrome c release from mitochondria did not change when BeWo cells were forced to fuse by addition of forskolin. Hence, it was suggested that externalization of phosphatidylserine is not triggered by execution caspases such as caspase 3, but rather by other mechanisms [36].

At the same time, van den Eijnde et al. [42] showed that in fusing myoblasts externalization of phosphatidylserine is spatially and temporarily restricted to certain cell–cell contact areas during fusion [42]. This raised the notion that PS externalization during fusion is spatially restricted to very small compartments of the plasma membrane. Such small areas between two fusing cells may not be accessible for antibodies or FITC-conjugated phosphatidylserine and thus may escape visualization by light microscopy.

6.4.2.2 Caspase 8 and Cytoskeletal Remodeling

Spectrin, band 4.1 and fodrin belong to the spectrin protein family of sub-membranous cytoskeletal proteins that possess binding sites for phosphatidylserine [61, 62]. The erythroid spectrin belongs to a larger family of cytoskeleton proteins with its non-erythroid homolog alpha-fodrin. Such proteins may be involved in sequestering this negatively charged phospholipid in the inner leaflet of the plasma membrane. Interestingly, target proteins of caspase activity include proteins of the sub-membranous cytoskeleton such as spectrin and fodrin [63]. In the villous trophoblast alpha-fodrin shows an inconsistent distribution in cytotrophoblasts with absence in the syncytiotrophoblast [53]. It was initially assumed that in the cytotrophoblast alpha-fodrin is cleaved by initiator caspases such as caspase 8 prior to fusion [53]. Recently, we have shown that caspase 8 is able to cleave alpha-fodrin and that alpha-fodrin expression is diminished in highly differentiated cytotrophoblasts, and entirely missing in the syncytiotrophoblast [53]. Thus, a direct contribution of active caspase 8 in cleavage of alpha-fodrin in fusing primary trophoblasts and BeWo cells was substantiated [26].

Immunohistochemical staining of villous trophoblast revealed a co-localization of active caspase 8 with vesicles of alpha-fodrin in newly formed sites of the syncytiotrophoblast [26]. This data suggested that alpha-fodrin is cleaved by active caspase 8 during the transition of a cytotrophoblast into the syncytiotrophoblast layer [26]. However, since the avidity and number of PS-binding sites in spectrin-like proteins is not enough to keep phosphatidylserine in the inner leaflet of a plasma membrane, cleavage of alpha-fodrin may not be sufficient to result in excessive externalization of phosphatidylserine [64].

As shown for alpha-fodrin, tight control of the cleavage of the sub-membranous cytoskeleton is an important process to prime a cell for the fusion process. The spectrin network maintains the curvature of the plasma membrane and respective degradation of spectrin affects membrane curvature and facilitates fusion, since an increasing curvature of a membrane increases fusogenicity [65]. The importance of the spectrin/fodrin network for fusion has been demonstrated by microinjection of anti-fodrin antibodies into bovine kidney epithelial cells [66]. The presence of the antibodies induced cell fusion in these non-fusogenic cells [66], pointing to a key role of the sub-membranous cytoskeleton in the fusion process.

6.4.3 Caspase 8 Regulation During and After Fusion

If the above interpretation of data is correct, then caspase 8 is turned into its active form already in the cytotrophoblast and is incorporated into the syncytiotrophoblast by fusion with this specific cytotrophoblast. For the multinucleated layer of the syncytiotrophoblast the incorporation of such hazardous material would have the risk of activating the apoptosis cascade. Hence, to prevent such a scenario, respective safety measures need to be in place:

1. The inactive proforms of caspase 8 seem to be present only in the cytotrophoblast layer, while in the syncytiotrophoblast these are absent. Activated caspase 8 that is incorporated into the syncytiotrophoblast by syncytial fusion may run out of inactive proforms within the syncytium, and thus further autocatalysis of this caspase is avoided.
2. In primary term trophoblasts it was shown that the level of trophoblast differentiation clearly impacts the activity of caspases [67]. In less differentiated (mononucleated) trophoblasts the activities of caspases 3, 6, 8 and 9 are significantly increased compared to the activities in more differentiated (multinucleated) cells [67]. Accordingly, specific mechanisms within the syncytiotrophoblast may inhibit active caspases but also activation of downstream caspases.
3. Members of the Bcl-2 family of proteins belong to those candidate proteins that may well be involved in regulating the progress of the apoptosis cascade in the syncytiotrophoblast. Anti-apoptotic members of this family such as Bcl-2 and Mcl-1 have been described to be expressed in the syncytiotrophoblast throughout gestation [68]. Thus, the syncytiotrophoblast may use anti-apoptotic Bcl-2 family members to down-regulate caspase activity after incorporation of active caspase 8 via syncytial fusion.
4. An active caspase 8 molecule may also be directly inhibited by inhibitors such as c-Flip (cellular FLICE inhibitory protein). C-Flip exists in two splice variants (short and long) and directly acts on and binds to caspase 8 leading to activation or inhibition of this protease [69]. In trophoblast derived cell lines such as BeWo, Jeg-3 and Jar cells c-Flip is expressed [70–72]. In cell lines c-Flip and caspase 8 are co-localized in the same cells. This is true for term villous trophoblast as well. In the cytotrophoblast layer caspase 8 co-localizes with c-Flip. However, c-Flip is present at specific sites in the syncytiotrophoblast as well, maybe at those sites where fusion will occur shortly [71].

6.5 Location Where Fusion Is Initiated in the Trophoblast

Several factors from maternal blood, the fetoplacental tissues as well as the trophoblast have been postulated to regulate cytotrophoblast fusion with the syncytiotrophoblast. However, the driving force initiating this event still awaits its detection. Thus, the question remains whether fusion is initiated by the syncytiotrophoblast, the cytotrophoblast, by both or by other cells such as Hofbauer cells, the placental macrophages.

6.5.1 Syncytiotrophoblast

Initiation of syncytial fusion by the syncytiotrophoblast requires dissolution of the plasma membrane at small and well defined sites. Such sites could presumably be those regions requiring fresh cytotrophoblast derived material. Indeed, degenerative changes within the syncytiotrophoblast were suggested to induce differentiation of the underlying cytotrophoblasts [73]. If the syncytiotrophoblast would route syncytial fusion, the respective cytotrophoblasts could be prepared to undergo fusion by leaving the cell cycle, adjusting differentiation towards fusion and adjusting the cellular expression pattern to the needs of the overlying syncytiotrophoblast. Hence, defined sites of the syncytiotrophoblast may release stimuli regulating the level of differentiation of underlying cytotrophoblasts.

However, experiments using trypsinization of villous explants to remove the syncytiotrophoblast revealed that under such conditions fusion of villous cytotrophoblasts still occurs [24, 33, 74]. This seems to contradict the hypothesis of initiating fusion by the syncytiotrophoblast. This only holds true if the syncytiotrophoblast-derived signal is a signal to promote differentiation and fusion. On the other hand, if this syncytial signal hinders fusion, then the reduction or loss of the signal may result in initiating fusion. From this point of view, the experiments listed above could well be explained by such a negative syncytial signal.

6.5.2 Cytotrophoblast

Initiation of fusion by the cytotrophoblast is difficult to anticipate without a close collaboration with the syncytiotrophoblast. If fusion is initiated and conducted by the cytotrophoblast without knowledge of the situation in the overlying syncytiotrophoblast, the latter could locally exceed but also exhaust in compounds derived from the cytotrophoblast. Hence, it is tempting to speculate that a cross-talk between the two layers exists to spatially and temporarily route syncytial fusion between a cytotrophoblast and the syncytiotrophoblast.

6.6 Concluding Remarks

Syncytial fusion of villous cytotrophoblasts with the syncytiotrophoblast needs to be tightly controlled and at the same time needs to take place throughout pregnancy. Only a continuous input of fresh cytotrophoblastic compounds assures the maintenance of the multinucleated syncytiotrophoblast and thus sustains the placental barrier. Different factors and players have been suggested to be essential for trophoblast fusion including cytokines, hormones, protein kinases, and transcription factors. More specifically, experiments on the syncytins 1 and 2 as well as on caspase 8 have opened the avenue to broaden the field of players in trophoblast fusion. However, a specific role for any of the suggested players is still not certain.

References

1. Chen EH, Olson EN (2004) Towards a molecular pathway for myoblast fusion in drosophila. Trends Cell Biol 14:452–460
2. Benirschke K, Kaufmann P, Baergen R (2006) Pathology of the human placenta, 5th edn. Springer, New York, NY
3. Midgley AR, Pierce GB Jr, Deneau GA et al (1963) Morphogenesis of syncytiotrophoblast in vivo: an autoradiographic demonstration. Science 141:349–350
4. Panigel M (1993) The origin and structure of extraembryonic tissues. In: Redman C, Sargent I, Starkey P, (eds) The human placenta. Blackwell Scientific Publications, London
5. Richart R (1961) Studies of placental morphogenesis. I. Radioautographic studies of human placenta utilizing tritiated thymidine. Proc Soc Exp Biol Med 106:829–831
6. Huppertz B, Kaufmann P, Kingdom JCP (2002) Trophoblast turnover in health and disease. Fetal Maternal Med Rev 13:17–32
7. Huppertz B, Kingdom JC (2004) Apoptosis in the trophoblast–role of apoptosis in placental morphogenesis. J Soc Gynecol Investig 11:353–362
8. Oren-Suissa M, Podbilewicz B (2007) Cell fusion during development. Trends Cell Biol 17:537–546
9. de Parseval N, Lazar V, Casella JF et al (2003) Survey of human genes of retroviral origin: identification and transcriptome of the genes with coding capacity for complete envelope proteins. J Virol 77:10414–10422
10. Rote NS, Chakrabarti S, Stetzer BP (2004) The role of human endogenous retroviruses in trophoblast differentiation and placental development. Placenta 25:673–683
11. Mi S, Lee X, Li X et al (2000) Syncytin is a captive retroviral envelope protein involved in human placental morphogenesis. Nature 403:785–789
12. Lee X, Keith JC Jr, Stumm N et al (2001) Downregulation of placental syncytin expression and abnormal protein localization in pre-eclampsia. Placenta 22:808–812
13. Blond JL, Lavillette D, Cheynet V et al (2000) An envelope glycoprotein of the human endogenous retrovirus HERV-W is expressed in the human placenta and fuses cells expressing the type D mammalian retrovirus receptor. J Virol 74:3321–3329
14. Frendo JL, Olivier D, Cheynet V et al (2003) Direct involvement of HERV-W env glycoprotein in human trophoblast cell fusion and differentiation. Mol Cell Biol 23:3566–3574
15. Yu C, Shen K, Lin M et al (2002) GCMa regulates the syncytin-mediated trophoblastic fusion. J Biol Chem 277:50062–50068
16. Chen CP, Chen LF, Yang SR et al (2008) Functional characterization of the human placental fusogenic membrane protein syncytin 2. Biol Reprod 79:815–823
17. Vargas A, Moreau J, Landry S et al (2009) Syncytin-2 plays an important role in the fusion of human trophoblast cells. J Mol Biol 392:301–318
18. Esnault C, Priet S, Ribet D et al (2008) A placenta-specific receptor for the fusogenic, endogenous retrovirus-derived, human syncytin-2. Proc Natl Acad Sci USA 105:17532–17537
19. Malassine A, Blaise S, Handschuh K et al (2007) Expression of the fusogenic HERV-FRD env glycoprotein (syncytin 2) in human placenta is restricted to villous cytotrophoblastic cells. Placenta 28:185–191
20. Malassine A, Frendo JL, Blaise S et al (2008) Human endogenous retrovirus-FRD envelope protein (syncytin 2) expression in normal and trisomy 21-affected placenta. Retrovirology 5:6
21. Butler TM, Elustondo PA, Hannigan GE et al (2009) Integrin-linked kinase can facilitate syncytialization and hormonal differentiation of the human trophoblast-derived BeWO cell line. Reprod Biol Endocrinol 7:51
22. Daoud G, Amyot M, Rassart E et al (2005) ERK1/2 and p38 regulate trophoblasts differentiation in human term placenta. J Physiol 566:409–423
23. Knerr I, Schubert SW, Wich C et al (2005) Stimulation of GCMa and syncytin via cAMP mediated PKA signaling in human trophoblastic cells under normoxic and hypoxic conditions. FEBS Lett 579:3991–3998
24. Baczyk D, Drewlo S, Proctor L et al (2009) Glial cell missing-1 transcription factor is required for the differentiation of the human trophoblast. Cell Death Differ 16:719–727
25. Gauster M, Siwetz M, Huppertz B (2009) Fusion of villous trophoblast can be visualized by localizing active caspase 8. Placenta 30:547–550
26. Gauster M, Siwetz M, Orendi K et al (2010) Caspases rather than calpains mediate remodelling of the fodrin skeleton during human placental trophoblast fusion. Cell Death Differ 17:336–345
27. Morrish DW, Bhardwaj D, Dabbagh LK et al (1987) Epidermal growth factor induces differentiation and secretion of human chorionic gonadotropin and placental lactogen in normal human placenta. J Clin Endocrinol Metab 65:1282–1290
28. Garcia-Lloret MI, Morrish DW, Wegmann TG et al (1994) Demonstration of functional cytokine-placental interactions: CSF-1 and GM-CSF stimulate human cytotrophoblast differentiation and peptide hormone secretion. Exp Cell Res 214:46–54

29. Yang M, Lei ZM, Rao Ch V (2003) The central role of human chorionic gonadotropin in the formation of human placental syncytium. Endocrinology 144:1108–1120
30. Crocker IP, Strachan BK, Lash GE et al (2001) Vascular endothelial growth factor but not placental growth factor promotes trophoblast syncytialization in vitro. J Soc Gynecol Investig 8:341–346
31. Li H, Dakour J, Guilbert LJ et al (2005) PL74, a novel member of the transforming growth factor-beta superfamily, is overexpressed in preeclampsia and causes apoptosis in trophoblast cells. J Clin Endocrinol Metab 90:3045–3053
32. Shi QJ, Lei ZM, Rao CV et al (1993) Novel role of human chorionic gonadotropin in differentiation of human cytotrophoblasts. Endocrinology 132:1387–1395
33. Leisser C, Saleh L, Haider S et al (2006) Tumour necrosis factor-alpha impairs chorionic gonadotrophin beta-subunit expression and cell fusion of human villous cytotrophoblast. Mol Hum Reprod 12:601–609
34. Lin C, Lin M, Chen H (2005) Biochemical characterization of the human placental transcription factor GCMa/1. Biochem Cell Biol 83:188–195
35. Baczyk D, Satkunaratnam A, Nait-Oumesmar B et al (2004) Complex patterns of GCM1 mRNA and protein in villous and extravillous trophoblast cells of the human placenta. Placenta 25:553–559
36. Das M, Xu B, Lin L et al (2004) Phosphatidylserine efflux and intercellular fusion in a BeWO model of human villous cytotrophoblast. Placenta 25:396–407
37. Savill J (1998) Apoptosis. Phagocytic docking without shocking. Nature 392:442–443
38. Lyden TW, Ng AK, Rote NS (1993) Modulation of phosphatidylserine epitope expression by BeWO cells during forskolin treatment. Placenta 14:177–186
39. Bevers EM, Williamson PL (2010) Phospholipid scramblase: an update. FEBS Lett 584:2724–2730
40. Rote NS, Wei BR, Xu C et al (2010) Caspase 8 and human villous cytotrophoblast differentiation. Placenta 31: 89–96
41. Huppertz B, Frank HG, Kingdom JC et al (1998) Villous cytotrophoblast regulation of the syncytial apoptotic cascade in the human placenta. Histochem Cell Biol 110:495–508
42. van den Eijnde SM, van den Hoff MJ, Reutelingsperger CP et al (2001) Transient expression of phosphatidylserine at cell-cell contact areas is required for myotube formation. J Cell Sci 114:3631–3642
43. Adler RR, Ng AK, Rote NS (1995) Monoclonal antiphosphatidylserine antibody inhibits intercellular fusion of the choriocarcinoma line, JAR. Biol Reprod 53:905–910
44. Martin I, Pecheur EI, Ruysschaert JM et al (1999) Membrane fusion induced by a short fusogenic peptide is assessed by its insertion and orientation into target bilayers. Biochemistry 38:9337–9347
45. Decout A, Labeur C, Goethals M et al (1998) Enhanced efficiency of a targeted fusogenic peptide. Biochim Biophys Acta 1372:102–116
46. Barnoy S, Glaser T, Kosower NS (1998) The calpain-calpastatin system and protein degradation in fusing myoblasts. Biochim Biophys Acta 1402:52–60
47. Bartoli M, Richard I (2005) Calpains in muscle wasting. Int J Biochem Cell Biol 37:2115–2133
48. Barnoy S, Maki M, Kosower NS (2005) Overexpression of calpastatin inhibits L8 myoblast fusion. Biochem Biophys Res Commun 332:697–701
49. Black S, Kadyrov M, Kaufmann P et al (2004) Syncytial fusion of human trophoblast depends on caspase 8. Cell Death Differ 11:90–98
50. Pop C, Salvesen GS (2009) Human caspases: activation, specificity, and regulation. J Biol Chem 284:21777–21781
51. Launay S, Hermine O, Fontenay M et al (2005) Vital functions for lethal caspases. Oncogene 24:5137–5148
52. Fernando P, Kelly JF, Balazsi K et al (2002) Caspase 3 activity is required for skeletal muscle differentiation. Proc Natl Acad Sci USA 99:11025–11030
53. Huppertz B, Frank HG, Reister F et al (1999) Apoptosis cascade progresses during turnover of human trophoblast: analysis of villous cytotrophoblast and syncytial fragments in vitro. Lab Invest 79:1687–1702
54. De Falco M, Fedele V, Cobellis L et al (2004) Immunohistochemical distribution of proteins belonging to the receptor-mediated and the mitochondrial apoptotic pathways in human placenta during gestation. Cell Tissue Res 318:599–608
55. Fong PY, Xue WC, Ngan HY et al (2006) Caspase activity is downregulated in choriocarcinoma: a cDNA array differential expression study. J Clin Pathol 59:179–183
56. Miehe U, Kadyrov M, Neumaier-Wagner P et al (2006) Expression of the actin stress fiber-associated protein CLP36 in the human placenta. Histochem Cell Biol 126:465–471
57. Gadella BM, Harrison RA (2000) The capacitating agent bicarbonate induces protein kinase A-dependent changes in phospholipid transbilayer behavior in the sperm plasma membrane. Development 127:2407–2420
58. Sessions A, Horwitz AF (1983) Differentiation-related differences in the plasma membrane phospholipid asymmetry of myogenic and fibrogenic cells. Biochim Biophys Acta 728:103–111
59. Mandal D, Mazumder A, Das P et al (2005) Fas-, caspase 8-, and caspase 3-dependent signaling regulates the activity of the aminophospholipid translocase and phosphatidylserine externalization in human erythrocytes. J Biol Chem 280:39460–39467

60. Ohtani T, Hatori M, Ito H et al (2000) Involvement of caspases in 5-FU induced apoptosis in an oral cancer cell line. Anticancer Res 20:3117–3121
61. Mombers C, Verkleij AJ, de Gier J et al (1979) The interaction of spectrin-actin and synthetic phospholipids. II. The interaction with phosphatidylserine. Biochim Biophys Acta 551:271–281
62. Sato SB, Ohnishi S (1983) Interaction of a peripheral protein of the erythrocyte membrane, band 4.1, with phosphatidylserine-containing liposomes and erythrocyte inside-out vesicles. Eur J Biochem 130:19–25
63. Janicke RU, Ng P, Sprengart ML et al (1998) Caspase-3 is required for alpha-fodrin cleavage but dispensable for cleavage of other death substrates in apoptosis. J Biol Chem 273:15540–15545
64. Gudi SR, Kumar A, Bhakuni V et al (1990) Membrane skeleton-bilayer interaction is not the major determinant of membrane phospholipid asymmetry in human erythrocytes. Biochim Biophys Acta 1023:63–72
65. Martens S, McMahon HT (2008) Mechanisms of membrane fusion: disparate players and common principles. Nat Rev Mol Cell Biol 9:543–556
66. Eskelinen S, Lehto VP (1994) Induction of cell fusion in cultured fibroblasts and epithelial cells by microinjection of EGTA, GTP gamma S and antifodrin antibodies. FEBS Lett 339:129–133
67. Yusuf K, Smith SD, Sadovsky Y et al (2002) Trophoblast differentiation modulates the activity of caspases in primary cultures of term human trophoblasts. Pediatr Res 52:411–415
68. Sakuragi N, Matsuo H, Coukos G et al (1994) Differentiation-dependent expression of the BCL-2 proto-oncogene in the human trophoblast lineage. J Soc Gynecol Investig 1:164–172
69. Chang DW, Xing Z, Pan Y et al (2002) C-FLIP(L) is a dual function regulator for caspase-8 activation and CD95-mediated apoptosis. Embo J 21:3704–3714
70. Dash PR, McCormick J, Thomson MJ et al (2007) Fas ligand-induced apoptosis is regulated by nitric oxide through the inhibition of fas receptor clustering and the nitrosylation of protein kinase cepsilon. Exp Cell Res 313: 3421–3431
71. Ka H, Hunt JS (2006) FLICE-inhibitory protein: expression in early and late gestation human placentas. Placenta 27:626–634
72. Rajashekhar G, Loganath A, Roy AC et al (2003) Resistance to fas-mediated cell death in BeWO and NJG choriocarcinoma cell lines: implications in immune privilege. Gynecol Oncol 91:89–100
73. Benirschke K, Kaufmann P, Baergen RN (2006) Pathology of the human placenta, 5th edn. Springer, New York, NY
74. Baczyk D, Dunk C, Huppertz B et al (2006) Bi-potential behaviour of cytotrophoblasts in first trimester chorionic villi. Placenta 27:367–374
75. Morrish DW, Bhardwaj D, Paras MT (1991) Transforming growth factor beta 1 inhibits placental differentiation and human chorionic gonadotropin and human placental lactogen secretion. Endocrinology 129:22–26
76. Jiang B, Kamat A, Mendelson CR (2000) Hypoxia prevents induction of aromatase expression in human trophoblast cells in culture: potential inhibitory role of the hypoxia-inducible transcription factor mash-2 (mammalian achaete-scute homologous protein-2). Mol Endocrinol 14:1661–1673
77. Jansson T (2001) Amino acid transporters in the human placenta. Pediatr Res 49:141–147
78. Kudo Y, Boyd CA (2002) Human placental amino acid transporter genes: expression and function. Reproduction 124:593–600
79. Kudo Y, Boyd CA (2004) RNA interference-induced reduction in CD98 expression suppresses cell fusion during syncytialization of human placental BeWO cells. FEBS Lett 577:473–477
80. Kudo Y, Boyd CA, Millo J et al (2003) Manipulation of CD98 expression affects both trophoblast cell fusion and amino acid transport activity during syncytialization of human placental BeWO cells. J Physiol 550:3–9
81. Frendo JL, Cronier L, Bertin G et al (2003) Involvement of connexin 43 in human trophoblast cell fusion and differentiation. J Cell Sci 116:3413–3421
82. Dalton P, Christian HC, Redman CW et al (2007) Membrane trafficking of CD98 and its ligand galectin 3 in BeWO cells–implication for placental cell fusion. Febs J 274:2715–2727
83. Huppertz B, Bartz C, Kokozidou M (2006) Trophoblast fusion: fusogenic proteins, syncytins and ADAMs, and other prerequisites for syncytial fusion. Micron 37:509–517
84. White L, Dharmarajan A, Charles A (2007) Caspase-14: a new player in cytotrophoblast differentiation. Reprod Biomed Online 14:300–307

Chapter 7
Macrophage Fusion and Multinucleated Giant Cells of Inflammation

Amy K. McNally and James M. Anderson

Abstract Macrophages undergo fusion with other macrophages to form the hallmark multinucleated giant cells of chronic inflammation. However, neither the existence of distinct morphological types of giant cells, the signaling pathways that induce their formation, the molecular mechanism(s) of macrophage fusion, nor the significance of macrophage multinucleation at chronic inflammatory sites are well understood. Our efforts have been focused on these unknowns, particularly as they relate to the foreign body-type giant cells that form on implanted biomaterials and biomedical devices. We have pursued the discoveries of human macrophage fusion factors (interleukin-4, interleukin-13, α-tocopherol) with emphasis on foreign body giant cells, and identified adhesion receptors and signaling intermediates, as well as an adhesion protein substrate (vitronectin) that supports macrophage fusion. Studies on the molecular mechanism of macrophage fusion have revealed it to be a mannose receptor-mediated phagocytic process with participation of the endoplasmic reticulum. Further phenotypic and functional investigations will foster new perspectives on these remarkable multinucleated cells and their physiological significances in multiple inflammatory processes.

7.1 Introduction

Multinucleated giant cells have long been regarded as hallmark indicators of chronic inflammatory processes. As early as 1868, Langhans reported unusual giant cells containing multiple peripherally-arranged nuclei in the granulomas of tuberculosis [1]. Other than this long-standing link with chronic inflammation, however, we know relatively little about these intriguing cells and even less about why they appear where and when they do.

In many cases where giant cells are observed, there is a definable pathological agent, such as in tuberculosis, in which the causative organism is a mycoplasma. Additional examples of known causes are persistent bacterial, viral, parasitic, or fungal infections. In other cases, giant cells arise where the chronic inflammatory cause is not precisely known, for example, in sarcoidosis, rheumatoid arthritis, and certain neoplasias [2]. Of particular interest for our research, giant cells appear where there is a non-phagocytosable foreign body in the form of an implanted biomedical device or biomaterial [3]. In fact, so-called foreign body giant cells have been observed to interface with vascular, cardiovascular, orthopedic, and breast prostheses for periods extending to 15 years [4] and to occupy as much as 25% of implant surface area [5]. Therefore, they are a prominent cell type on biomaterials and have been widely linked to the biodegradation of certain biomedical polymers in vivo.

A.K. McNally (✉)
Department of Pathology, Wolstein Research Building, Case Western Reserve University, Cleveland, OH 44106, USA
e-mail: akm2@case.edu

T. Dittmar, K.S. Zänker (eds.), *Cell Fusion in Health and Disease*, Advances in Experimental Medicine and Biology 713, DOI 10.1007/978-94-007-0763-4_7, © Springer Science+Business Media B.V. 2011

From a cell biological perspective, the most interesting and striking feature of giant cells is that they are actually multinucleated macrophages, formed by macrophage fusion with other macrophages [6, 7]. Other well known examples of cell–cell fusion, such as myoblast fusion, sperm/ovum fusion, or osteoclast formation, are clearly function-driven aspects of normal physiology. In contrast, and because macrophages in their mononuclear form appear to be effective in other immune and inflammatory scenarios, the biological basis of macrophage multinucleation at various sites of chronic inflammation remains only speculative. For example, "frustrated phagocytosis" has been suggested as a driving force for multinucleation [8], which could potentially serve to combine phagocytic forces that are otherwise ineffective. An alternative possibility is that macrophage multinucleation might function to sequester a nonphagocytosable foreign body in order to protect host tissues from the adverse consequences of an on-going chronic inflammatory response. In a polarized cell type, as has been proposed by Vignery [9], both of these situations could be the case. Nevertheless, the single common denominator in these otherwise pathologically distinguishable scenarios appears to be the persistent, i.e. unresolvable by phagocytosis, presence of foreign microorganisms or materials. Beyond this, the precise molecular mechanism of macrophage fusion has not been elucidated, and the potential physiological significance of multinucleation itself, so key to understanding these cells, is as yet unclear.

7.2 Morphological Types of Multinucleated Giant Cells

A further dimension to these unknowns stems from the existence of morphological "variants" of multinucleated giant cells, of which there are two major recognized types. These have also long been observed, and yet their potential differences have been largely, and even surprisingly, overlooked. However, in order to advance our understanding of multinucleated macrophage biology, we must begin to view distinct morphological types as more than vague "variants" of the same thing. In keeping with structure/function relationships in biological systems, our in vitro findings with human macrophage fusion support this view and raise new questions on the potential significances of *types* of multinucleation.

The type of giant cell originally observed by Langhans is consistently circular or ovoid in shape with a limited number of nuclei, often arranged in a characteristic circular or semi-circular "horseshoe" pattern (Fig. 7.1a). These multinucleated cells vary somewhat in diameter but seldom exceed

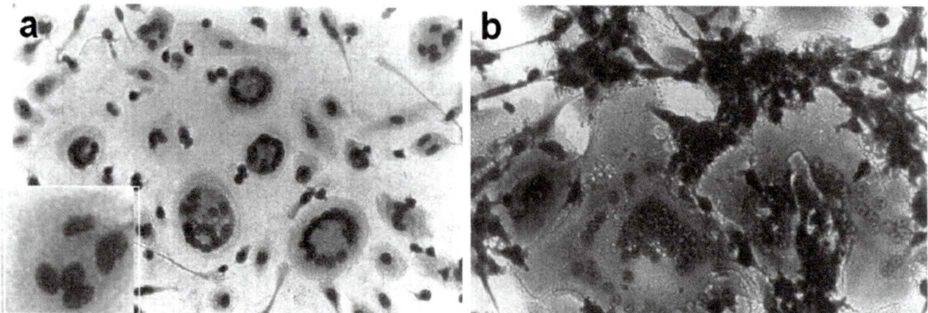

Fig. 7.1 Micrographic images of (a) LGC or (b) FBGC giant cell morphological types induced with IFN-γ + IL-3 or IL-4, respectively. The *inset* at *lower left* in (**a**) is an enlargement of the multinucleated cell seen at the *top right corner* of the same image. It demonstrates a random nuclear arrangement under LGC-inducing conditions, illustrating how classification of giant cells by morphology alone has generated confusion in the literature

50 microns, nor do they normally contain more than about 10–20 nuclei, which indicates a limited potential for macrophage fusion. Langhans-type giant cells (LGC) are commonly seen in association with granulomas due to chronic microbial infections [10].

Quite unmistakably distinct from LGC are the foreign body-type giant cells (FBGC) which are found interfacing with foreign materials such as surgical sutures or implanted biomedical devices (Fig. 7.1b). These exhibit an irregularly-shaped cytoplasm which is highly variable in size. Numbers of nuclei in FBGC are also widely variable and range from ten to many tens to even hundreds within a cytoplasm that may exceed one millimeter in diameter [5, 11]. This suggests that, unlike LGC, a mechanism to restrict degrees of fusion does not operate in FBGC. In further contrast to LGC, FBGC exhibit no definable patterns of nuclear arrangments, with multiple nuclei randomly scattered throughout the extensive cytoplasm of these irregularly-shaped cells.

There also exists a gray area between these two readily identifiable giant cell types, which is occupied by multinucleated cells with relatively few nuclei (usually around 3–10) and no particular pattern of nuclear arrangement (for example, the multinucleated cell in the top right corner of Fig. 7.1a, which is enlarged in the inset at lower left). This is the cell type that is responsible for apparent confusion in the literature because: (1) the classification of giant cells is as yet insufficiently based almost completely on morphology, and (2) macrophage fusion leading to giant cell formation is a morphological continuum, of which we can glimpse only part. Although their random nuclear arrangements usually place these cells in the foreign body giant cell category, this may, in fact, be misleading. Obviously, they could be precursors to larger and more highly multinucleated FBGC. However, they may actually be LGC precursors, i.e in the early stages of formation, post several fusion events but prior to the circular or semi-circular arrangement of nuclei. Finally, they could represent a non-Langhans-, non-foreign body-type giant cell. Further studies focused on the phenotypes and functions of morphologic types of giant cells are rquired to illuminate these unknowns.

7.3 Differential Signaling Pathways for Macrophage Multinucleation

In our in vitro studies with human monocyte-derived macrophages, we were able to demonstrate that the FBGC and LGC morphologies clearly arise under the influences of very different cytokines. As depicted in Fig. 7.2, FBGC can be induced by interleukin (IL)-4 [11] or IL-13 [12], or, as we later discovered, by a non-cytokine moiety, α-tocopherol [13]. Alternatively, LGC generation is mediated by interferon (IFN)-γ plus a macrophage maturation factor [11, 14]. The FBGC or LGC generated in our parallel in vitro systems were morphologically indistinguishable from those observed adherent to implanted biomaterials or in association with infectious granulomas, respectively. Both IL-4 and IFN-γ had been previously linked to macrophage fusion [14–17]. It was difficult to compare these investigations, however, because of considerable variations in cell sources and culture conditions. We demonstrated that, under otherwise identical culture conditions, these two distinct cytokines induced two morphologically distinguishable types of giant cells from human blood monocyte-derived macrophages [11]. This indicated, for the first time, that the occurrence of distinct types of giant cells at chronic inflammatory sites may represent different host responses to diverse inflammatory stimuli.

7.3.1 Interleukin-4 and Interleukin-13

Our early efforts to establish an in vitro system of human macrophage fusion that would duplicate FBGC morphology were aided by a study from McInnis and Rennick [15], who, working with IL-4-treated mouse bone marrow macrophages, were able to achieve a fusion rate of approximately 10%. However few, the FBGC that they generated were morphologically very similar to the

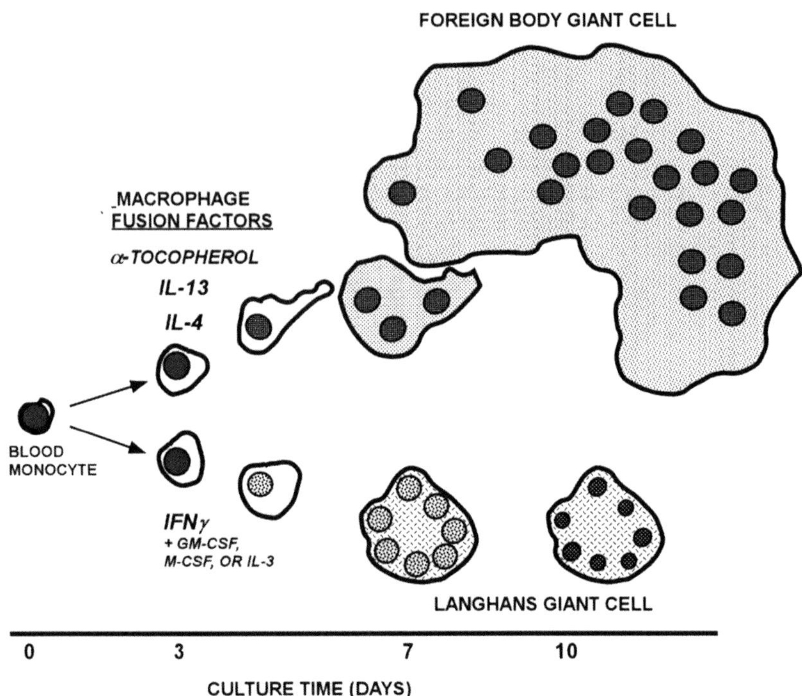

Fig. 7.2 Morphologically distinct types of giant cells are induced by differential cytokine signals from human blood monocyte-derived macrophages in vitro. IL-4, IL-13, or α-tocopherol promotes the formation of foreign body-type giant cells (FBGC) on implanted biomaterials. FBGC vary greatly in cytoplasmic areas and numbers of randomly-arranged nuclei. In contrast, Langhans-type giant cells (LGC) are induced by IFN-γ in the presence of a macrophage maturation factor such as GM-CSF, M-CSF, or IL-3. LGC are relatively much smaller and are characterized by a circular or semi-circular arrangement of nuclei within an ovoid cytoplasm. At longer culture times, LGC develop pyknotic nuclei, indicating apoptosis, whereas pyknotic nuclei are not observed in FBGC

FBGC observed on retrieved biomedical materials. Other investigators had, up to that point, also reported varying degrees of human macrophage fusion using INF-γ [14, 16] or supernatant from lectin-stimulated mononuclear leukocytes [17–19], but these cultures mainly appeared to produce "polykaryons" which bore no resemblance to giant cells observed in vivo [16] or Langhans-type multinucleated cells [14, 17], which are not oberved interfacing with implanted biomaterials [3]. In two of these studies, each attempted to extend the finding of McInnis and Rennick from mouse to human monocytes/macrophages. They were each unsuccessful, concluding that IL-4 was not a human macrophage fusion factor and that the effect observed by McInnis and Rennick was a species-specific one [14, 17]. However, these investigators had added IL-4 directly to human monocytes in culture, which actually inhibits monocyte adhesion [20]. Instead, our approach provided a period of monocyte-to-macrophage development prior to the addition of IL-4. Thus, we were able to demonstrate that IL-4 is indeed a potent human macrophage fusion factor [11], and that the process of IL-4-induced FBGC formation requires a degree of macrophage development prior to the induction of fusion. We initially found that this could be achieved by the inclusion of macrophage maturation factors, such as granulocyte-macrophage colony-stimulating factor (GM-CSF), macrophage (M)-CSF, or IL-3 in the cultures [11]. However, mauration factors do not, by themselves, induce fusion and are not required for

IL-4-induced fusion per se. We later learned that if monocytes/macrophages were cultured on chemically supportive surfaces, such as arginine-glycine-aspartate (RGD)-modified cell culture polystyrene, the addition of IL-4 only was sufficient to induce FBGC formation [21]. At the time, IL-13 was also emerging as alternative macrophage activation cytokine with multiple activities that were similar to IL-4. Therefore, we tested IL-13 under the same culture conditions that had been established for IL-4 and discovered that it, too, was a potent human macrophage fusion factor that, by itself, could induce the formation of FBGC in a manner indistinguishable from that of IL-4 [22]. In vivo studies confirmed a role for IL-4 in FBGC formation on biomaterials [23]. Our more recent in vivo efforts with athymic (nude) mice suggest that Th2 lymphocytes are not the source of IL-4/IL-13 [24]. Other possibilities include mast cells, eosinophils, basophils, natural killer (NK) lymphocytes, and NKT lymphocytes [25, 26]. These avenues remain to be addressed.

7.3.2 *Interferon-γ*

In parallel cultures and under identical conditions, we tested IFN-γ as a fusion factor compared to IL-4. We found that IFN-γ induced fusion leading to the formation of LGC only [11]. Unlike FBGC formation, IFN-γ-mediated fusion does require the cocommitant presence of a macrophage maturation factor, which can be either granulocyte-macrophage colony-stimulating factor (GM-CSF), macrophage-CSF, or IL-3 [11]. As illustrated in Fig. 7.2, LGC appear to have a limited lifespan, as nuclear pyknosis occurs with continued culture times, whereas FBGC cultures do not develop pyknotic nuclei (our unpublished observations). In the presence of IL-4, apoptosis does not occur [27], and the resultant FBGC exhibit morphologically normal nuclei.

7.3.3 *α-Tocopherol*

It was also intriguing that vitamin E incorporated in a poly(etherurethane) biomaterial induced increases in FBGC formation yet decreases in biomaterial degradation in vivo ([28] and unpublished data). This was the first evidence to disconnect IL-4-induced macrophage fusion from biomedical polymer degradation. In vitro studies to directly pursue this revealed that vitamin E (90% α-tocopherol) moderately induced macrophage fusion and increased IL-4-induced FBGC formation. However, the purified α-tocopherol isomer alone most remarkably induced macrophage fusion, leading to cultures of confluent FBGC below normal plasma tocopherol concentrations [13]. This was not the case with the structurally similar antioxidants probucol or Trolox, suggesting that the α-tocopherol effects on FBGC formation were independent of its antioxidant activity. In this regard, multiple activities of α-tocopherol have been described that are independent of its antioxidant properties, including effects on phospholipase A_2, protein kinase C (PKC), adhesion, and diacylglycerol kinase activity ([29] and see below). This study revealed that α-tocopherol is yet a third as well as the most potent human macrophage fusion factor, inducing striking multinucleated giant cells of the foreign body-type.

7.4 Mechanisms of Adhesion that Support FBCG Formation

Based on our collective experiences, monocyte-derived macrophage fusion leading to FBGC or LGC formation requires adhesion success. This is highly dependent on culture material surface chemistry and/or adsorbed blood proteins [21, 30]. Culture materials that do not support initial monocyte adhesion and its ensuing macrophage morphological development (cytoplasmic expansion) beyond monocyte adhesion cannot support giant cell formation. Apparently, this requires the engagement of select integrins with appropriate adsorbed protein ligands to initiate activation of specific adhesion signals. Our investigations on adhesion mechanisms that support FBGC formation have focused on specific components of adhesion success. These include the adsorption of appropriate blood proteins or the provision of a supportive culture material surface, the identification of relevant adhesion receptors (integrins), integrin-mediated signaling (adhesion kinases), cytoskeletal responses (microfilaments and microtubules), and the assembly of focal adhesion structures (podosomes). In addition, we have pursued mechanisms of material-dependent adhesion failure, or anoikis (apoptosis) as a means of intervention in the process of FBGC formation on biomaterials.

7.4.1 Adhesion Receptors (Integrins)

Initial efforts to identify adhesion receptors that support FBGC formation revealed that both β1 and β2 integrins mediated adhesion during IL-4-induced FBGC formation [31]. Initial monocyte adhesion required functional β2 integrins, whereas, during the IL-4 induction of macrophage fusion, an additional dependence on functional β1 integrins was acquired. Of note, we did not find a functional role for β3 integrins in FBGC formation, nor did we detect β3 integrin in our culture system. This indicated that FBGC adhesion differs from that of osteoclasts, which utilize αVβ3 integrins for adhesion to bone [32]. Subsequent studies focused on the identities of the α integrin partners to these heterodimeric adhesion receptors [33]. Immunoprecipitation of fusing macrophage/FBGC lysates with anti-β1 integrin and immunoblotting revealed the presence of α5 and αV, as well as α2 and α3. As expected, αM and αX immunoprecipitated with β2 but not with β1. We did not detect α4 or several other β integrin partners. Immunocytochemistry coupled with confocal microscopy indicated that α5 and αX are poorly expressed on day 0. However, following the induction of macrophage fusion by IL-4 on day 3, they were each readily detectable in fusing macrophages/FBGC on day 7. In contrast, αM and αV were present throughout the culture period, with very strong αM and αX expression on day 7. We also demonstrated expression and co-localization of α3, α5, or αV with β1 on fusing macrophages/FBGC at this time point as well as strong co-localization of αM and αX with β2 at cell–cell fusion interfaces. Therefore, IL-4-induced FBGC are characterized by the expression of αMβ2 and αXβ2 > α5β1 and αVβ1 > α3β1 and α2β1. Thus, monocytes/macrophages and FBGC express a select group of adhesion receptors with potential for binding to specific blood proteins that may adsorb to biomaterials and to extracellular matrix proteins, including complement C3, fibrin(ogen), fibronectin, Factor X, vitronectin, certain collagens, laminin, and perhaps others as new ligands for these receptors become known.

7.4.2 RGD and Vitronectin

Initially, we discovered that arginine-glycine-aspartate (RGD)-modified culture polystyrene supported optimal monocyte-to-macrophage development in the absence of any other macrophage maturation factors [21], suggesting that an adhesion protein(s) containing this prototypical cell attachment sequence [34] is critical for this morphological progression. Combined with our later integrin receptor findings, we directly addressed the identification of relevant ligand(s) for these adhesion receptors. The approach was to adsorb potential integrin protein ligands to polystyrene culture surfaces. We thereby found that IL-4-induced FBGC formation did not proceed in vitro on material adsorbed with complement C3bi, fibrinogen, plasma fibronectin, cell-derived (fibroblast) fibronectin, collagen types I or IV, or laminin [30]. Surprisingly, these proteins also completely restricted macrophage adhesion and development and FBGC formation on RGD-modified surfaces, our optimal FBGC substrate material. In striking contrast, FBGC formation readily occurred on adsorbed vitronectin, which contains the RGD cell attachment sequence, and this effect of vitronectin was comparable to FBGC formation on adsorbed RGD peptide. Therefore, although fusing macrophages/FBGC express several β1 and β2 integrins and thereby would appear to possess broad ligand binding potential, they selectively utilize specific adhesive proteins to support macrophage adhesion and fusion leading to FBGC formation. Although fibrinogen, fibronectin, collagens, and laminin are well known to support the adhesion of multiple other cell types, they do not support macrophage development leading to FBGC formation. These findings indicate that the optimal cell binding RGD adhesive sequence for FBGC formation is presented in the blood-derived protein vitronectin, and that the propensity for vitronectin adsorption may be one mechanism for the material surface chemistry dependency of FBGC formation on biomaterials.

7.4.3 Adhesion Kinases

Integrin activation is a process in which conformational changes that increase integrin ligand binding affinity occur [35]. In phagocytic cells, activation leads to integrin clustering, which promotes intracellular signaling pathways that collectively control cytoskeletal rearrangements and formation of adhesion structures. These, in turn, support cell mobility, survival, and synthetic abilities. The integrin signaling proline-rich tyrosine kinase-2 (PYK2) is a member of the focal adhesion kinase (FAK) family and is highly expressed in macrophages [36]. PYK2 is co-localized with paxillin, talin, vinculin, and αMβ2 integrin in cell adhesion structures termed podosomes (see below). PYK2 is tyrosine phosphorylated upon macrophage adhesion and has been functionally linked to integrin-mediated regulation of cell spreading and migration [36]. We have demonstrated strong signals for FAK and PYK2 in whole cell lysates of IL-4-induced fusing macrophages/FBGC, the expression of which each increases during macrophage development and FBGC formation [37]. Further, FBGC adhesion was abrogated by the tyrosine kinase inhibitor genistein and by the phosphatidylinositol-3-kinase inhibitors wortmannin and LY294002 [31].

7.4.4 Microfilaments and Microtubules

Our early inhibitor studies demonstrated that F-actin was necessary for IL-4- or IL-13-induced FBGC formation because cytochalasins, which disrupt actin microfilaments, inhibited macrophage fusion in a concentration-dependent manner [38]. Importantly, the concentrations of cytochalasins that interfered with fusion did not decrease macrophage adhesion, cytoplasmic spreading, or motility but did prevent internalization of yeast via mannose receptor-mediated phagocytosis. This indicated that the mechanism of fusion is related to phagocytosis (see below, Molecular Mechanism of Macrophage Fusion). Furthermore, nocodazole restricted macrophage fusion in a concentration-dependent manner, pointing to an additional role for microtubules in this phenomenon [39].

7.4.5 Podosomes

Podosomes are specialized macrophage adhesion structures located at the cell/substrate interface where actin microfilaments terminate at the cell membrane [40]. We have noted that podosomes are a striking feature of FBGC, which display dense peripheral rings of these adhesion structures [12]. Surrounding the actin microfilament core, podosomes are specifically associated with β2 integrins in the membrane and are composed of gelsolin, paxillin, talin, vinculin, and other proteins, such as the actin-bundling protein, fascin-1 [41]. Beta integrin subunit binding by talin is the last common step in integrin activation and is believed to represent the center of converging integrin activation signals leading to formation of adhesion structures [35]. Adhesive structural proteins (gelsolin, paxillin, talin, fascin-1) are detectable during macrophage development and strongly up-regulated in fusing macrophages/FBGC on RGD-modified surfaces in vitro [37].

7.4.6 Adhesion Failure (Anoikis)

The alternative to adhesion success is anoikis, which is defined as apoptosis specifically due to adhesion failure [42]. Anoikis is a normal biological mechanism for the control and regulation of cell proliferation and tissue development. Cell death signaling by this mechanism is believed to be regulated by the cytoskeleton. Our in vitro observations have been that monocytes initially adhere quite well to most surfaces. However, those that fail to maintain adhesion and undergo macrophage morphological development, i.e. cytoplasmic spreading, apparently do not survive and/or cannot form FBGC. Thus, whether adhesion signals that promote integrin clustering, adhesion kinase activation, cytoplasmic spreading, and adhesion structure (podosome) formation are initiated and maintained is evidently determined by the material adherence substrate. Exploiting this natural cellular phenomenon to better understand the material surface chemistry dependence of macrophage survival and FBGC formation on biomaterials, we have evaluated apoptosis on several types of modified materials by measuring early and late events in apoptosis. As initial studies with differentially-modified polystyrenes had indicated, macrophage development and FBGC formation are highly material surface-dependent [43]. We found that a mechanism for this phenomenon is the biomaterial surface chemistry-dependent induction of apoptosis [44–46]. In addition, we found that activation of caspase-3, an intermediate indicator of apoptosis signaling, in inflammatory cells under shear stress leads to cell detachment [47]. It is of particular interest that caspases are known to cleave gelsolin, an important component of podosomes, and thereby disrupt adhesive interactions, leading to apoptosis/anoikis [42].

7.5 Molecular Mechanism of Macrophage Fusion

7.5.1 A Role for Mannose Receptors (MR)

Upon the establishment of our in vitro system of IL-4-induced FBGC formation, we could begin to address the moleular mechanism of macrophage-macrophage fusion. Inasmuch as IL-4 was reported to most stongly induce MR on macrophages, a finding which fostered the concept of "alternative activation" [48], we tested the effects of previously described inhibitors of MR activity on IL-4-induced FBGC formation [49]. Patterns of inhibition were consistent with participation of MR in the mechanism of fusion, and MR were found to be specifically up-regulated by IL-4 in our culture system and concentrated at fusion interfaces [49]. The formation of LGC, induced by a cytokine-enriched supernatant from lectin-stimulated mononuclear leukocytes, was also inhibited by α-mannan [43], indicating that similar mechanisms of fusion may operate to induce both types of giant cells. At that time, the MR was already well known for its key role in innate immunity via the clearance of microorganisms bearing terminal mannose oligosaccharides from mammalian tissues. Our data revealed a novel function for this endocytic/phagocytic receptor in the formation of multinucleated macrophages.

7.5.2 A Phagocytic Mechanism for Fusion with Participation of the Endoplasmic Reticulum

We extended these findings as well as our previously demonstrated requirements for F-actin in macrophage fusion with similarities to phagocytosis [38] and found that macrophage fusion leading to FBGC formation exhibits multiple additional features of phagocytosis [39]. Exploiting multiple pharmacological inhibitors, we discovered critical roles for vacuolar-type ATPase, microtubules (see above) and the endoplasmic reticulum (ER) in macrophage fusion. Further, we found a specific requirement for the calcium-independent phospholipase A_2 (iPLA_2), but not calcium-dependent PLA_2 (cPLA_2), secretory PLA_2 (sPLA_2), cyclooxygenase, or lipoxygenase in the mechanism of fusion. Immunocytochemistry confirmed iPLA_2 expression and absence of cPLA_2 or sPLA_2 expression in macrophages/FBGCs. As markers of ER-mediated phagocytosis, calnexin and calregulin were detectable on non-permeabilized fusing macrophages and also concentrated at fusion interfaces where they co-localized with actin in permeabilized macrophages/FBGCs. Furthermore, ER markers co-localized with concanavalin A reactivity, which is a marker of potential MR ligand, on non-permeabilized fusing macrophages, suggesting that the ER may present relevant MR ligand at fusion interfaces. These data demonstrated for the first time that the mechanism of macrophage fusion leading to formation of multinucleated giant cells occurs by ER-mediated phagocytosis. We believe that these findings reveal a mechanism by which such extensive degrees of cell–cell fusion and cytoplasmic spreading, characteristic features of these remarkable multinucleated cells, can take place.

7.5.3 Diacyl Glycerol (DG)-Dependent and -Independent PKCs

To further investigate the macrophage fusion signaling pathways that promote and support FBGC formation, we investigated the participation of PKC in the IL-4-induced fusion of human monocyte-derived macrophage in vitro [50]. The PKC inhibitors H-7, calphostin C, and GF109203X attenuated macrophage fusion, whereas H-8, which is more selective for PKA and PKG, did not. Macrophage fusion was also prevented by the phospholipase C inhibitor, Et-18-OCH_3, the PKC isoform inhibitors GO6983 or rottlerin and by peptide inhibitors for PKC (20–28), PKCβ, or PKCζ but not by HBDDE or peptide inhibitors for PKCε or PKA. In cultures of fusing macrophages/FBGCs, we detected only PKCα, β, δ, and ζ by immunoprecipitation and immunoblotting, and we also observed strong expression of these isoforms by immunocytochemistry. Our collective results suggest that the γ, ε, η, μ, θ, or ι PKC isoforms are not required in the mechanism of IL-4-induced macrophage fusion; whether PKCα is required is unclear. However, new evidence is provided that FBGC formation is specifically supported by PKCβ, PKCδ, and PKCζ in combined diacylglycerol-dependent (PKCβ and PKCδ) and -independent (PKCζ) signaling pathways. Importantly, inhibition of PKCβ and PKCδ do not affect macrophage development or cytoplasmic spreading during monocyte-to-macrophage differentiation but interfere with FBGC formation at the point of macrophage fusion. In contrast, inhibition of PKCζ has more drastic effects on the morphological progression by restricting early events that must support macrophage development and cytoplasmic spreading. Whether PKCζ is also required for macrophage fusion itself cannot be determined from the present studies. The collective data from studies of PKCβ, δ, and ζ suggest that these isoforms play prominent roles in phagocytosis signaling as well as in migration and macrophage differentiation [51–54].

7.5.4 Diacylglycerol Kinase

Consistent with the reported activation of diacylglycerol kinase by α-tocopherol, we found that the diacylglycerol kinase inhibitor R59022 completely abrogates FBGC formation [13]. R59022 inhibition of IL-4-induced FBGC formation is reversed by α-tocopherol, suggesting that FBGC formation induced by both fusion factors requires diacylglycerol kinase activation. This study suggests a novel role for diacylglycerol kinase in the mechanism of macrophage fusion/FBGC formation at sites of chronic inflammation. Diacylglycerol kinase converts diacylglycerol to phosphatidic acid, which may act as a fusogenic lipid in membranes [55]. Consistent with this, we discovered that propranolol, which also promotes phosphatidate formation, greatly enhances IL-4-induced FBGC formation (unpublished data).

7.5.5 *Matrix Metalloproteinase (MMP)-13*

A role for MMPs in IL-4-induced macrophage fusion was indicated by pharmacological inhibition with actinonin, which inhibits aminopeptidases such as collagenase MMPs and blocks approximately 60% of macrophage fusion at 50 μM [56]. In addition, CL-82198, reported to be a selective inhibitor of MMP-13 that does not affect MMP-1 or MMP-9, strongly inhibits fusion by approximately 60% between 1 and 5 μM. Epigallocatechin gallate, which inhibits MMP-2, MMP-9, and MMP-12, and NNGH, which inhibits MMP-3 and MMP-12 do not restrict macrophage fusion at the relevant concentrations [56]. Finally, we confirmed that MMP-13 is detectable in lysates of fusing macrophages/FBGC (unpublished finding).

7.6 Phenotypic and Functional Profiles of Multinucleated Giant Cells

New perspectives continue to emerge on "classical" versus "alternative" macrophage activation, phenotype acquisition, and switching [57–60]. The concept of alternative macrophage activation was introduced to distinguish Th2 lymphokine (IL-4)-activated from Th1 lymphokine (IFN-γ)- or classically activated macrophages [48, 61]. IL-4 (as well as its relative IL-13) has evolved as a prototypical alternative macrophage activation signal, as its effects on monocytes/macrophages are largely antagonistic to those of IFN-γ. Therefore, IL-4 promotes a macrophage phenotypic profile that is distinct from the classically activated macrophage generated by the pro-inflammatory cytokine IFN-γ.

For example, MR-mediated phagocytosis is strongly induced by IL-4 and inhibited by IFN-γ, which instead supports IgG-mediated phagocytosis. The production of pro-inflammatory cytokines (IL-1, IL-6, tumor necrosis factor-α) is induced by IFN-γ but inhibited by IL-4, which instead induces anti-inflammatory cytokines (IL-1 receptor antagonist and IL-10), thereby promoting wound healing and matrix deposition. Reactive oxygen and nitrogen species are induced by IFN-γ, but IL-4 inhibits these activities and instead induces arginase-1 (in the mouse), which mediates collagen deposition and tissue repair.

Collectively, the classically activated macrophage exhibits capacities for cellular immunity, microbicide, and tissue damage, whereas alternatively activated macrophages exhibit enhanced capacities for humoral immunity, allergic responses, and repair processes such as fibrosis, matrix remodeling, phagocytosis, and, notably, down-modulation of inflammation [62–65]. Thus, it is reasonable to predict, in keeping with current views of structure/function relationships in biological systems, that morphologically distinct types of giant cells induced by IL-4 (or IL-13) and IFN-γ will prove to exhibit distinguishable phenotypes and possess distinct functional capacities.

The FBGC that form on implanted materials have been widely believed to directly bring about the degradation of certain implanted biomaterials, negatively impacting their efficacy and biocompatibility. This perspective stems from early scanning electron micrographic analyses, which revealed pitting and cracking on poly(etherurethane) biomaterial surfaces in FBGC "footprints", i.e. areas from which adherent FBGC had been removed [66]. This seemed to indicate, and it was thus was inferred, that polymer degradation resulted from concentration of phagocytic oxidative and microbicidal activities within an acidic closed compartment or microenvironment at the FBGC/biomaterial interface.

Paradoxically, this perspective appears to be incompatible with our subsequent findings on FBGC formation. Of three discovered human macrophage fusion factors, IL-4, IL-13, and α-tocopherol [11, 13, 22], each is well and widely known, not to promote, but to down-modulate so-called "pro-inflammatory" and oxidative activities of macrophages. IL-4 and IL-13 are cytokines which each exert inhibitory effects on respiratory burst activity, the expression and secretion of pro-inflammatory cytokines, monocyte adhesion to endothelium, cytotoxic activities, and chronic destructive experimental arthritis [62, 67]. Instead, they induce mannose receptor expression, wound healing, angiogenesis, and tissue remodeling [61, 63, 65, 68].

Our findings with α-tocopherol, which is not a cytokine but the major component of vitamin E, further oppose the view that FBGC are the perpetraters of oxidative damage to biomedical polymers. In addition to its well known antioxidant properties and host tissue protective effects [69], α-tocopherol exerts non-antioxidant effects on macrophages, including the activation of diacylglycerol kinase related to production of phosphatidic and lysophosphatidic acids, actin polymerization, chemotaxis, cellular migration, and cell survival [29, 70–73]. Interestingly, α-tocopherol induces connective tissue growth factor (CTGF), which may promote connective tissue fibrosis at sites of chronic tissue injury and wound healing [74]. Accordingly, we have identified CTGF in cultures of fusing macrophages/FBGC (unpublished data). This may indicate that CTGF is synthesized and secreted by macrophages/FBGC at sites of biomaterial implantation.

Resolution of this apparent paradox is possible, however, if one supposes that biomaterial-adherent monocyte-derived macrophages initially exhibit a pro-inflammatory and oxidative phenotype with capacities for biomaterial surface degradation. Subsequently, this phenotype could undergo down-modulation by IL-4- and/or IL-13 signaling in a process of alternative macrophage activation that is accompanied by fusion of the adherent macrophages. The resultant biomaterial-adherent FBGC would exhibit capacities for wound healing, angiogenesis, and/or tissue remodeling, and yet the "footprints" of pro-inflammatory and oxidative activities would remain on the biomaterial surface and appear to have been mediated by the FBGC. If so, and if one could accelerate IL-4-induced alternative macrophage activation and FBGC formation, biomaterial degradation should be reduced. This is consistent with our observed decreased biomaterial degradation coupled with increased FBGC formation on vitamin E-modified poly(etherurethane) in vivo (unpublished data). Whether decreased biomaterial degradation resulted from the antioxidant activity of vitamin E and/or to vitamin E (α-tocopherol)-mediated increases in FBGC formation remains an intriguing question.

Accordingly, our in vitro studies on cytokine production by biomaterial-adherent monocytes, macrophages, and IL-4-induced FBGC support the concept of a time-dependent phenotypic switch. Initially, we found a cytokine switch from a pro-inflammatory to an alternative macrophage activation phenotype and a dissociation between pro-inflammatory cytokine production and FBGC formation [75]. In additional cytokine array studies aimed at determining the direct influences of lymphocytes in co-culture with monocytes/macrophages, we discovered that temporal patterns of cytokine production switch from an initial pro-inflammatory phenotype with IFN-γ production [76]. Further, when acute inflammatory cells (polymorphonuclear leukocytes) are included in initial co-culture with monocytes/macrophages (from day 0 to day 3), they exert a negative influence on subsequent IL-4-induced FBGC formation [77].

7.6.1 FBGC Versus Osteoclasts

Regarding the above-mentioned functional differences between giant cell types, it is now clear that osteoclasts differ from FBGC by several significant criteria. Most obviously, although both cell types are multinucleated and of monocyte-derived macrophage origin, osteoclasts are a feature of normal, non-inflammatory physiology [32]. FBGC, on the other hand, arise only under conditions of chronic inflammation [2, 3]. In addition and as discussed, FBGC are induced by IL-4 both in vitro and in vivo, but osteoclast formation and mature osteoclast function are inhibited by IL-4 [78]. Conversely, osteoclasts are formed under the influences of receptor activator of nuclear factor-κB ligand (RANKL) and tumor necrosis factor-α, the latter of which does not support FBGC formation [11]. In terms of adhesion, osteoclasts adhere to bone via αVβ3 and αVβ5 integrin interactions with osteopontin and bone sialoprotein [32]. Adhesion mechanisms that operate in FBGC do not include β3 integrins but, as outlined above, appear to involve αMβ2, αXβ2, αVβ1, perhaps additional β1 integrins, and adsorbed vitronectin [30, 31, 33]. The role of β5 integrins in FBGC adhesion has not yet been addressed. Phenotypic studies to probe for the expression of recognized osteoclast markers in IL-4-induced FBGC indicate that these multinucleated giant cells do not express tartrate-resistant acid phosphatase or calcitonin receptors (manuscript in preparation), further differentiating FBGC from osteoclasts.

7.7 Summary

Our perspectives on giant multinucleated cells of inflammation have evolved considerably in the last two decades. Significant progress has been made in determining macrophage fusion factors, signaling pathways, adhesion receptors/proteins, and fusion mechanisms for FBGC formation. Importantly and based on their generation from clearly opposing signaling pathways, the long-observed morphological distinctions between FBGC and LGC now extend to potential phenotype and function. Macrophage multinucleation is apparently a much more complex phenomenon than the coincidental existence of giant cell "variants". Further studies on these intriguing cells, directed at establishing new phenotypic classifications between types of giant cells, will broaden our perspectives of chronic inflammatory processes and perhaps reveal additional complexities therein. Ongoing investigations will ultimately reveal the physiological significances and roles played by these remarkable cells in inflammation.

Acknowledgments Supported by the National Institute of Biomedical Imaging and Bioengineering, National Institutes of Health.

References

1. Langhans T (1868) Über Riesenzellen mit wandständigen Kernen in Tuberkeln und die fibröse Form des Tuberkels. Arch Pathol Anat 42:382–404
2. Chambers TJ, Spector WG (1982) Inflammatory giant cells. Immunobiology 161:283–289
3. Anderson JM (2000) Multinucleated giant cells. Curr Opin Hematol 7:40–47
4. Anderson JM (1988) Inflammatory response to implants. ASAIO Trans 34:101–107
5. Zhao OH, Anderson JM, Hiltner A et al (1992) Theoretical analysis on cell size distribution and kinetics of foreign-body giant cell formation in vivo on polyurethane elastomers. J Biomed Mat Res 26:1019–1038
6. Murch AR, Grounds MD, Marshall CA et al (1982) Direct evidence that inflammatory multinucleate giant cells form by fusion. J Pathol 137:177–180
7. Sutton JS, Weiss L (1966) Transformation of monocytes in tissue culture into macrophages, epithelioid cells, and multinucleated giant cells. An electron microscope study. J Cell Biol 28:303–332
8. Henson PM, Henson JE, Fittschen C et al (1988) Phagocytic cells: Degranulation and secretion. In: Gallin JI, Snyderman R (eds) Inflammation: Basic principles and clinical correlates, edn, Raven, New York, NY
9. Vignery A, Niven-Fairchild T, Ingbar DH et al (1989) Polarized distribution of Na+,K+-ATPase in giant cells elicited in vivo and in vitro. J Histochem Cytochem 37:1265–1271
10. Brodbeck WG, Anderson JM (2009) Giant cell formation and function. Curr Opin Hematol 16:53–57
11. McNally AK, Anderson JM (1995) Interleukin-4 induces foreign body giant cells from human monocytes/macrophages. Differential lymphokine regulation of macrophage fusion leads to morphological variants of multinucleated giant cells. Am J Pathol 147:1487–1499

12. DeFife KM, Jenney CR, Colton E et al (1999) Cytoskeletal and adhesive structural polarizations accompany IL-13-induced human macrophage fusion. J Histochem Cytochem 47:65–74
13. McNally AK, Anderson JM (2003) Foreign body-type multinucleated giant cell formation is potently induced by alpha-tocopherol and prevented by the diacylglycerol kinase inhibitor R59022. Am J Pathol 163:1147–1156
14. Enelow RI, Sullivan GW, Carper HT et al (1992) Induction of multinucleated giant cell formation from in vitro culture of human monocytes with interleukin-3 and interferon-gamma: comparison with other stimulating factors. Am J Respir Cell Mol Biol 6:57–62
15. McInnes A, Rennick DM (1988) Interleukin 4 induces cultured monocytes/macrophages to form giant multinucleated cells. J Exp Med 167:598–611
16. Weinberg JB, Hobbs MM, Misukonis MA (1984) Recombinant human gamma-interferon induces human monocyte polykaryon formation. Proc Natl Acad Sci USA 81:4554–4557
17. Takashima T, Ohnishi K, Tsuyuguchi I et al (1993) Differential regulation of formation of multinucleated giant cells from concanavalin A-stimulated human blood monocytes by IFN-gamma and IL-4. J Immunol 150:3002–3010
18. Most J, Neumayer HP, Dierich MP (1990) Cytokine-induced generation of multinucleated giant cells in vitro requires interferon-gamma and expression of LFA-1. Eur J Immunol 20:1661–1667
19. Orentas RJ, Reinlib L, Hildreth JE (1992) Anti-class II MHC antibody induces multinucleated giant cell formation from peripheral blood monocytes. J Leukoc Biol 51:199–209
20. Elliott MJ, Gamble JR, Park LS et al (1991) Inhibition of human monocyte adhesion by interleukin-4. Blood 77:2739–2745
21. Anderson JM, Defife K, McNally A et al (1999) Monocyte, macrophage and foreign body giant cell interactions with molecularly engineered surfaces. J Mat Sci 10:579–588
22. DeFife KM, Jenney CR, McNally AK et al (1997) Interleukin-13 induces human monocyte/macrophage fusion and macrophage mannose receptor expression. J Immunol 158:3385–3390
23. Kao WJ, McNally AK, Hiltner A et al (1995) Role for interleukin-4 in foreign-body giant cell formation on a poly(etherurethane urea) in vivo. J Biomed Mat Res 29:1267–1275
24. Rodriguez A, Macewan SR, Meyerson H et al (2009) The foreign body reaction in T-cell-deficient mice. J Biomed Mat Res 90:106–113
25. Gessner A, Mohrs K, Mohrs M (2005) Mast cells, basophils, and eosinophils acquire constitutive IL-4 and IL-13 transcripts during lineage differentiation that are sufficient for rapid cytokine production. J Immunol 174: 1063–1072
26. O'Connor GM, Hart OM, Gardiner CM (2006) Putting the natural killer cell in its place. Immunology 117:1–10
27. Brodbeck WG, Shive MS, Colton E et al (2002) Interleukin-4 inhibits tumor necrosis factor-alpha-induced and spontaneous apoptosis of biomaterial-adherent macrophages. J Lab Clin Med 139:90–100
28. Schubert MA, Wiggins MJ, DeFife KM et al (1996) Vitamin E as an antioxidant for poly(etherurethane urea): in vivo studies. Student Research Award in the Doctoral Degree Candidate Category, Fifth World Biomaterials Congress (22nd Annual Meeting of the Society for Biomaterials), Toronto, Canada, May 29-June 2, 1996. J Biomed Mat Res 32:493–504
29. Azzi A, Stocker A (2000) Vitamin E: non-antioxidant roles. Progress Lipid Res 39:231–255
30. McNally AK, Jones JA, Macewan SR et al (2008) Vitronectin is a critical protein adhesion substrate for IL-4-induced foreign body giant cell formation. J Biomed Mat Res 86:535–543
31. McNally AK, Anderson JM (2002) Beta1 and beta2 integrins mediate adhesion during macrophage fusion and multinucleated foreign body giant cell formation. Am J Pathol 160:621–630
32. Teitelbaum SL (2007) Osteoclasts: what do they do and how do they do it? Am J Pathol 170:427–435
33. McNally AK, Macewan SR, Anderson JM (2007) alpha subunit partners to beta1 and beta2 integrins during IL-4-induced foreign body giant cell formation. J Biomed Mat Res 82:568–574
34. Ruoslahti E (1996) RGD and other recognition sequences for integrins. Annu Rev Cell Dev Biol 12:697–715
35. Ratnikov BI, Partridge AW, Ginsberg MH (2005) Integrin activation by talin. J Thromb Haemost 3:1783–1790
36. Duong LT, Rodan GA (2000) PYK2 is an adhesion kinase in macrophages, localized in podosomes and activated by beta(2)-integrin ligation. Cell Motil Cytoskeleton 47:174–188
37. MacEwan SR, McNally A, Anderson JM (2007) Focal adhesion kinase, proline-rich tyrosine kinase-2, thrombospondin, and fascin-1 expression in adherent macrophages and foreign body giant cells. Society for Biomaterials Annual Meeting, Chicago, IL, April 18–21,703
38. DeFife KM, Jenney CR, Colton E et al (1999) Disruption of filamentous actin inhibits human macrophage fusion. Faseb J 13:823–832
39. McNally AK, Anderson JM (2005) Multinucleated giant cell formation exhibits features of phagocytosis with participation of the endoplasmic reticulum. Exp Mol Pathol 79:126–135
40. Marx J (2006) Cell biology. Podosomes and invadopodia help mobile cells step lively. Science 312:1868–1869
41. Calle Y, Burns S, Thrasher AJ et al (2006) The leukocyte podosome. Eur J Cell Biol 85:151–157
42. Frisch SM, Screaton RA (2001) Anoikis mechanisms. Curr Opin Cell Biol 13:555–562

43. McNally A (1994) Mechanisms of monocyte/macrophage adhesion and fusion on different surfaces. PhD Thesis, Case Western Reserve University, Cleveland, OH
44. Brodbeck WG, Colton E, Anderson JM (2003) Effects of adsorbed heat labile serum proteins and fibrinogen on adhesion and apoptosis of monocytes/macrophages on biomaterials. J Mat Sci 14:671–675
45. Brodbeck WG, Patel J, Voskerician G et al (2002) Biomaterial adherent macrophage apoptosis is increased by hydrophilic and anionic substrates in vivo. Proc Natl Acad Sci USA 99:10287–10292
46. Jones JA, Dadsetan M, Collier TO et al (2004) Macrophage behavior on surface-modified polyurethanes. J Biomat Sci 15:567–584
47. Shive MS, Brodbeck WG, Anderson JM (2002) Activation of caspase 3 during shear stress-induced neutrophil apoptosis on biomaterials. J Biomed Mat Res 62:163–168
48. Stein M, Keshav S, Harris N et al (1992) Interleukin 4 potently enhances murine macrophage mannose receptor activity: a marker of alternative immunologic macrophage activation. J Exp Med 176:287–292
49. McNally AK, DeFife KM, Anderson JM (1996) Interleukin-4-induced macrophage fusion is prevented by inhibitors of mannose receptor activity. Am J Pathol 149:975–985
50. McNally AK, Macewan SR, Anderson JM (2008) Foreign body-type multinucleated giant cell formation requires protein kinase C beta, delta, and zeta. Exp Mol Pathol 84:37–45
51. Carnevale KA, Cathcart MK (2003) Protein kinase C beta is required for human monocyte chemotaxis to MCP-1. J Biol Chem 278:25317–25322
52. Larsen EC, DiGennaro JA, Saito N et al (2000) Differential requirement for classic and novel PKC isoforms in respiratory burst and phagocytosis in RAW 264.7 cells. J Immunol 165:2809–2817
53. Larsson C (2006) Protein kinase C and the regulation of the actin cytoskeleton. Cell Signal 18:276–284
54. Liu Q, Ning W, Dantzer R et al (1998) Activation of protein kinase C-zeta and phosphatidylinositol 3′-kinase and promotion of macrophage differentiation by insulin-like growth factor-I. J Immunol 160:1393–1401
55. Harsh DM, Blackwood RA (2001) Phospholipase A(2)-mediated fusion of neutrophil-derived membranes is augmented by phosphatidic acid. Biochem Biophys Res Commun 282:480–486
56. Jones JA, McNally AK, Chang DT et al (2008) Matrix metalloproteinases and their inhibitors in the foreign body reaction on biomaterials. J Biomed Mat Res 84:158–166
57. Mantovani A, Sica A, Locati M (2007) New vistas on macrophage differentiation and activation. Eur J Immunol 37:14–16
58. Mantovani A, Sica A, Sozzani S et al (2004) The chemokine system in diverse forms of macrophage activation and polarization. Trends Immunol 25:677–686
59. Porcheray F, Viaud S, Rimaniol AC et al (2005) Macrophage activation switching: an asset for the resolution of inflammation. Clin Exp Immunol 142:481–489
60. Taylor PR, Martinez-Pomares L, Stacey M et al (2005) Macrophage receptors and immune recognition. Annu Rev Immunol 23:901–944
61. Goerdt S, Politz O, Schledzewski K et al (1999) Alternative versus classical activation of macrophages. Pathobiology 67:222–226
62. Gordon S (2003) Alternative activation of macrophages. Nat Rev Immunol 3:23–35
63. Gratchev A, Kzhyshkowska J, Utikal J et al (2005) Interleukin-4 and dexamethasone counterregulate extracellular matrix remodelling and phagocytosis in type-2 macrophages. Scand J Immunol 61:10–17
64. Moestrup SK, Moller HJ (2004) CD163: a regulated hemoglobin scavenger receptor with a role in the anti-inflammatory response. Ann Med 36:347–354
65. Mosser DM (2003) The many faces of macrophage activation. J Leukoc Biol 73:209–212
66. Zhao Q, Topham N, Anderson JM et al (1991) Foreign-body giant cells and polyurethane biostability: in vivo correlation of cell adhesion and surface cracking. J Biomed Mat Res 25:177–183
67. Malefyt RW (1999) Role of interleukin-10, interleukin-4, and interleukin-13 in resolving inflammatory responses. In: Gallin JL, Snyderman R (eds) Inflammation: Basic principles and clinical correlates, edn, Lippincott Williams & Wilkins, Philadelphia, PA
68. Kodelja V, Muller C, Tenorio S et al (1997) Differences in angiogenic potential of classically vs alternatively activated macrophages. Immunobiology 197:478–493
69. Brigelius-Flohe R, Traber MG (1999) Vitamin E: function and metabolism. Faseb J 13:1145–1155
70. Chan SS, Monteiro HP, Schindler F et al (2001) Alpha-tocopherol modulates tyrosine phosphorylation in human neutrophils by inhibition of protein kinase C activity and activation of tyrosine phosphatases. Free Rad Res 35: 843–856
71. Koh JS, Lieberthal W, Heydrick S et al (1998) Lysophosphatidic acid is a major serum noncytokine survival factor for murine macrophages which acts via the phosphatidylinositol 3-kinase signaling pathway. J Clin Invest 102: 716–727
72. Lee IK, Koya D, Ishi H et al (1999) d-Alpha-tocopherol prevents the hyperglycemia induced activation of diacylglycerol (DAG)-protein kinase C (PKC) pathway in vascular smooth muscle cell by an increase of DAG kinase activity. Diabetes Res Clin Pract 45:183–190

73. Topham MK, Prescott SM (1999) Mammalian diacylglycerol kinases, a family of lipid kinases with signaling functions. J Biol Chem 274:11447–11450
74. Leask A, Holmes A, Abraham DJ (2002) Connective tissue growth factor: a new and important player in the pathogenesis of fibrosis. Curr Rheumatol Rep 4:136–142
75. Jones JA, Chang DT, Meyerson H et al (2007) Proteomic analysis and quantification of cytokines and chemokines from biomaterial surface-adherent macrophages and foreign body giant cells. J Biomed Mat Res 83:585–596
76. Chang DT, Colton E, Matsuda T et al (2009) Lymphocyte adhesion and interactions with biomaterial adherent macrophages and foreign body giant cells. J Biomed Mat Res 91:1210–1220
77. Kirk JT, McNally AK, Anderson JM (2010) Polymorphonuclear leukocyte inhibition of monocytes/macrophages in the foreign body reaction. J Biomed Mat Res 94:683–687
78. Moreno JL, Kaczmarek M, Keegan AD et al (2003) IL-4 suppresses osteoclast development and mature osteoclast function by a STAT6-dependent mechanism: irreversible inhibition of the differentiation program activated by RANKL. Blood 102:1078–1086

Chapter 8
Molecular Mechanisms of Myoblast Fusion Across Species

Adriana Simionescu and Grace K. Pavlath

Abstract Skeletal muscle development, growth and regeneration depend on the ability of progenitor myoblasts to fuse to one another in a series of ordered steps. Whereas the cellular steps leading to the formation of a multinucleated myofiber are conserved in several model organisms, the molecular regulatory factors may vary. Understanding the common and divergent mechanisms regulating myoblast fusion in *Drosophila melanogaster* (fruit fly), *Danio rerio* (zebrafish) and *Mus musculus* (mouse) provides a better insight into the process of myoblast fusion than any of these models could provide alone. Deciphering the mechanisms of myoblast fusion from simpler to more complex organisms is of fundamental interest to skeletal muscle biology and may provide therapeutic avenues for various diseases that affect muscle.

8.1 Introduction

Skeletal muscle is composed of multinucleated myofibers, which are postmitotic. Myofibers form by the fusion of mononucleated progenitor myoblasts with one another in a series of ordered steps, including differentiation, elongation, migration, adhesion, membrane alignment and finally membrane union. Each of these steps is highly regulated by a variety of molecules. In this chapter, we compare and contrast molecular regulators of myoblast fusion in three model organisms, *Drosophila melanogaster* (fruit fly), *Danio rerio* (zebrafish) and *Mus musculus* (mouse), in order to gain a deeper insight into the general mechanisms regulating myoblast fusion.

8.2 Current Model Systems for Studying Myoblast Fusion

Currently, three model organisms are utilized for studying the process of myoblast fusion. These are the fruit fly (*Drosophila melanogaster*), an invertebrate, as well as the zebrafish (*Danio rerio*) and the mouse (*Mus musculus*), both vertebrates. Each of these model systems offers unique advantages for studying myoblast fusion. In this chapter, we utilize both vertebrate and invertebrate model systems to highlight the current knowledge about the process of myoblast fusion. Below, we emphasize the advantages of each model system for studying myoblast fusion, as well as specifics about the myoblast fusion process in each organism, with an attempt to compare and contrast this process and the molecules regulating it in these three model organisms.

G.K. Pavlath (✉)
Department of Pharmacology, Emory University, Atlanta, GA 30322, USA
e-mail: gpavlat@emory.edu

T. Dittmar, K.S. Zänker (eds.), *Cell Fusion in Health and Disease*, Advances in Experimental Medicine and Biology 713, DOI 10.1007/978-94-007-0763-4_8, © Springer Science+Business Media B.V. 2011

8.3 General Aspects of Myoblast Fusion in *Drosophila melanogaster*

The *Drosophila* model system offers many advantages for studying embryonic myoblast fusion. In contrast to vertebrates, the somatic musculature of *Drosophila* is much less complex, muscle development takes less time, and each resulting muscle is composed of a single, multinucleated myofiber, as opposed to a bundle of myofibers. Also, the small size and short life cycle of *Drosophila* facilitate generating a large number of flies for genetic manipulations. Together with the small size of the *Drosophila* genome, these properties have led to the widespread use of genetic screens and the discovery of genetic mutants with defects in myoblast fusion. The genes responsible for these defects were subsequently identified, and new molecular pathways were characterized. Studying these genetic mutants at the ultrastructural level by transmission electron microscopy enabled a deeper understanding of the membrane dynamics occurring during alignment and breakdown, leading to the formation of a mature, multinucleated muscle. In addition, cell-type specific labeling and live imaging of actin

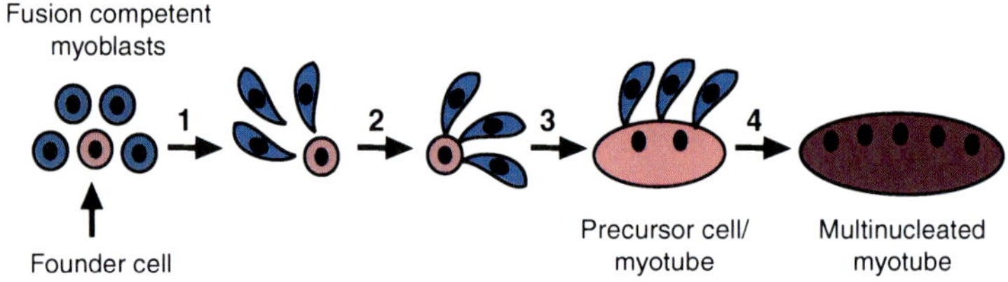

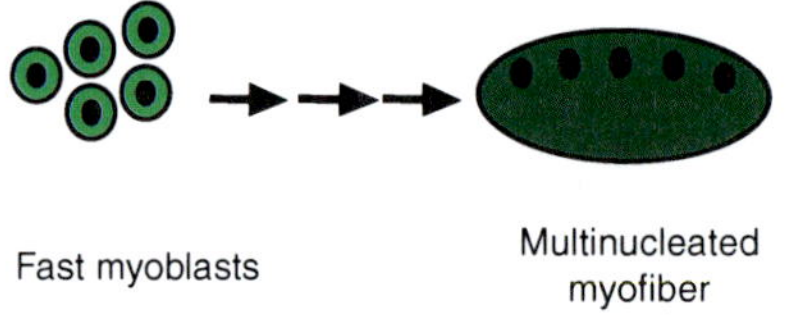

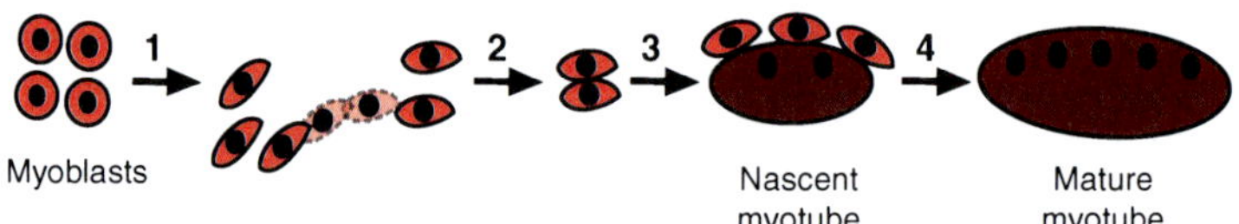

Fig. 8.1 Myoblast fusion in *Drosophila melanogaster*, *Danio rerio* and *Mus musculus*. (**a**) During embryonic myoblast fusion in the fruit fly (*Drosophila melanogaster*), recognition (*1*) and adhesion (*2*) between fusion competent myoblasts and the founder cell occurs, leading to the formation of a binucleated precursor cell/myotube (*3*). Subsequently, more fusion competent myoblasts fuse with the precursor cell/myotube, giving rise to a multinucleated myotube (*4*). (**b**) During embryonic myoblast fusion in the zebrafish (*Danio rerio*), fast myoblasts differentiate and fuse with one another, giving rise to multinucleated myofibers. (**c**) During embryonic and adult myoblast fusion in the mouse (*Mus musculus*), differentiated myoblasts elongate and migrate towards other myoblasts (*1*). Subsequently, myoblasts adhere to one another (*2*) and fuse, first giving rise to a nascent myotube (*3*), and then more myoblasts fuse with the nascent myotube, giving rise to a mature myotube (*4*)

in these mutants facilitated studies of the cellular dynamics of myoblast fusion in vivo. Together, these techniques have provided insights into the process of myoblast fusion in *Drosophila* at various levels.

During embryonic myoblast fusion in *Drosophila*, two types of myoblasts, the founder cell (FC) and the fusion competent myoblast (FCM) fuse with one another [1, 2] (Fig. 8.1a). Thus, fusion is specific, directional and asymmetric, as it only occurs between FCs and FCMs, and not between two equivalent myoblasts. Initially, each FC "seeds" a distinct muscle, and gives the future muscle its characteristics, including location, orientation, insertion points and size. As surrounding FCMs fuse with the FC, the nuclei of the newly incorporated FCMs adopt the same pattern of gene expression as the FC [2]. Two current models exist for embryonic myoblast fusion in *Drosophila* [3–5]. The first model proposes two distinct steps in the formation of the multinucleated myotube [4, 5]. In the first step of fusion, each FC fuses with several FCMs to give rise to a bi- or trinucleate precursor cell, and in the second step, further fusion events occur with the precursor cell to give rise to the mature muscle. This model implies that some mutants may not progress at all beyond the bi- or trinucleate precursor cell stage, and suggests that a distinct subset of genes is necessary for each of the two steps. However, all fusion mutants identified are capable of occasional fusions, suggesting a second model known as the two-phase model, in which the same genes are necessary throughout the entire fusion process [3, 5]. Key to this model is the fact that rare and limited fusion events occur in the first phase of fusion, and more frequent fusion events occur in the second phase [3]. The transition between the first and second phase in this model may be due to either limiting factors whose levels are tightly controlled during fusion and may increase in the second phase or to the spatial arrangements of FCs and FCMs in vivo. Thus, in the first phase, few FCMs are close to FCs, resulting in limited fusion events, whereas in the second phase, FCMs that were previously unable to contact FCs, will migrate towards the FCs and fuse with them, increasing the frequency of fusion. Although attempts to merge these two models have been made [4], further data are required.

Myoblast fusion in *Drosophila* continues during pupal development, and is mediated by myoblasts derived from adult muscle precursors, which are set aside during embryonic development by asymmetric division of FCs [2, 6]. These myoblasts then undergo either de novo fusion or fusion with existing larval muscles [6]. However, much less is understood about the process of myoblast fusion during these stages compared to embryonic myoblast fusion. Therefore, we focus on the molecular mechanisms regulating embryonic myoblast fusion in *Drosophila* in this chapter.

8.4 Molecules Regulating Myoblast Fusion in *Drosophila melanogaster*

As myoblast fusion in *Drosophila* occurs between two distinct types of myoblasts, the FC and the FCM, not surprisingly, some molecules regulating fusion are asymmetrically distributed between the two cells (Fig. 8.2). Initially, transmembrane proteins mediate the recognition (indicated by filopodia extension) and adhesion steps. Following adhesion, downstream signaling pathways are activated, which involve cytoplasmic adaptors and actin cytoskeletal regulators, leading to actin cytoskeletal remodeling. Actin foci form at the site of adhesion between two cells [7]. An important determinant of the extent of myoblast fusion is the presence, size and number of actin foci. Some mutants in which fusion defects are observed display long-term persistence of actin foci, indicating that actin foci dissolution is necessary to enable membrane fusion.

Some of the proteins regulating myoblast fusion in *Drosophila* have been studied in multiple organisms (Table 8.1) and show conserved functions. However, for other proteins, although homologs exist in higher organisms, their role in fusion has not been investigated (Table 8.1). Further studies on the role of these later proteins in myoblast fusion in higher organisms will determine whether they too are functionally conserved, or whether novel functions emerge. Below, we discuss the major molecules that regulate myoblast fusion in *Drosophila*.

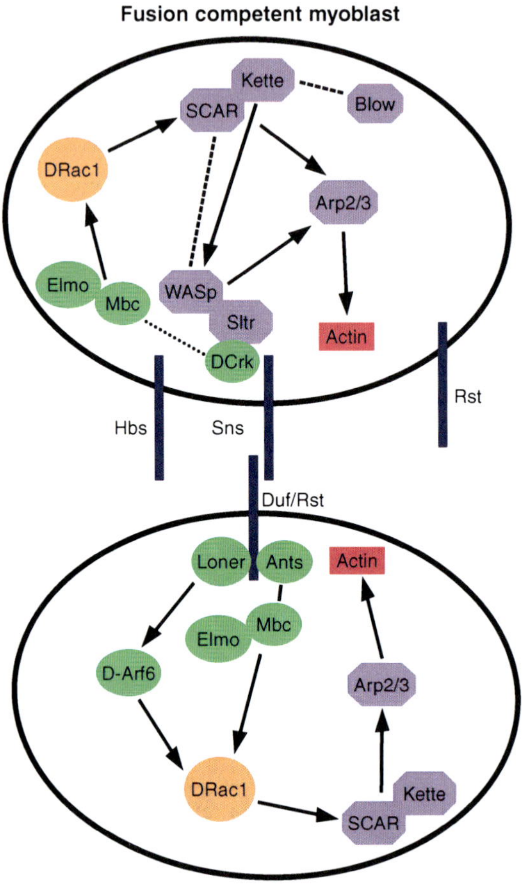

Fig. 8.2 Molecular mechanisms of myoblast fusion in *Drosophila melanogaster*. During embryonic myoblast fusion in *Drosophila melanogaster*, some transmembrane proteins (*bars*) interact with one another, resulting in recruitment of cytoplasmic proteins (*ovals*), leading to activation of *Drosophila* Rac1 (DRac1). Actin cytoskeletal rearrangements mediated by other cytoplasmic proteins (*octagons*) occur at the site of fusion by both DRac1-dependent and independent pathways. *Solid arrows* = known regulators; *solid lines* = known biochemical interactions; *dashed lines* = genetic interactions; *dotted lines* = potential biochemical interactions. Ants = Antisocial; Arf6 = Adenosine diphosphate ribosylation factor 6; Arp2/3 = Actin related protein 2/3; Blow = Blown fuse; Crk = CT10 regulator of kinase; D = prefix, *Drosophila* protein; Duf = Dumbfounded; Elmo = Engulfment and cell motility; Hbs = Hibris; Mbc = Myoblast city; Rac1 = Ras-related C3 botulinum toxin substrate 1; Rst = Roughest; SCAR = Suppressor of cyclic AMP receptor; Sltr = Solitary; Sns = Sticks and stones; WASp = Wiskott-Aldrich syndrome protein

8.4.1 Recognition and Adhesion

In order for myoblast fusion to occur, transmembrane proteins must first mediate recognition and adhesion between the FC and the FCM. The major molecules currently known to mediate recognition and adhesion (Fig. 8.2), as well as their mode of interaction, are discussed in this section.

FCs express two transmembrane proteins that belong to the Immunoglobulin superfamily, Dumbfounded/Kin of irre C (Duf/Kirre) and Roughest/Irregular chiasm C (Rst/IrreC) (Fig. 8.2), which

Table 8.1 Conserved proteins with known and putative roles in myoblast fusion in *Drosophila melanogaster*, *Danio rerio* and *Mus musculus*[a]

Function	*Drosophila melanogaster*	*Danio rerio*	*Mus musculus*	References
Recognition and adhesion	Duf/Kirre	Kirrel	Kirrel family[b]	[8, 9, 43, 44]
	Rst/IrreC	Kirrel	Kirrel family[b]	[8, 43, 44]
	Sns	Nephrin[b]	Nephrin	[10, 13, 44]
	Hbs	Nephrin[b]	Nephrin	[11–13, 44]
Cytoplasmic adaptor	DCrk	Crk/CrkL	?[c]	[27, 45]
Actin cytoskeletal regulator	Mbc	Dock1,5	Dock1,5	[23, 24, 45, 119, 120]
	Loner	?[c]	Brag2/ARF-GEP$_{100}$	[21, 120]
	D-Arf6	?[c]	Arf6	[21, 111]
	DRac1	Rac1	Rac1	[16, 43, 69, 121]
	Kette	?[c]	Nap1	[7, 34, 35, 56]
	WASp	?[c]	N-WASP	[28, 30–32]
	Sltr/D-WIP	?[c]	WIP	[28, 31]

[a]Homologous proteins that have a role in myoblast fusion in two out of the three organisms.
[b]Homologous protein whose fusion role has not been studied.
[c]Homolog not known, or protein has not been studied in skeletal muscle.

act redundantly on FCs during myoblast fusion [8, 9]. Deletion of *duf* does not cause obvious muscle defects, and loss of *rst* causes only mild muscle defects. However, in the absence of both genes, FCMs extend filopodia in random orientations, with no preference for the FC itself, and fusion is blocked. As ectopic Duf or Rst mediate recruitment of FCMs, both proteins function as FCM attractants [8, 9]. These observations suggest that Duf or Rst expression in FCs is essential for FCM attraction towards the FC. Although Rst is also expressed in FCMs, its role there is not essential for attraction towards ectopic sites of Duf or Rst [8].

FCMs express two transmembrane proteins that belong to the Immunoglobulin superfamily, Sticks and stones (Sns) and Hibris (Hbs) (Fig. 8.2), which are not found on FCs [10–12]. In *sns* mutants, fusion does not occur beyond the bi- or trinucleate precursor cell [10, 13]. Although embryos that lack *hbs* do not have an overt muscle phenotype [12], they exhibit a partial fusion block, resulting in a number of unfused myoblasts [11]. Moreover, in *sns hbs* double mutants, only mononucleated cells exist, indicating that *sns* and *hbs* function redundantly for the formation of the bi- or trinucleate precursor cell [13]. The role of Hbs itself during myoblast fusion appears to be both positive and negative. Hbs overexpression in *sns* mutant embryos drives fusion beyond the precursor cell stage [13], however, when Hbs is overexpressed in wild type embryos, a block in fusion occurs [11, 12]. One potential explanation proposed is that Hbs and Sns interact, and that Hbs negatively regulates Sns by various mechanisms [11, 13]. However, fully understanding the role of Hbs in myoblast fusion requires further studies.

As fusion occurs only between the FC and the FCM, Duf and Rst on the FC may interact with Sns and Hbs on the FCM to mediate recognition and adhesion. Duf and Sns colocalize at points of contact between the FC and the FCM [14], suggesting that they may interact in vivo. In addition, in embryos deficient for both Duf and Rst, Sns localization is disrupted [14], suggesting that Sns localization depends on Duf and Rst. In vitro studies using the Schneider line 2 (S2) cell line [15], a non-adherent cell line, enable insights into the potential interactions between these transmembrane proteins, as well as with their intracellular partners, by expression of various recombinant proteins in these cells. These studies further support the idea that Duf/Rst and Sns may function as a receptor ligand/pair for mediating FC-FCM adhesion [14] (Fig. 8.2).

8.4.2 Cytoplasmic Adaptors and Actin Cytoskeletal Regulators

The initial recognition and adhesion between the FC and the FCM, mediated by ligand/receptor pairs, is followed by downstream signaling, likely involving the binding of intracellular proteins to the cytoplasmic tails of these transmembrane proteins. These downstream pathways lead to actin cytoskeletal remodeling, which enables the final step of membrane fusion. Many of these proposed signaling pathways were determined by in vitro studies in S2 cells, and in vivo evidence exists only for some of them. Below, we discuss the major molecules involved in regulating myoblast fusion downstream of the Duf/Rst and Sns receptor/ligand pair (Fig. 8.2).

8.4.2.1 Signaling Pathways Activating DRac1

Downstream of Duf and Rst, two intracellular pathways have been described, which converge at *Drosophila* Ras-related C3 botulinum toxin substrate 1 (DRac1) (Fig. 8.2), a Rho GTPase with a role in actin cytoskeletal remodeling. DRac1 was implicated in myoblast fusion due to its mutant phenotype: expression of a dominant negative DRac1 leads to delayed myoblast fusion initially, and then excessive fusion [16], whereas expression of a constitutively active DRac1 also leads to a block in fusion [16, 17].

In the first pathway leading to DRac1 activation is Antisocial/Rolling pebbles 7 (Ants/Rols7) (Fig. 8.2), a cytoplasmic adaptor protein that localizes to FCs and is required for myoblast fusion at a stage beyond the initial attraction and adhesion between the FC and the FCM [18–20]. In *ants* mutant embryos, FCMs still extend filopodia towards the FCs, but fusion beyond the bi- or trinucleate precursor does not occur [18, 19]. Both in vivo and in vitro studies suggest that Ants acts downstream of Duf/Rst in FCs [18, 20–22] (Fig. 8.2).

The next important player is Myoblast city (Mbc) (Fig. 8.2), a cytoplasmic protein required for myoblast fusion [23, 24]. Mbc likely has a role in very early fusion stages, as the initial recognition and adhesion steps between FCs and FCMs do not occur in *mbc* mutants [17]. In vitro, Mbc interacts with Ants [18], suggesting a potential role for Mbc in FCs. Additionally, Mbc interacts in vitro with Engulfment and Cell Motility (Elmo) [25, 26] (Fig. 8.2), another protein essential for myoblast fusion [26]. In the *Drosophila* eye Mbc and Elmo act together as a non-conventional GEF for DRac1 [26]. However, the connection between Mbc/Elmo/DRac1 and actin polymerization during embryonic myoblast fusion in *Drosophila* is currently not understood. In vitro interactions between Mbc and the cytoplasmic adaptor protein *Drosophila* CT10 regulator of kinase (DCrk) protein [25, 27], and between DCrk and Sns [28], suggest that Mbc may regulate actin cytoskeletal dynamics through a pathway involving DCrk and Sns in FCMs (Fig. 8.2). While DCrk may regulate actin cytoskeletal dynamics potentially through Mbc, Mbc can also direct fusion in the absence of DCrk binding [29]. Therefore, the exact role of DCrk in embryonic myoblast fusion in *Drosophila* remains to be determined.

In the second pathway leading to DRac1 activation is Loner (Fig. 8.2), a GEF, which functions independently of Ants [21]. Loner is localized to FCs, and is required for myoblast fusion likely after the recognition and adhesion steps [21], as in *loner* mutants, FCMs extend filopodia towards FCs, but do not fuse with them. As indicated by both in vivo and in vitro studies, Loner functions downstream of Duf/Rst in FCs [21]. Furthermore, Loner functions as a GEF for the small GTPase *Drosophila* adenosine diphosphate ribosylation factor 6 (D-Arf6), a class III GTPase [21] (Fig. 8.2). D-Arf6 is also required for myoblast fusion, as the expression of a dominant negative D-Arf6 in FCs leads to a disruption in myoblast fusion [21]. As the phenotype of a dominant negative D-Arf6 in FCs is similar to, but less severe than that of loss of Loner, D-Arf6 is likely an essential mediator of myoblast fusion downstream of Loner [21] (Fig. 8.2). DRac1 is no longer localized at fusion sites in *loner* mutants, suggesting that the Loner/D-Arf6 pathway may be required for the proper localization of DRac1 in FCs [21] (Fig. 8.2), enabling actin cytoskeletal regulation during embryonic myoblast fusion in *Drosophila*.

8.4.2.2 Signaling Pathways Directly Upstream of Arp2/3

Most pathways regulating actin cytoskeletal dynamics in FCs and FCMs appear to converge at the actin-related protein 2/3 (Arp2/3) (Fig. 8.2). Below we detail the regulation of actin cytoskeletal rearrangements that occurs during embryonic myoblast fusion, modulated by Arp2/3 and its upstream regulators. Given the complexities of, and some controversy in, the regulation of the actin cytoskeleton upstream of Arp2/3 during embryonic myoblast fusion, only one possible model is presented in Fig. 8.2.

Arp2/3, which is required for myoblast fusion [7, 30], is activated by two actin nucleation-promoting factors (NPFs), suppressor of cyclic AMP receptor (SCAR) and Wiskott-Aldrich syndrome protein (WASp) (Fig. 8.2). In the complete absence of either SCAR or WASp in vivo, a severe fusion block occurs, suggesting that both SCAR and WASp are required for myoblast fusion [7, 28, 30–32]. The two NPFs also appear to regulate each other during the fusion process. Thus, *scar* and *wasp* interact genetically [30], however how SCAR and WASP function together during myoblast fusion is controversial [30, 33].

Kette regulates the activity of SCAR to modulate Arp2/3-dependent actin polymerization. However, the actual mechanism by which Kette regulates SCAR is controversial, as both positive and negative regulation has been proposed [7, 34]. Kette is expressed in both the FC and the FCM (Fig. 8.2) [35], and localizes mostly to the cytoplasm, and some to the membrane [34]. Kette is required for myoblast fusion, as *kette* mutants exhibit a block in myoblast fusion after the formation of the precursor cell [35]. In addition, *kette* genetically interacts with *blow* (Fig. 8.2), which encodes Blown Fuse (Blow), a cytoplasmic protein restricted to the FCM [35, 36]. Like Kette, Blow is required for fusion beyond the precursor cell [17, 35]. Additionally, Kette regulates the activity of WASP (Fig. 8.2) to mediate Arp2/3-dependent actin cytoskeletal rearrangements both positively and negatively [32, 34].

Finally, the cytoplasmic protein Solitary (Sltr)/D-WASp-interacting protein (D-WIP), is expressed in the FCM but not the FC [28, 31]. Sltr is required for fusion beyond the binucleate precursor cell [28, 31]. Sltr positively regulates Arp2/3-dependent actin polymerization by interacting with WASp and recruiting it to fusion sites, indicating that Sltr and WASp may function together to mediate the fusion process [28, 31]. Furthermore, Sltr is localized to fusion sites by Sns [28, 31]. These data suggest that Sltr is part of a pathway that involves Sns, WASp and Arp 2/3 to regulate actin dynamics. In vitro studies support specific interactions among molecules in this pathway, and further identify DCrk as a direct link between Sns and Sltr [28, 31] (Fig. 8.2). Together, these data suggest a role for Sltr in FCMs for regulating actin dynamics.

8.5 General Aspects of Myoblast Fusion in *Danio rerio*

Zebrafish models are primarily used to study embryonic myoblast fusion. The zebrafish offers many advantages for studying embryonic myoblast fusion, including the rapid timing of embryonic development, the large number of progeny, and the utilization of large-scale genetic screens. In addition, the optical transparency of the zebrafish embryo and larvae enable detailed in vivo analyses of the mutants identified through genetic screens. Also, morpholino injections can be easily utilized to disrupt signaling pathways with potential roles in fusion.

Fewer details are known about embryonic myoblast fusion in *Danio rerio* compared to the other two model organisms presented in this chapter. During embryonic myoblast fusion in zebrafish, two types of muscle precursor cells, slow and fast myoblasts, exist, and have different properties [37–43]. Slow myoblasts are fusion incompetent, whereas fast myoblasts are capable of fusing with one another. Initially, slow myoblasts differentiate and give rise to a superficial layer of mononucleated slow-twitch fibers; subsequently, fast myoblasts fuse with one another to give rise to deeper syncytial fast-twitch fibers (Fig. 8.1b) [38–40, 43]. The zebrafish slow fibers are specialized for slow-force, long duration contractions, due to their slow myosin isoforms and oxidative metabolism. In contrast, zebrafish fast fibers are specialized for high-force, short duration contractions, due to their fast myosin isoforms and glycolytic metabolism. The existence of different fiber types in the vertebrate *Danio rerio* confers a complexity that does not exist in the invertebrate *Drosophila melanogaster*.

8.6 Molecules Regulating Myoblast Fusion in *Danio rerio*

Since fast myoblasts fuse to one another during embryonic myoblast fusion in *Danio rerio*, most molecules regulating their fusion are likely expressed on all fast myoblasts, conferring an apparent symmetry to this process in *Danio rerio* (Fig. 8.1b). This apparent symmetry is in direct contrast to *Drosophila melanogaster*, where fusion occurs between two distinct myoblast types (Fig. 8.1a).

8.6.1 Recognition and Adhesion

In order for fusion to occur, muscle cells must first recognize and adhere to one another. Far fewer molecules are known to regulate myoblast fusion in *Danio rerio* than in the other two model organisms discussed in this chapter. The molecules that mediate fusion of fast myoblasts to one another in *Danio rerio* (Table 8.1) were discovered by homology searches of the *Drosophila* fusion genes in the *Danio rerio* genome database. Subsequently, morpholino studies indicated the importance of these molecules during various steps of the fusion process in *Danio rerio*.

Nephrin is the *Danio rerio* homolog of *Drosophila* Sns and Hbs proteins. *Nephrin* morphants exhibit smaller muscles and several disorganized myosepta, consistent with a muscle defect [44]. However, whether the zebrafish nephrin regulates fusion of fast myoblasts has not specifically been studied. Kirrel, the *Danio rerio* homolog of *Drosophila* Duf and Rst proteins, is required for myoblast fusion in zebrafish, as *kirrel* morphants exhibit large clusters of unfused fast myoblasts [43]. Kirrel localization at cell–cell contacts during zebrafish myoblast fusion further suggests a role for Kirrel in the fusion process [43].

8.6.2 Cytoplasmic Adaptors and Actin Cytoskeletal Regulators

Homologs of cytoplasmic proteins that regulate the cytoskeleton during myoblast fusion in *Drosophila* also exist in zebrafish. For example, Rac1 is required for zebrafish myoblast fusion [43]. *Rac1* morphants exhibit a defect in myoblast fusion, and expression of a constitutively active human Rac1, highly homologous to the zebrafish Rac1, leads to hyperfusion. As this hyperfusion phenotype does not occur in *kirrel* morphants, Rac likely acts downstream of Kirrel during myoblast fusion [43].

A few other cytoplasmic proteins with a role in the fusion of fast myoblasts to one another are currently known. The Dedicator of cytokinesis 1 and 5 (Dock 1 and 5) proteins, the *Danio rerio* homologs of the *Drosophila* Mbc protein, are required for myoblast fusion in zebrafish [45]. *Dock 1* and *dock 5* morphants exhibit a large number of unfused myoblasts, which elongate to form mononucleated fibers. In addition, the adaptor proteins Crk and CT10 regulator of kinase-like (CrkL), homologs of the *Drosophila* DCrk protein, are also required for myoblast fusion and are known to interact with Dock proteins [45]. *Crk* and *crkl* morphants exhibit a phenotype similar to that of *dock* morphants, while Crk or CrkL overexpression enhances fusion. Whether a pathway exists in which the proteins mentioned above act from the membrane to the actin cytoskeleton to mediate myoblast fusion in zebrafish is unknown. Future studies are needed to expand our knowledge of the molecular players regulating myoblast fusion in *Danio rerio*.

8.7 General Aspects of Myoblast Fusion in *Mus musculus*

Mouse models offer a number of advantages for the study of myoblast fusion not found in the other two lower organisms discussed earlier in this chapter. The 2-year life span of mice affords the opportunity to study fusion in vertebrates over a greater age range, starting with embryonic development and continuing well into adulthood and old age. In addition, as in any vertebrate, individual skeletal muscles are composed of variable proportions of fast-twitch and slow-twitch myofibers, which determine the force and duration of muscle contraction. This complexity in muscle types results in the unique ability to study myoblast fusion in different types of adult muscles in the context of various physiologic stimuli, such as exercise, hypertrophy and regeneration.

From the genetic standpoint, transgenic technologies are well developed in mice, allowing tissue-specific and developmental-specific expression of transgenes or knockout of endogenous genes. Thus, myoblast fusion can be analyzed in the context of human disease mutations, as many mouse models of human neuromuscular disease exist. Furthermore, DNA plasmids expressing stimulatory or inhibitory molecules can easily be electroporated into adult muscles to manipulate different signaling pathways involved in myoblast fusion. From the biochemical standpoint, many antibodies against various mouse proteins exist that are useful for addressing questions related to fusion at the single-cell level.

Cell culture techniques are also well developed in mice, as opposed to the other two model organisms discussed earlier, allowing in vitro studies of myoblast fusion. The advantages of in vitro studies are the ability to: (1) carefully control the environment of the cells; (2) easily manipulate cellular components with drugs, siRNA or DNA constructs; and (3) study myoblast fusion using time-lapse microscopy. Much of the work pertaining to myoblast fusion in *Mus musculus* derives from in vitro studies utilizing both primary muscle cells and established cell lines in which individual steps regulating the fusion process can be easily dissected and analyzed. These in vitro studies demonstrate that multinucleated myotubes, which are the equivalent of immature myofibers found in vivo, form in a series of ordered steps. Initially, myoblasts differentiate into elongated cells that migrate towards one other and undergo recognition and adhesion. Several differentiated myoblasts then fuse to one another to form a small nascent myotube with a few myonuclei. Additional differentiated myoblasts fuse with the nascent myotubes during subsequent rounds of fusion to generate a mature myotube that contains many myonuclei (Fig. 8.1c). Currently, many studies couple in vitro experiments using primary muscle cells with in vivo studies of adult regenerative myogenesis in mice to obtain a more comprehensive view of the regulatory networks governing myoblast fusion.

Myofiber formation during embryogenesis occurs in distinct overlapping phases involving different classes of myoblasts [46–51]. During primary myogenesis, embryonic myoblasts fuse to form primary myofibers that are necessary to establish the basic muscle pattern. A second wave of myogenesis involves fusion of fetal myoblasts either with each other to form secondary myofibers that surround the primary myofibers, or with primary myofibers. At the end of secondary myogenesis, late in development, a basal lamina surrounds each myofiber and a third type of muscle precursor cell, the adult satellite cell, can be morphologically identified lying between the basal lamina and the myofiber plasma membrane. Myoblasts derived from satellite cells are responsible for postnatal muscle growth and regeneration [52].

The best studied model of myoblast fusion in the mouse is that occurring during adult regenerative myogenesis; similar techniques have not been developed to study regeneration in any of the model systems presented earlier. In response to trauma, myofibers degenerate and muscle regeneration is induced. Satellite cells proliferate in response to injury, and their progeny myoblasts differentiate into myocytes, which fuse with one another or with existing myofibers to restore normal tissue architecture [52]. Unfortunately, current technologies do not exist for directly visualizing myoblast fusion in vivo in mice. However, much information derives from morphologic and biochemical measurements of myofiber formation and growth in response to localized injury in mice as an indirect readout of myoblast fusion. Thus, the formation of regenerated myofibers is easily identified histologically by the presence of centrally located nuclei within the myofiber and can be quantified both by measurement of myofiber number/field and myofiber size. In addition, during muscle regeneration, developmental isoforms of several proteins are re-expressed and then eventually replaced by adult isoforms when regeneration is completed. The efficiency of muscle regeneration under different conditions can also be quantified by analysis of the timecourse of expression of these various developmental isoforms.

8.8 Molecules Regulating Specific Processes During Myoblast Fusion in *Mus musculus*

While myoblast heterogeneity in both embryonic and adult mice exists in terms of protein expression as well as proliferative and fusion capacities [49, 53], specific subsets of myoblasts that seed the formation of myofibers, such as the FC in *Drosophila* myogenesis have not been identified. Some similarities are observed in the types of molecules regulating myoblast fusion in mice and in the other two lower organisms discussed so far suggesting functional conservation (Table 8.1). However, new types of molecules that have not been identified to date in lower organisms such as secreted molecules and transcription factors also regulate fusion in mice (Table 8.2). Also, a number of pathways demonstrated to control myoblast fusion in mice are not found in lower organisms and may have evolved in higher eukaryotes to regulate the plasticity of muscle growth and regeneration, allowing maintenance of muscle mass over an extended lifespan and the growth of muscles containing multiple myofibers. Below we discuss the molecular factors regulating myoblast fusion in *Mus musculus* in the context of the cellular events from the initial elongation step to the final step of cell fusion with nascent myofibers, resulting in mature myofibers.

8.8.1 Elongation and Membrane Alterations

As myoblasts differentiate in vitro, they elongate and migrate towards other differentiated myoblasts to form groups of aligned cells [54–56] (Fig. 8.1c). Elongation is a permanent change and likely a complex interplay among multiple types of proteins as alterations in integrins, matrix remodeling enzymes and molecules that affect the cytoskeletal network result in defects in myoblast elongation (Table 8.2). Elongation is followed by extension of lamellopodia and filopodia, cell extensions composed of actin filaments, which make contact with neighboring muscle cells [56–60]. While the role of these cell extensions in the fusion process is unknown, these structures are sites for the localization of adhesion molecules [59, 61] and signaling molecules [58, 61]. Filopodia are reminiscent of axon growth cones and may respond to chemoattractants produced by other cells and be necessary to recognize other myoblasts capable of fusing.

8.8.2 Migration

Migration is necessary to achieve cell–cell contact during myogenesis (Fig. 8.1c), which is required both to trigger differentiation [62] and to allow differentiated myoblasts to fuse with one another and with nascent myotubes. Some regulatory factors that influence myoblast migration modulate the velocity or direction of cell migration, whereas others regulate the clearance of the extracellular matrix at the leading edge of migrating cells, thus facilitating cell motility [63–66]. Myotube formation and growth in vitro are enhanced by both positive [CD164, interleukin 4 (IL4), mannose receptor (MR) and mouse odorant receptor 23 (MOR23)] and negative [prostacyclin] regulators of cell

Table 8.2 Molecules regulating myoblast fusion in *Mus musculus*

Process	Molecule	Type	Location	References
Elongation and membrane alterations	EB3	Microtubule binding protein	Cytoplasm	[122]
	Kindlin-2	Integrin-associated adaptor protein	Cytoplasm	[123]
	MT1-MMP	Extracellular matrix protease	Membrane	[54]
	Non-muscle myosin 2A	Structural protein	Cytoplasm	[55]
	RhoE	RhoA inhibitor	Cytoplasm	[115]
Migration	CD164	CXCR4-associated sialomucin	Membrane	[124]
	Interleukin 4	Cytokine	Extracellular	[65, 66]
	Mannose receptor	Collagen clearance	Membrane	[64]
	MOR23	G-protein coupled receptor	Membrane	[63]
	Prostacyclin	Prostaglandin	Extracellular	[67]
Adhesion	M-cadherin	Cadherin	Membrane	[69, 109, 111]
	MOR23	G-protein coupled receptor	Membrane	[63]
	NCAM	Ig superfamily protein	Membrane	[125]
	Nephrin	Ig superfamily protein	Membrane	[44]
	VCAM	Ig superfamily protein	Membrane	[126]
Actin cytoskeletal dynamics	Arf6	GTPase for Rac1	Cytoplasm	[21, 111]
	Brag2/ARF-GEP$_{100}$	GEF for Arf6/Rac1	Cytoplasm	[120]
	cAMP	Lamellipodium formation	Cytoplasm	[58]
	Cdc42	GTPase for N-WASP	Cytoplasm	[121]
	Dock1,5	GEF for Rac1	Cytoplasm	[119, 120]
	Filamin C	Actin cross-linking protein	Cytoplasm	[127]
	Nap1	GEF for WAVE complex formation	Cytoplasm	[56]
	N-WASP	Actin nucleation promoting factor	Cytoplasm	[28]
	Phospholipase D	Enzyme for production of phosphatidic acid	Cytoplasm	[111]
	Protein kinase A	Lamellipodium formation	Cytoplasm	[58]
	Rac1	GTPase for Arp2/3	Cytoplasm	[69, 121]
	Trio	GEF for Rac1, RhoA, RhoG	Cytoplasm	[69]
	WIP	Actin binding protein	Cytoplasm	[28]
Integrin signaling	β1 integrin	Integrin	Membrane	[83]
	Focal adhesion kinase	Tyrosine kinase	Cytoplasm	[84]
	Talin 1, 2	Integrin-associated adaptor protein	Cytoplasm	[128]

Table 8.2 (continued)

Process	Molecule	Type	Location	References
Fusion with nascent myotubes	EHD2	Myoferlin-associated protein	Cytoplasm	[129]
	Follistatin	Glycoprotein	Extracellular	[97, 99]
	Growth hormone	Peptide hormone	Extracellular	[105]
	NFATc2	Transcription factor	Nucleus	[85]
	Interleukin 4	Cytokine	Extracellular	[65]
	Mannose receptor	Collagen clearance	Membrane	[64]
	mTOR	Kinase	Cytoplasm	[100, 102]
	Myoferlin	Phospholipid binding protein	Membrane	[130]
	Prostaglandin F2α	Prostaglandin	Extracellular	[104]
	SHP-2	Phosphatase for c-src	Cytoplasm	[107]
Process not categorized	Calpain 3	Cysteine protease	Cytoplasm	[113, 114]
	Caveolin-3	Structural component of caveolae	Membrane	[116]
	CD9	Transmembrane 4 superfamily protein	Membrane	[88]
	CD81	Transmembrane 4 superfamily protein	Membrane	[88]
	cGMP	Cyclic nucleotide	Cytoplasm	[96, 97]
	c-src	Tyrosine kinase	Cytoplasm	[107]
	FOXO1a	Transcription factor	Nucleus	[91, 93, 131]
	GRP94	Chaperone glycoprotein	Endoplasmic reticulum	[89, 90, 132]
	Nitric oxide	Soluble gas	Extracellular	[96–98]
	Rho/Rock	Kinase for FOXO1a	Cytoplasm	[92]
	Smn	Component of spliceosome	Nucleus	[94]

migration (Table 8.2). Whereas positive migratory factors promote cell fusion by increasing the probability of myoblasts being close to one another, negative migratory factors may enhance cell fusion by acting as a "brake" on migrating cells to facilitate cell–cell contact and adhesion [67]. Thus, the net balance between these two classes of migratory regulators would be critical for the formation and growth of myofibers.

8.8.3 Recognition and Adhesion

In order for fusion to occur, myoblasts must recognize and adhere to one another (Fig. 8.1c) [68], and these two processes are regulated by multiple classes of molecules in *Mus musculus* (Table 8.2). This molecular diversity may allow not only more specificity in myoblast recognition and adhesion, but also activation of specific intracellular signaling pathways, such as Rac1 [69], cAMP [63] and tyrosine kinases [70]. Also, dynamic clustering and dispersion of lipid rafts appears to be necessary for regulating the accumulation of adhesion-complex proteins at presumptive fusion sites in vitro [59]. The recognition and adhesion of myoblasts prior to fusion likely involves multiple adhesion molecules, but the interplay between such molecules is largely unknown.

Following adhesion, alignment occurs through the parallel apposition of the membranes of elongated myoblasts with other myoblasts or with nascent myotubes [71–73]. By transmission electron microscopy, coated vesicles are observed in close proximity to the aligned plasma membranes where membrane union occurs in small regions as in *Drosophila* [71]. During development [74] or muscle regeneration [75], unilamellar vesicles are also observed in close apposition to the fusing membranes of muscle cells. The function of these vesicles is unknown.

8.8.4 Actin Cytoskeletal Dynamics and Integrin Signaling

Extensive cytoskeletal reorganization occurs before and after fusion [76]. Visualization of the actin cytoskeleton in fusing mouse myoblasts in vitro reveals similar dynamic changes as in *Drosophila* developmental myoblast fusion in vivo [56, 60, 77, 78]. Given the number of proteins regulating myoblast fusion that impinge on actin reorganization (Table 8.2), processes dependent on the actin cytoskeleton must play fundamental roles in myoblast fusion. Indeed, latrunculin B, an inhibitor of actin polymerization, inhibits fusion of mouse myoblasts in vitro [28, 56, 79]. Following cell–cell adhesion, the structure of the actin cytoskeleton at the contact site of fusing myoblasts is highly regulated by a complex signaling cascade [80]. Various GTPases and guanine nucleotide exchange factors (GEFs) are activated, which in turn impinge on Wiskott-Aldrich Syndrome Protein (WASP) and Wasp family verprolin-homologous protein (WAVE) proteins. Activation of both WASP and WAVE is critical for the Arp2/3 complex to initiate actin polymerization. In the absence of proper actin cytoskeletal remodeling, F-actin structures accumulate at the plasma membrane of apposed mouse myoblasts and are correlated with a decrease in myoblast fusion as seen in *Drosophila* [56].

Integrins and integrin signaling also play roles in modulating myoblast fusion in mice (Table 8.2). Integrins are heterodimeric transmembrane receptors comprised of an α and a β chain that bind to the extracellular matrix or to cell-surface ligands as well as to various intracellular proteins and regulate numerous downstream signaling pathways [81]. Skeletal muscle expresses many integrin subunits that are regulated during myogenesis [82]. Integrins could function in myoblasts by regulating the formation or expression of protein complexes necessary for fusion or by relaying signals for remodeling of the actin cytoskeleton [83, 84].

8.8.5 Cell Fusion with Nascent Myotubes

Differentiated myoblasts fuse with one another to form small, nascent myotubes with few nuclei and subsequently, additional myoblasts fuse with the nascent myotube, leading to the mature multinucleated cell characterized by increased myonuclear number and cell size (Fig. 8.1c) [85]. The requirement for distinct molecules at different stages of myoblast fusion was first suggested by experiments in C2C12 cells, a muscle cell line. Treatment of C2C12 cells with wheat germ agglutinin (WGA) resulted in normal formation of nascent myotubes, but an inhibition of mature myotube formation [86], suggesting that specific cell surface molecules were required in later steps of fusion. Subsequently, a number of molecules with diverse cellular locations and functions were discovered to be required only for fusion of myoblasts with nascent myotubes both in vitro and in vivo but not for myoblast-myoblast fusion (Table 8.2). Why myoblast-myotube fusion should require unique molecules compared to myoblast-myoblast fusion is unknown but may be related to specific challenges inherent with myoblast fusion to a multinucleated cell. Additionally, these molecules could represent a fine-tuning mechanism for controlling the ultimate number of nuclei within a myotube/myofiber. Furthermore, these molecules may direct myoblast fusion to specific sites along the myotube or with specific myotubes. Finally, these molecules could specifically control growth of regenerating myofibers rather than allowing new myofibers to form, and thus constitute a means of controlling the number of regenerated myofibers after muscle injury.

8.9 Other Molecules Regulating Myoblast Fusion in *Mus musculus*

For a number of molecules the exact step in the fusion process that they regulate in *Mus musculus* is unknown (Table 8.2). These molecules are localized to diverse cellular compartments suggesting that they likely control different processes during myoblast fusion. A few molecules are discussed below to illustrate differential functions arising from this cellular compartmentalization.

Two membrane proteins with a role in myoblast fusion are the Transmembrane 4 superfamily molecules CD9 and CD81. Transmembrane 4 superfamily molecules can associate with each other and with other cell surface molecules, as well as with cytoplasmic signaling molecules, suggesting a role in cell–cell adhesion [87]. During myogenesis, both CD9 and CD81 form complexes with several different β1 integrins, and blocking antibodies against CD9 and CD81 greatly inhibit myotube formation in C2C12 myoblasts in vitro [88]. These results suggest that CD9 and CD81 regulate β1 integrin signaling, which is important for myoblast fusion [83].

Another important molecule is glucose-regulated protein 94 (GRP94), a muscle-specific protein that functions as a molecular chaperone in the endoplasmic reticulum. GRP94 knockdown in C2C12 myoblasts inhibits fusion, whereas GRP94 overexpression accelerates fusion [89], demonstrating that GRP94 is important in regulating myoblast fusion. Recent studies indicate that GRP94 is a molecular chaperone necessary for the synthesis of insulin-like growth factor 2 (IGF-2), which, like insulin-like growth factor 1 (IGF-1), can regulate proliferation and differentiation during myogenesis, but how IGF-2 specifically plays a role in the fusion process is unknown [90].

Few nuclear factors have been identified with a regulatory role in myoblast fusion. The transcription factor Forkhead box gene, group O 1a (FOXO1a) is expressed in skeletal muscle and translocates to the nucleus at the onset of differentiation. Ectopic expression of a mutant form of FOXO1a lacking the transactivation domain severely inhibits fusion of primary mouse muscle cells, whereas expression of a non-phosphorylated FOXO1a enhances the rate and extent of fusion [91]. The Rho/ROCK pathway likely acts upstream of FOXO1a. ROCK can phosphorylate FOXO1a in vitro, and the addition of a ROCK inhibitor to differentiating C2C12 myoblasts leads to nuclear accumulation of FOXO1a and accelerated myoblast fusion [92]. Interestingly, FOXO1a regulates transcription of cyclic GMP-dependent kinase 1, which in turn phosphorylates FOXO1a, abolishing its DNA binding activity

[93]. This negative feedback loop involving FOXO1a and cyclic GMP-dependent kinase 1 may help control the rate of myoblast fusion.

Another nuclear factor associated with fusion is Survival motor neuron protein (Smn), a component of the spliceosome that may help regulate RNA splicing. Smn knockdown leads to a large number of unfused myoblasts [94], suggesting Smn may regulate splicing of RNA transcripts for proteins that promote myoblast fusion, but this hypothesis has not been directly tested.

8.10 Integrated Pathways Regulating Myoblast Fusion in *Mus musculus*

A small percentage of the molecules identified with a role in myoblast fusion in *Drosophila melanogaster*, *Danio rerio,* or *Mus musculus*, mainly those that regulate cell adhesion and actin cytoskeletal rearrangements, has been studied across species, and for the most part display conserved functions (Table 8.1). A greater number of molecules have been identified to date that regulate myoblast fusion in *Mus musculus* than in either *Drosophila melanogaster* or *Danio rerio*, suggesting increased complexity in the molecular mechanisms and/or redundancy among molecules regulating fusion across species. Surprisingly few of these molecules in *Mus musculus* can be placed into larger integrative pathways. The three largest integrative pathways regulating myoblast fusion, which have been studied in *Mus musculus* but not in *Drosophila melanogaster* or in *Danio rerio,* are discussed below.

8.10.1 Molecules that Enhance Follistatin Expression

The first pathway centers around follistatin, a secreted protein that inhibits the activity of myostatin, a negative regulator of skeletal muscle hypertrophy (Fig. 8.3a) [95]. Follistatin expression can be modulated by several different mechanisms during myogenesis. One mechanism is through the transient increase in cGMP that occurs at the onset of myoblast fusion in primary mouse myoblasts resulting in the production of nitric oxide by nitric oxide synthase [96, 97]. This increase in cGMP stimulates transcription of follistatin through the transcription factors myoD, cAMP response element binding protein (CREB) and nuclear factor of activated T cells c2 (NFATc2), known mediators of myogenesis. Pharmacologic enhancement of nitric oxide or cGMP levels [96–98] enhance cell fusion, whereas decreased levels of these two molecules diminish cell fusion [96, 97] due to changes in the production of follistatin. A second mechanism of enhancing production of follistatin is by treatment of myoblasts with deacetylase inhibitors. Deacetylase inhibitors increase transcription of follistatin through a MyoD/CREB/NFATc2-dependent pathway and cause myotube hypertrophy by increasing myoblast-myotube fusion [99]. Whether follistatin produced in response to nitric oxide/cGMP signaling also specifically stimulates myoblast-myotube fusion is unknown.

Recently, studies on the role of mammalian target of rapamycin (mTOR) (Fig. 8.3a), a serine/threonine protein kinase [100, 101], in myoblast fusion have provided a third mechanism of regulating follistatin levels [102]. Early studies demonstrated mTOR activity is necessary for the secretion of an unidentified factor that promotes fusion of myoblasts with nascent myotubes [100]; this mTOR-regulated fusion-promoting factor was recently identified as follistatin [102]. Given the importance of follistatin in myoblast fusion, further pathways are likely to intersect with the regulation of its expression.

8.10.2 Molecules Upstream and Downstream of NFATc2

The second pathway centers around nuclear factor of activated T cells c2 (NFATc2), a transcription factor that plays a central role in orchestrating fusion of myoblasts with nascent myotubes (Fig. 8.3b)

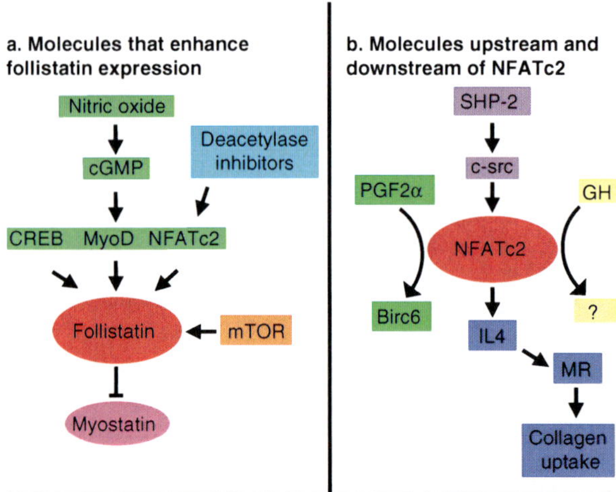

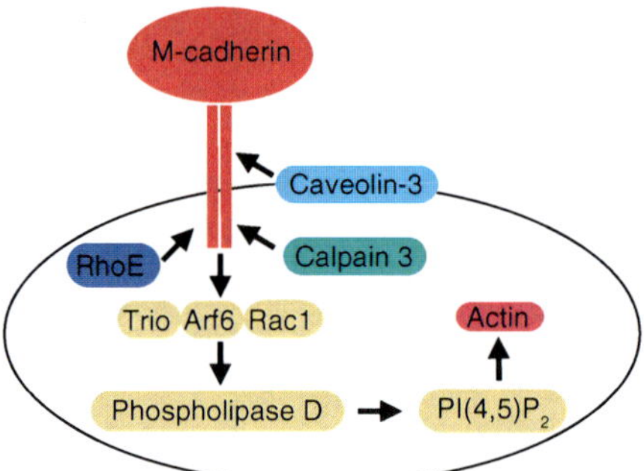

Fig. 8.3 Integrated pathways regulating adult myoblast fusion in *Mus musculus*. (**a**) Molecules that enhance follistatin expression are shown. Follistatin expression is increased by two cellular pathways: nitric oxide (via cGMP and its targets, CREB, MyoD and NFATc2) and mTOR. In addition, treatment of myoblasts with deacetylase inhibitors enhances expression of follistatin. Follistatin binds to and inhibits myostatin, a negative regulator of muscle growth. (**b**) Molecules upstream and downstream of NFATc2 are shown. NFATc2 is required for the fusion-promoting activity of PGF2α, GH and SHP-2 (via c-src). The mechanism by which GH acts is unknown, however PGF2α activates expression of Birc6, an anti-apoptotic protein, and SHP-2 stimulates secretion of IL4. IL4 enhances expression of MR, a molecule that regulates myoblast migration likely through collagen clearance. (**c**) Upstream and downstream molecules regulating M-cadherin signaling are shown. Both expression and localization of M-cadherin are regulated by caveolin-3, calpain 3 and RhoE. M-cadherin exists in a complex with three cytoplasmic proteins (Trio, Arf6 and Rac1), which regulate actin cytoskeletal rearrangements through the activation of phospholipase D and the production of PI(4,5)P_2. See text for more details. Arf6 = Adenosine diphosphate ribosylation factor 6; Birc6 = Baculoviral IAP repeat-containing 6; cGMP = Cyclic guanosine monophosphate; CREB = cAMP response element binding protein; c-src = Rous sarcoma oncogene cellular homolog; GH = Growth hormone; IL4 = interleukin 4; M-cadherin = Muscle cadherin; MR = Mannose receptor; mTOR = Mammalian target of rapamycin; MyoD = Myogenic differentiation antigen; NFATc2 = Nuclear factor of activated T cells c2; PGF2α = Prostaglandin F2α; PI(4,5)P_2 = Phosphatidylinositol 4,5-bisphosphate; Rac1 = Ras-related C3 botulinum toxin substrate 1; SHP-2 = SH2 domain-containing tyrosine phosphatase

[85, 103]. NFATc2 null mice are able to normally form regenerating myofibers after injury, but these myofibers grow at a slower rate and never reach the same final myofiber size of wild type mice. Similarly, NFATc2 null myoblasts can only form small myotubes in vitro due to a defect in the recruitment and/or fusion of myogenic cells with nascent myotubes [85]. Importantly, NFATc2 regulates expression of the cytokine IL4 by muscle cells [65]. In the absence of IL4 and IL4 receptor α, myoblast fusion with nascent myotubes is also defective in vitro and in vivo [65]. IL4 likely promotes myoblast fusion in part by regulating expression of the mannose receptor (MR), a cell surface endocytic C-type lectin. Like IL4 null myoblasts, MR null myoblasts form smaller myotubes in vitro and myofibers in vivo [64]. Specifically, MR null myoblasts display a reduction in general motility, as well as an impairment in directed migration towards unknown factors released by fusing muscle cells in culture. Collagen uptake is also decreased in MR null muscle cells, supporting a role for this receptor in helping to clear extracellular matrix from the leading edge of migrating cells, thus facilitating migration during the fusion process.

Molecules upstream of NFATc2 are also key players in regulating myoblast fusion with nascent myotubes (Fig. 8.3b). NFATc2 is required for the increase in myonuclear number due to prostaglandin $F_{2\alpha}$ (PGF2α) [104] as well as to growth hormone (GH) [105] in vitro. PGF2α-mediated activation of NFATc2 leads to expression of baculoviral IAP repeat-containing protein 6 (Birc6), an anti-apoptotic molecule that increases the pool of myoblasts available for fusion [106]. However, GH does not activate NFATc2 signaling nor secretion of IL4, rather GH may act in a parallel pathway to NFATc2. The mechanism by which GH promotes myoblast fusion with nascent myotubes is currently unknown. Additionally, NFATc2 may also be downstream of the tyrosine kinase c-src [107]. The tyrosine phosphatase SHP-2 stimulates c-src, resulting in activation of NFAT and subsequent fusion of myoblasts with myotubes in vitro [107]. Similar to NFATc2 null muscle cells [85], IL4 is decreased in SHP-2 null muscle cells in vitro and in vivo, supporting the idea that NFATc2 is a target for positive regulation by SHP-2 during fusion. However, whether the small size of muscles in SHP-2 null mice in vivo is a consequence of disrupting NFATc2 activation solely as opposed to other NFAT family members needs to be determined.

8.10.3 Upstream and Downstream Molecules Regulating M-Cadherin Signaling

The third integrative pathway centers around the calcium-dependent adhesion molecule M-cadherin (Fig. 8.3c). M-cadherin regulates myoblast fusion in vitro [69, 108, 109], but M-cadherin null mice do not have defects in skeletal muscle formation or regeneration, suggesting potential compensation by other cell adhesion molecules such as N-cadherin in vivo [110]. M-cadherin-dependent cell–cell adhesion mediates Rac1 GTPase activation via the Rho-GEF Trio in vitro [69]. M-cadherin, Rac1 and Trio exist in a multiprotein complex at the time of fusion together with ARF6, a GTPase [111]. Upon ARF6 knockdown, the M-cadherin/Rac1/Trio complex does not form. ARF6 regulates myoblast fusion through activation of phospholipase D and production of phosphatidylinositol 4,5-bisphosphate ($PI(4,5)P_2$) [111], a molecule that regulates actin cytoskeletal reorganization at the plasma membrane, vesicle trafficking and membrane curvature [112], suggesting a role for these processes in myoblast fusion.

M-cadherin expression and localization can be modulated by several factors (Fig. 8.3c). Primary myoblasts from calpain 3 null mice give rise to myotubes containing an increased number of myonuclei in vitro [113], likely due to the increased levels of membrane-associated M-cadherin [114]. In addition, RhoE is important both for M-cadherin expression as well as for M-cadherin accumulation at cell–cell contacts in vitro [115]. Finally, caveolin-3 regulates both expression and localization of M-cadherin. Myoblasts from caveolin-3 overexpressing mice fail to form myotubes and to upregulate M-cadherin expression early in differentiation, whereas myoblasts from caveolin-3 null mice show enhanced fusion and fail to downregulate M-cadherin in vitro [116]. In addition, caveolin-3 appears to regulate the localization of M-cadherin in caveolae membranes at later stages of differentiation; this localization into caveolae membranes is hypothesized to inhibit M-cadherin-mediated signaling and hence, myoblast fusion. Additional experiments are needed to understand the exact mechanisms by which calpain 3, RhoE and caveolin-3 regulate M-cadherin levels and/or localization.

8.11 Conclusions and Future Directions

Myoblast fusion is a complex, highly regulated process characterized by molecular diversity in both lower and higher organisms. While *Drosophila melanogaster*, *Danio rerio* and *Mus musculus* each offer unique advantages for studying myoblast fusion, the information gleaned from studying all of these organisms together provides a greater understanding of myoblast fusion than would be gained from studying any single organism.

Further research in multiple areas detailed below is needed to fully elucidate the molecular regulation of myoblast fusion. (1) A detailed understanding of the spatial and temporal coordination of myoblast fusion is lacking. Fusion does not occur between all cells, nor does it always take place along the entire membrane of a cell. Studies of how muscle cells recognize each other rather than other cells in the tissue and how they decide when and where on the cell membrane to fuse would be beneficial to understanding myoblast fusion. (2) Myoblast heterogeneity is observed in *Mus musculus*, which raises the question of whether some mammalian myoblasts could be the equivalent of the founder cell and fusion competent myoblast in *Drosophila melanogaster*. Further studies are needed to elucidate the functional consequences of this heterogeneity for myoblast fusion in mice. (3) The actual membrane fusogen molecules regulating the final step of myoblast fusion are not known in any organism. Fusogens identified in other cell types could give clues as to the types of molecules controlling this final step [117, 118]. (4) Methods for single cell imaging in muscles of live mice do not exist. Developing such sensitive methods of imaging would provide an unprecedented level of insight into vertebrate myoblast fusion in different physiologic settings such as muscle regeneration or hypertrophy. (5) Various molecules are likely important for myoblast fusion in different physiologic settings. An emphasis should be placed on identifying these regulatory molecules and pathways. MicroRNAs are likely to play critical roles in myoblast fusion as they regulate diverse processes in multiple cells, but little work has been done to identify regulatory networks for microRNAs in myoblast fusion. Such studies could lead to the identification of new regulatory molecules, as well as provide insight into the control of known regulatory pathways during myoblast fusion. Furthermore, large-scale screens in higher organisms using small molecule or siRNA libraries are also likely to identify new molecules with a role in myoblast fusion.

Further research in these various areas will enable a deeper understanding of the molecular mechanisms by which myoblast fusion occurs. Various cell transplantation therapies for muscular disorders are under development for use in humans and are being tested in mice using different types of muscle cells, but the efficacy of these therapies is still low. Studies of myoblast fusion may provide improved therapeutic strategies for diseases that affect skeletal muscle.

Acknowledgments GKP is supported by grants AR047314, AR051372, AR052730 and NS069234 from the National Institutes of Health.

References

1. Bate M (1990) The embryonic development of larval muscles in Drosophila. Development 110:791–804
2. Baylies MK, Bate M, Ruiz Gomez M (1998) Myogenesis: a view from Drosophila. Cell 93:921–927
3. Beckett K, Baylies MK (2007) 3D analysis of founder cell and fusion competent myoblast arrangements outlines a new model of myoblast fusion. Dev Biol 309:113–125

4. Onel SF, Renkawitz-Pohl R (2009) FuRMAS: triggering myoblast fusion in Drosophila. Dev Dyn 238: 1513–1525
5. Richardson B, Beckett K, Baylies M (2008) Visualizing new dimensions in Drosophila myoblast fusion. Bioessays 30:423–431
6. Roy S, VijayRaghavan K (1999) Muscle pattern diversification in Drosophila: the story of imaginal myogenesis. BioEssays 21:486–498
7. Richardson BE, Beckett K, Nowak SJ et al (2007) SCAR/WAVE and Arp2/3 are crucial for cytoskeletal remodeling at the site of myoblast fusion. Development 134:4357–4367
8. Strunkelnberg M, Bonengel B, Moda LM et al (2001) rst and its paralogue kirre act redundantly during embryonic muscle development in Drosophila. Development 128:4229–4239
9. Ruiz-Gomez M, Coutts N, Price A et al (2000) Drosophila dumbfounded: a myoblast attractant essential for fusion. Cell 102:189–198
10. Bour BA, Chakravarti M, West JM et al (2000) Drosophila SNS, a member of the immunoglobulin superfamily that is essential for myoblast fusion. Genes Dev 14:1498–1511
11. Artero RD, Castanon I, Baylies MK (2001) The immunoglobulin-like protein Hibris functions as a dose-dependent regulator of myoblast fusion and is differentially controlled by Ras and Notch signaling. Development 128:4251–4264
12. Dworak HA, Charles MA, Pellerano LB et al (2001) Characterization of Drosophila hibris, a gene related to human nephrin. Development 128:4265–4276
13. Shelton C, Kocherlakota KS, Zhuang S et al (2009) The immunoglobulin superfamily member Hbs functions redundantly with Sns in interactions between founder and fusion-competent myoblasts. Development 136: 1159–1168
14. Galletta BJ, Chakravarti M, Banerjee R et al (2004) SNS: Adhesive properties, localization requirements and ectodomain dependence in S2 cells and embryonic myoblasts. Mech Dev 121:1455–1468
15. Rogers SL, Rogers GC (2008) Culture of Drosophila S2 cells and their use for RNAi-mediated loss-of-function studies and immunofluorescence microscopy. Nat Protoc 3:606–611
16. Luo L, Liao YJ, Jan LY et al (1994) Distinct morphogenetic functions of similar small GTPases: Drosophila Drac1 is involved in axonal outgrowth and myoblast fusion. Genes Dev 8:1787–1802
17. Doberstein SK, Fetter RD, Mehta AY et al (1997) Genetic analysis of myoblast fusion: blown fuse is required for progression beyond the prefusion complex. J Cell Biol 136:1249–1261
18. Chen EH, Olson EN (2001) Antisocial, an intracellular adaptor protein, is required for myoblast fusion in Drosophila. Dev Cell 1:705–715
19. Rau A, Buttgereit D, Holz A et al (2001) rolling pebbles (rols) is required in Drosophila muscle precursors for recruitment of myoblasts for fusion. Development 128:5061–5073
20. Menon SD, Chia W (2001) Drosophila rolling pebbles: a multidomain protein required for myoblast fusion that recruits D-Titin in response to the myoblast attractant Dumbfounded. Dev Cell 1:691–703
21. Chen EH, Pryce BA, Tzeng JA et al (2003) Control of myoblast fusion by a guanine nucleotide exchange factor, loner, and its effector ARF6. Cell 114:751–762
22. Menon SD, Osman Z, Chenchill K et al (2005) A positive feedback loop between Dumbfounded and Rolling pebbles leads to myotube enlargement in Drosophila. J Cell Biol 169:909–920
23. Rushton E, Drysdale R, Abmayr SM et al (1995) Mutations in a novel gene, myoblast city, provide evidence in support of the founder cell hypothesis for Drosophila muscle development. Development 121:1979–1988
24. Erickson MR, Galletta BJ, Abmayr SM (1997) Drosophila myoblast city encodes a conserved protein that is essential for myoblast fusion, dorsal closure, and cytoskeletal organization. J Cell Biol 138:589–603
25. Ishimaru S, Ueda R, Hinohara Y et al (2004) PVR plays a critical role via JNK activation in thorax closure during Drosophila metamorphosis. EMBO J 23:3984–3994
26. Geisbrecht ER, Haralalka S, Swanson SK et al (2008) Drosophila ELMO/CED-12 interacts with Myoblast city to direct myoblast fusion and ommatidial organization. Dev Biol 314:137–149
27. Galletta BJ, Niu XP, Erickson MR et al (1999) Identification of a Drosophila homologue to vertebrate Crk by interaction with MBC. Gene 228:243–252
28. Kim S, Shilagardi K, Zhang S et al (2007) A critical function for the actin cytoskeleton in targeted exocytosis of prefusion vesicles during myoblast fusion. Dev Cell 12:571–586
29. Balagopalan L, Chen MH, Geisbrecht ER et al (2006) The CDM superfamily protein MBC directs myoblast fusion through a mechanism that requires phosphatidylinositol 3,4,5-triphosphate binding but is independent of direct interaction with DCrk. Mol Cell Biol 26:9442–9455
30. Berger S, Schafer G, Kesper DA et al (2008) WASP and SCAR have distinct roles in activating the Arp2/3 complex during myoblast fusion. J Cell Sci 121:1303–1313
31. Massarwa R, Carmon S, Shilo BZ et al (2007) WIP/WASp-based actin-polymerization machinery is essential for myoblast fusion in Drosophila. Dev Cell 12:557–569

32. Schafer G, Weber S, Holz A et al (2007) The Wiskott-Aldrich syndrome protein (WASP) is essential for myoblast fusion in Drosophila. Dev Biol 304:664–674
33. Gildor B, Massarwa R, Shilo BZ et al (2009) The SCAR and WASp nucleation-promoting factors act sequentially to mediate Drosophila myoblast fusion. EMBO Rep 10:1043–1050
34. Bogdan S, Klambt C (2003) Kette regulates actin dynamics and genetically interacts with Wave and Wasp. Development 130:4427–4437
35. Schroter RH, Lier S, Holz A et al (2004) kette and blown fuse interact genetically during the second fusion step of myogenesis in Drosophila. Development 131:4501–4509
36. Schroter RH, Buttgereit D, Beck L et al (2006) Blown fuse regulates stretching and outgrowth but not myoblast fusion of the circular visceral muscles in Drosophila. Differentiation 74:608–621
37. Baxendale S, Davison C, Muxworthy C et al (2004) The B-cell maturation factor Blimp-1 specifies vertebrate slow-twitch muscle fiber identity in response to Hedgehog signaling. Nat Genet 36:88–93
38. Devoto SH, Melancon E, Eisen JS et al (1996) Identification of separate slow and fast muscle precursor cells in vivo, prior to somite formation. Development 122:3371–3380
39. Roy S, Wolff C, Ingham PW (2001) The u-boot mutation identifies a Hedgehog-regulated myogenic switch for fiber-type diversification in the zebrafish embryo. Genes Dev 15:1563–1576
40. Blagden CS, Currie PD, Ingham PW et al (1997) Notochord induction of zebrafish slow muscle mediated by Sonic hedgehog. Genes Dev 11:2163–2175
41. Du SJ, Devoto SH, Westerfield M et al (1997) Positive and negative regulation of muscle cell identity by members of the hedgehog and TGF-beta gene families. J Cell Biol 139:145–156
42. Ochi H, Westerfield M (2007) Signaling networks that regulate muscle development: lessons from zebrafish. Dev Growth Differ 49:1–11
43. Srinivas BP, Woo J, Leong WY et al (2007) A conserved molecular pathway mediates myoblast fusion in insects and vertebrates. Nat Genet 39:781–786
44. Sohn RL, Huang P, Kawahara G et al (2009) A role for nephrin, a renal protein, in vertebrate skeletal muscle cell fusion. Proc Natl Acad Sci USA 106:9274–9279
45. Moore CA, Parkin CA, Bidet Y et al (2007) A role for the Myoblast city homologues Dock1 and Dock5 and the adaptor proteins Crk and Crk-like in zebrafish myoblast fusion. Development 134:3145–3153
46. Kelly AM, Zacks SI (1969) The histogenesis of rat intercostal muscle. J Cell Biol 42:135–153
47. Harris AJ, Duxson MJ, Fitzsimons RB et al (1989) Myonuclear birthdates distinguish the origins of primary and secondary myotubes in embryonic mammalian skeletal muscles. Development 107:771–784
48. Ross JJ, Duxson MJ, Harris AJ (1987) Formation of primary and secondary myotubes in rat lumbrical muscles. Development 100:383–394
49. Biressi S, Molinaro M, Cossu G (2007) Cellular heterogeneity during vertebrate skeletal muscle development. Dev Biol 308:281–293
50. Biressi S, Tagliafico E, Lamorte G et al (2007) Intrinsic phenotypic diversity of embryonic and fetal myoblasts is revealed by genome-wide gene expression analysis on purified cells. Dev Biol 304:633–651
51. Hutcheson DA, Zhao J, Merrell A et al (2009) Embryonic and fetal limb myogenic cells are derived from developmentally distinct progenitors and have different requirements for beta-catenin. Genes Dev 23: 997–1013
52. Charge SB, Rudnicki MA (2004) Cellular and molecular regulation of muscle regeneration. Physiol Rev 84: 209–238
53. Sherwood RI, Christensen JL, Conboy IM et al (2004) Isolation of adult mouse myogenic progenitors: functional heterogeneity of cells within and engrafting skeletal muscle. Cell 119:543–554
54. Ohtake Y, Tojo H, Seiki M (2006) Multifunctional roles of MT1-MMP in myofiber formation and morphostatic maintenance of skeletal muscle. J Cell Sci 119:3822–3832
55. Swailes NT, Colegrave M, Knight PJ et al (2006) Non-muscle myosins 2A and 2B drive changes in cell morphology that occur as myoblasts align and fuse. J Cell Sci 119:3561–3570
56. Nowak SJ, Nahirney PC, Hadjantonakis AK et al (2009) Nap1-mediated actin remodeling is essential for mammalian myoblast fusion. J Cell Sci 122:3282–3293
57. Yoon S, Molloy MJ, Wu MP et al (2007) C6ORF32 is upregulated during muscle cell differentiation and induces the formation of cellular filopodia. Dev Biol 301:70–81
58. Mukai A, Hashimoto N (2008) Localized cyclic AMP-dependent protein kinase activity is required for myogenic cell fusion. Exp Cell Res 314:387–397
59. Mukai A, Kurisaki T, Sato SB et al (2009) Dynamic clustering and dispersion of lipid rafts contribute to fusion competence of myogenic cells. Exp Cell Res 315:3052–3063
60. Stadler B, Blattler TM, Franco-Obregon A (2010) Time-lapse imaging of in vitro myogenesis using atomic force microscopy. J Microsc 237:63–69
61. Abramovici H, Gee SH (2007) Morphological changes and spatial regulation of diacylglycerol kinase-zeta, syntrophins, and Rac1 during myoblast fusion. Cell Motil Cytoskeleton 64:549–567

62. Krauss RS, Cole F, Gaio U et al (2005) Close encounters: regulation of vertebrate skeletal myogenesis by cell-cell contact. J Cell Sci 118:2355–2362
63. Griffin CA, Kafadar KA, Pavlath GK (2009) MOR23 promotes muscle regeneration and regulates cell adhesion and migration. Dev Cell 17:649–661
64. Jansen KM, Pavlath GK (2006) Mannose receptor regulates myoblast motility and muscle growth. J Cell Biol 174:403–413
65. Horsley V, Jansen KM, Mills ST et al (2003) IL-4 acts as a myoblast recruitment factor during mammalian muscle growth. Cell 113:483–494
66. Lafreniere JF, Mills P, Bouchentouf M et al (2006) Interleukin-4 improves the migration of human myogenic precursor cells in vitro and in vivo. Exp Cell Res 312:1127–1141
67. Bondesen BA, Jones KA, Glasgow WC et al (2007) Inhibition of myoblast migration by prostacyclin is associated with enhanced cell fusion. FASEB J 21:3338–3345
68. Knudsen KA (1985) The calcium-dependent myoblast adhesion that precedes cell fusion is mediated by glycoproteins. J Cell Biol 101:891–897
69. Charrasse S, Comunale F, Fortier M et al (2007) M-cadherin activates Rac1 GTPase through the Rho-GEF trio during myoblast fusion. Mol Biol Cell 18:1734–1743
70. Li H, Lemay S, Aoudjit L et al (2004) SRC-family kinase Fyn phosphorylates the cytoplasmic domain of nephrin and modulates its interaction with podocin. J Am Soc Nephrol 15:3006–3015
71. Lipton BH, Konigsberg IR (1972) A fine-structural analysis of the fusion of myogenic cells. J Cell Biol 53: 348–364
72. Rash JE, Fambrough D (1973) Ultrastructural and electrophysiological correlates of cell coupling and cytoplasmic fusion during myogenesis in vitro. Dev Biol 30:166–186
73. Wakelam MJ (1985) The fusion of myoblasts. Biochem J 228:1–12
74. Kalderon N, Gilula NB (1979) Membrane events involved in myoblast fusion. J Cell Biol 81:411–425
75. Robertson TA, Grounds MD, Mitchell CA et al (1990) Fusion between myogenic cells in vivo: an ultrastructural study in regenerating murine skeletal muscle. J Struct Biol 105:170–182
76. Fulton AB, Prives J, Farmer SR et al (1981) Developmental reorganization of the skeletal framework and its surface lamina in fusing muscle cells. J Cell Biol 91:103–112
77. Duan R, Gallagher PJ (2009) Dependence of myoblast fusion on a cortical actin wall and nonmuscle myosin IIA. Dev Biol 325:374–385
78. Swailes NT, Knight PJ, Peckham M (2004) Actin filament organization in aligned prefusion myoblasts. J Anat 205:381–391
79. O'Connor RS, Steeds CM, Wiseman RW et al (2008) Phosphocreatine as an energy source for actin cytoskeletal rearrangements during myoblast fusion. J Physiol 586:2841–2853
80. Kurisu S, Takenawa T (2009) The WASP and WAVE family proteins. Genome Biol 10:226
81. Hynes RO (2002) Integrins: bidirectional, allosteric signaling machines. Cell 110:673–687
82. Gullberg D (2003) Cell biology: the molecules that make muscle. Nature 424:138–140
83. Schwander M, Leu M, Stumm M et al (2003) Beta1 integrins regulate myoblast fusion and sarcomere assembly. Dev Cell 4:673–685
84. Quach NL, Biressi S, Reichardt LF et al (2009) Focal adhesion kinase signaling regulates the expression of caveolin 3 and beta1 integrin, genes essential for normal myoblast fusion. Mol Biol Cell 20:3422–3435
85. Horsley V, Friday BB, Matteson S et al (2001) Regulation of the growth of multinucleated muscle cells by an NFATC2-dependent pathway. J Cell Biol 153:329–338
86. Muroya S, Takagi H, Tajima S et al (1994) Selective inhibition of a step of myotube formation with wheat germ agglutinin in a murine myoblast cell line, C2C12. Cell Struct Funct 19:241–252
87. Maecker HT, Todd SC, Levy S (1997) The tetraspanin superfamily: molecular facilitators. FASEB J 11: 428–442
88. Tachibana I, Hemler ME (1999) Role of transmembrane 4 superfamily (TM4SF) proteins CD9 and CD81 in muscle cell fusion and myotube maintenance. J Cell Biol 146:893–904
89. Gorza L, Vitadello M (2000) Reduced amount of the glucose-regulated protein GRP94 in skeletal myoblasts results in loss of fusion competence. FASEB J 14:461–475
90. Wanderling S, Simen BB, Ostrovsky O et al (2007) GRP94 is essential for mesoderm induction and muscle development because it regulates insulin-like growth factor secretion. Mol Biol Cell 18:3764–3775
91. Bois PR, Grosveld GC (2003) FKHR (FOXO1a) is required for myotube fusion of primary mouse myoblasts. EMBO J 22:1147–1157
92. Nishiyama T, Kii I, Kudo A (2004) Inactivation of Rho/ROCK signaling is crucial for the nuclear accumulation of FKHR and myoblast fusion. J Biol Chem 279:47311–47319
93. Bois PR, Brochard VF, Salin-Cantegrel AV et al (2005) FoxO1a-cyclic GMP-dependent kinase I interactions orchestrate myoblast fusion. Mol Cell Biol 25:7645–7656

94. Shafey D, Cote PD, Kothary R (2005) Hypomorphic Smn knockdown C2C12 myoblasts reveal intrinsic defects in myoblast fusion and myotube morphology. Exp Cell Res 311:49–61
95. Lee SJ, McPherron AC (2001) Regulation of myostatin activity and muscle growth. Proc Natl Acad Sci USA 98:9306–9311
96. Lee KH, Baek MY, Moon KY et al (1994) Nitric oxide as a messenger molecule for myoblast fusion. J Biol Chem 269:14371–14374
97. Pisconti A, Brunelli S, Di Padova M et al (2006) Follistatin induction by nitric oxide through cyclic GMP: a tightly regulated signaling pathway that controls myoblast fusion. J Cell Biol 172:233–244
98. Long JH, Lira VA, Soltow QA et al (2006) Arginine supplementation induces myoblast fusion via augmentation of nitric oxide production. J Muscle Res Cell Motil 27:577–584
99. Iezzi S, Di Padova M, Serra C et al (2004) Deacetylase inhibitors increase muscle cell size by promoting myoblast recruitment and fusion through induction of follistatin. Dev Cell 6:673–684
100. Park IH, Chen J (2005) Mammalian target of rapamycin (mTOR) signaling is required for a late-stage fusion process during skeletal myotube maturation. J Biol Chem 280:32009–32017
101. Ge Y, Wu AL, Warnes C et al (2009) mTOR regulates skeletal muscle regeneration in vivo through kinase-dependent and kinase-independent mechanisms. Am J Physiol Cell Physiol 297:C1434–1444
102. Sun Y, Ge Y, Drnevich J et al (2010) Mammalian target of rapamycin regulates miRNA-1 and follistatin in skeletal myogenesis. J Cell Biol 189:1157–1169
103. Pavlath GK, Horsley V (2003) Cell fusion in skeletal muscle–central role of NFATC2 in regulating muscle cell size. Cell Cycle 2:420–423
104. Horsley V, Pavlath GK (2003) Prostaglandin F2(alpha) stimulates growth of skeletal muscle cells via an NFATC2-dependent pathway. J Cell Biol 161:111–118
105. Sotiropoulos A, Ohanna M, Kedzia C et al (2006) Growth hormone promotes skeletal muscle cell fusion independent of insulin-like growth factor 1 up-regulation. Proc Natl Acad Sci USA 103:7315–7320
106. Jansen KM, Pavlath GK (2008) Prostaglandin F2alpha promotes muscle cell survival and growth through upregulation of the inhibitor of apoptosis protein BRUCE. Cell Death Differ 15:1619–1628
107. Fornaro M, Burch PM, Yang W et al (2006) SHP-2 activates signaling of the nuclear factor of activated T cells to promote skeletal muscle growth. J Cell Biol 175:87–97
108. Zeschnigk M, Kozian D, Kuch C et al (1995) Involvement of M-cadherin in terminal differentiation of skeletal muscle cells. J Cell Sci 108 (Pt 9):2973–2981
109. Charrasse S, Comunale F, Grumbach Y et al (2006) RhoA GTPase regulates M-cadherin activity and myoblast fusion. Mol Biol Cell 17:749–759
110. Hollnagel A, Grund C, Franke WW et al (2002) The cell adhesion molecule M-cadherin is not essential for muscle development and regeneration. Mol Cell Biol 22:4760–4770
111. Bach AS, Enjalbert S, Comunale F et al (2010) ADP-ribosylation factor 6 regulates mammalian myoblast fusion through phospholipase D1 and phosphatidylinositol 4,5-bisphosphate signaling pathways. Mol Biol Cell 21:2412–2424
112. Donaldson JG (2008) Arfs and membrane lipids: sensing, generating and responding to membrane curvature. Biochem J 414:e1–2
113. Kramerova I, Kudryashova E, Tidball JG et al (2004) Null mutation of calpain 3 (p94) in mice causes abnormal sarcomere formation in vivo and in vitro. Hum Mol Genet 13:1373–1388
114. Kramerova I, Kudryashova E, Wu B et al (2006) Regulation of M-cadherin -{beta}-catenin complex by calpain 3 during terminal stages of myogenic differentiation. Mol Cell Biol 26:8437–8447
115. Fortier M, Comunale F, Kucharczak J et al (2008) RhoE controls myoblast alignment prior fusion through RhoA and ROCK. Cell Death Differ 15:1221–1231
116. Volonte D, Peoples AJ, Galbiati F (2003) Modulation of myoblast fusion by caveolin-3 in dystrophic skeletal muscle cells: implications for Duchenne muscular dystrophy and limb-girdle muscular dystrophy-1C. Mol Biol Cell 14:4075–4088
117. Dupressoir A, Marceau G, Vernochet C et al (2005) Syncytin-A and syncytin-B, two fusogenic placenta-specific murine envelope genes of retroviral origin conserved in Muridae. Proc Natl Acad Sci USA 102: 725–730
118. Dupressoir A, Vernochet C, Bawa O et al (2009) Syncytin-A knockout mice demonstrate the critical role in placentation of a fusogenic, endogenous retrovirus-derived, envelope gene. Proc Natl Acad Sci USA 106: 12127–12132
119. Laurin M, Fradet N, Blangy A et al (2008) The atypical Rac activator Dock180 (Dock1) regulates myoblast fusion in vivo. Proc Natl Acad Sci USA 105:15446–15451
120. Pajcini KV, Pomerantz JH, Alkan O et al (2008) Myoblasts and macrophages share molecular components that contribute to cell-cell fusion. J Cell Biol 180:1005–1019
121. Vasyutina E, Martarelli B, Brakebusch C et al (2009) The small G-proteins Rac1 and Cdc42 are essential for myoblast fusion in the mouse. Proc Natl Acad Sci USA 106:8935–8940

122. Straube A, Merdes A (2007) EB3 regulates microtubule dynamics at the cell cortex and is required for myoblast elongation and fusion. Curr Biol 17:1318–1325
123. Dowling JJ, Vreede AP, Kim S et al (2008) Kindlin-2 is required for myocyte elongation and is essential for myogenesis. BMC Cell Biol 9:36
124. Bae GU, Gaio U, Yang YJ et al (2008) Regulation of myoblast motility and fusion by the CXCR4-associated sialomucin, CD164. J Biol Chem 283:8301–8309
125. Suzuki M, Angata K, Nakayama J et al (2003) Polysialic acid and mucin type o-glycans on the neural cell adhesion molecule differentially regulate myoblast fusion. J Biol Chem 278:49459–49468
126. Rosen GD, Sanes JR, LaChance R et al (1992) Roles for the integrin VLA-4 and its counter receptor VCAM-1 in myogenesis. Cell 69:1107–1119
127. Dalkilic I, Schienda J, Thompson TG et al (2006) Loss of FilaminC (FLNc) results in severe defects in myogenesis and myotube structure. Mol Cell Biol 26:6522–6534
128. Conti FJ, Monkley SJ, Wood MR et al (2009) Talin 1 and 2 are required for myoblast fusion, sarcomere assembly and the maintenance of myotendinous junctions. Development 136:3597–3606
129. Doherty KR, Demonbreun AR, Wallace GQ et al (2008) The endocytic recycling protein EHD2 interacts with myoferlin to regulate myoblast fusion. J Biol Chem 283:20252–20260
130. Doherty KR, Cave A, Davis DB et al (2005) Normal myoblast fusion requires myoferlin. Development 132: 5565–5575
131. Kamei Y, Miura S, Suzuki M et al (2004) Skeletal muscle FOXO1 (FKHR) transgenic mice have less skeletal muscle mass, down-regulated Type I (slow twitch/red muscle) fiber genes, and impaired glycemic control. J Biol Chem 279:41114–41123
132. Ostrovsky O, Eletto D, Makarewich C et al (2009) Glucose regulated protein 94 is required for muscle differentiation through its control of the autocrine production of insulin-like growth factors. Biochim Biophys Acta 1803:333–341

Chapter 9
Cell-Fusion-Mediated Reprogramming: Pluripotency or Transdifferentiation? Implications for Regenerative Medicine

Daniela Sanges*, Frederic Lluis*, and Maria Pia Cosma

Abstract Cell–cell fusion is a natural process that occurs not only during development, but as has emerged over the last few years, also with an important role in tissue regeneration. Interestingly, in-vitro studies have revealed that after fusion of two different cell types, the developmental potential of these cells can change. This suggests that the mechanisms by which cells differentiate during development to acquire their identities is not irreversible, as was considered until a few years ago. To date, it is well established that the fate of a cell can be changed by a process known as reprogramming. This mainly occurs in two different ways: the differentiated state of a cell can be reversed back into a pluripotent state (pluripotent reprogramming), or it can be switched directly to a different differentiated state (lineage reprogramming). In both cases, these possibilities of obtaining sources of autologous somatic cells to maintain, replace or rescue different tissues has provided new and fundamental insights in the stem-cell-therapy field. Most interestingly, the concept that cell reprogramming can also occur in vivo by spontaneous cell fusion events is also emerging, which suggests that this mechanism can be implicated not only in cellular plasticity, but also in tissue regeneration. In this chapter, we will summarize the present knowledge of the molecular mechanisms that mediate the restoration of pluripotency in vitro through cell fusion, as well as the studies carried out over the last 3 decades on lineage reprogramming, both in vitro and in vivo. How the outcome of these studies relate to regenerative medicine applications will also be discussed.

9.1 Cell–Cell Fusion Methodologies

Membrane fusion is fundamental to the life of eukaryotic cells. Cellular trafficking and compartmentalization, intercellular communication, cell division, and many other physiological events are all dependent on this basic process. Fusion between two cells also occurs in a wide range of developmental and pathological processes [1].

In 1975, a biophysical discussion agreed upon the definition of the fusion process as the mixing of entrapped contents between two membrane-enclosed aqueous compartments that involves the mixing of the membrane contents, but with little escape of the entrapped contents [2].

Even though the process of cell–cell fusion is a physiological process during mammalian development [3], artificial fusion has also been used to merge together two cell types to generate a third

---

M.P. Cosma (✉)
Institució Catalana de Recerca i Estudis Avançats (ICREA), Barcelona, Spain; Center for Genomic Regulation (CRG), 08003 Barcelona, Spain
e-mail: pia.cosma@crg.eu

*These two authors contribute equally to this work.

T. Dittmar, K.S. Zänker (eds.), *Cell Fusion in Health and Disease*, Advances in Experimental Medicine and Biology 713, DOI 10.1007/978-94-007-0763-4_9, © Springer Science+Business Media B.V. 2011

cell type that would display hybrid characteristics different from both of the original cells. The development of monoclonal antibodies by Kohler and Milstein [4], for example, relied on the formation of "hybridomas" that were created by fusing antibody-producing cells with cancer cells. Following this idea, induced fusion has also been used to study other processes, such as the plasticity of cells of different origins. In this case, fusion has been induced mainly by two different methods, as now described.

9.1.1 PEG-Mediated Fusion

Polyethylene glycol (PEG) is an oligomer or polymer of ethylene oxide. PEG has been used to fuse cells from a long time [5]. However, in the 1970s, very little was known about the mechanisms by which PEG induced fusion. It was shown that the action of PEG in promoting cell–cell fusion was not due to effects such as surface absorption, crosslinking or solubilization, but that the major effect of PEG for membrane merging was due to volume exclusion, which induces an osmotic force that brings the membranes into close contact, resulting in the membrane dehydration necessary to induce fusion [6].

9.1.2 Electrofusion

One key aspect of membrane surfaces is their surface charge. Electrofusion consists of the application of pulsed electric fields [7] that allow membrane permeabilization (a reversible process without any dramatic membrane rupturing if controlled parameters are used). Thus, electropermeabilized cells brought into contact are fusogenic [8]. In subsequent experiments, when cells were first brought in contact and then an electric pulse was applied, this resulted in an increase in hybrid formation, suggesting that fusion takes place when the two cell surfaces in contact are electropermeabilized [9].

9.2 Somatic Cell Reprogramming

Cell–cell fusion has been extensively used in more recent years to study the plasticity of differentiated cells, a concept that can be strictly related to the capacity of adult cells to undergo reprogramming.

Somatic cell reprogramming can be referred to as the transition from one cell type into another. There are two major types of reprogramming: (i) reprogramming of differentiated cells into pluripotent cells; and (ii) lineage reprogramming of differentiated cells into different differentiated cells.

Reprogrammed pluripotent cells show:

(1) demethylation and reactivation of genes that are essential for pluripotency, such as *Oct4, Nanog* and *Sox2*;
(2) silencing of somatic markers;
(3) reactivation of the silent X chromosome;
(4) potential to form teratomas and to differentiate in tissues that are derived from the three germ layers after injection into nude mice;
(5) potential to generate chimeras, meaning that the reprogrammed cells can give rise to different tissues of the body.

In the lineage reprogramming that consists of the transition between specialized cellular identities, the new reprogrammed cells will acquire the features of the cells to which they convert to. In another

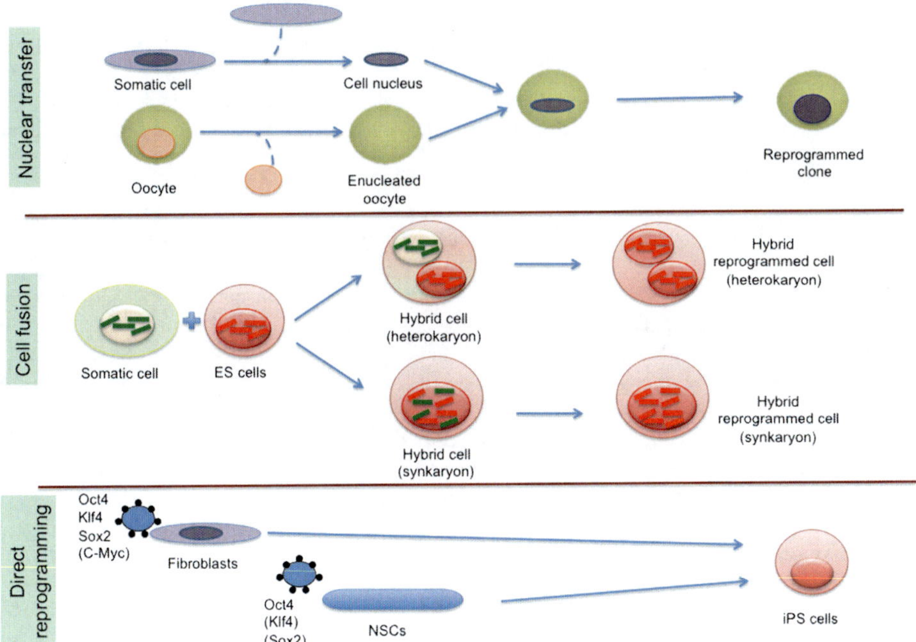

Fig. 9.1 Strategies to induce pluripotent reprogramming. There are several strategies used to induce reprogramming. In nuclear transfer, the somatic nucleus is transferred into an enucleated oocyte to yield a totipotent cell with the genetic material from the somatic cell (cloning). Cell fusion between ESCs and somatic cells results in the generation of hybrid cells where the somatic nucleus is reprogrammed into a pluripotent state. Generation of heterokaryons with two separated nuclei or synkaryons with one nucleus is possible. Finally, in the direct reprogramming strategy, the transduction of three transcription factors, Oct4, Klf4, Sox2 (with or without c-Myc) in a somatic cell induces the formation of iPSCs, which are pluripotent cells. It is also possible to generate iPSCs by transducing NSCs with only Oct4

words, if fibroblasts are to become muscle cells, they will need to express muscle cell genes and silence the expression of their fibroblast-specific genes.

To date, there are several strategies that have been used to generate pluripotent cells from somatic cells (Fig. 9.1).

9.2.1 Somatic Cell Nuclear Transfer

The process of somatic cell nuclear transfer (SCNT) involves the transfer of the nucleus of a somatic cell into an enucleated oocyte, with the goal of generating a totipotent cell. Then, under the correct conditions, an adult organism can develop. This technique is also known as cloning.

The cloning of animal cells was first performed when Briggs and King created frog embryo clones in 1952 [10]. They took fertilized frog eggs, removed the chromosomes/spindle apparatus, and replaced these with nuclei from older embryos. The net result was clones of adult frogs that were identical to the originating frog cell nuclei.

More recently, the nuclear reprogramming capacity of oocytes has been demonstrated in mammals by the production of cloned sheep, cows, mice and pigs [11–14]. However, all of the epigenetic marks that have been examined in cloned embryos show abnormalities [15] the success rate of this cloning is very low, and there are high levels of early and late embryonic lethality. This has been suggested to be due to inaccurate and incomplete reprogramming, including incorrect DNA methylation of the somatic donor nucleus [15–19].

However, this process of cloning, and especially if carried out in humans, is fraught with fears that such events could lead to reproductive cloning: the formation of an embryo that can become an adult after implantation in the uterus.

9.2.2 Nuclear Reprogramming of Somatic Cells by Cell–Cell Fusion

Similar to the production of cloned animals from somatic nuclear transfer, the state of a somatic nucleus can be reprogrammed to that of a pluripotent stem cell by hybridization with pluripotent cells. This is due to the observation that in the majority of hybrids, the phenotype of the less-differentiated cell-fusion partner is dominant over the phenotype of the more-differentiated cell-fusion partner.

As early as the 1970s, cell–cell fusion was used to study nuclear reprogramming. In one study, mouse embryonic carcinoma cells (ECCs) were fused with primary thymocytes, and the resultant hybrid cells showed properties of pluripotent carcinoma cells [20]. More recently, different types of somatic cells have been fused with embryonic stem cells (ESCs), and both mouse (mESCs) and human (hESCs) ESCs have been shown to have the ability to reprogram somatic cells [21, 22]. Moreover, fusion between specialized cells can lead to the formation of hybrids that acquire one of the two distinct phenotypes, as was seen, for example, by fusing murine muscle cells with human primary diploid cells derived from many different embryonic lineages [23–25].

As the aim of this chapter is to summarize the information relating to cell-fusion-mediated reprogramming, the knowledge in this field is detailed below.

9.2.3 Direct Reprogramming of Somatic Cells

The first evidence that adult somatic cells can be reprogrammed into cells with ESC-like characteristics was reported by Takahaski and Yamanaka [26], when they transduced the four genes encoding for the Oct4, Klf4, Sox2 and c-Myc factors into fetal and adult mouse fibroblasts, using retrovirus infection. The overexpression of the transduced trangenes produced some cells with an ESC-like phenotype, which are known as induced pluripotent stem cells (iPSCs). Soon after, it was shown that it was possible to reprogram human fibroblasts with these same four factors, or with the combination of the Oct4, Sox2, Nanog and LIN28 factors [27]. Subsequently, other cells, such as B lymphocytes, hepatocytes, gastric epithelial cells and others, have been successfully reprogrammed by the same combinations of factors, and also by a subset of these factors, or with new "blends" of different factors [28, 29].

c-Myc is a potent oncogene that is implicated in cell proliferation, DNA replication, cell growth and metastasis formation [30–32]. When iPSCs were generated using c-Myc, Oct4, Sox2 and Klf4, about 15% of the mice derived from these iPSCs developed tumors within 4 months. Subsequent studies have reported that both human and mouse adult fibroblasts can be reprogrammed using only three genes, with c-Myc being omitted here [33, 34]. This observation has stimulated many studies towards the identification of the "essential" reprogramming factors.

Klf4 is an abundant transcript in ESCs, although Klf4 knock-down does not lead to an obvious phenotype; this is probably due to functional redundancy with other Klf family members in ESCs [35, 36]. It has been shown that Klf4 can be replaced by the orphan nuclear receptor Esrrb, which can act in conjunction with Oct4 and Sox2 to mediate reprogramming. Esrrb-reprogrammed cells share similar marker expression and epigenetic signatures with respect to ESCs [37].

Oct4 and Sox2 are probably the most important factors for the induction of reprogramming; however, even these can be replaced. Both of these factors are required for the maintenance of ESC pluripotency and for self-renewal [38, 39]. A lack of Oct4 in embryos impairs their ability to develop the inner cell mass [40], while Sox2 loss-of-function results in defective epiblasts and differentiation into trophoblast cells [41]. To date, it has been shown that for the reprogramming of somatic cells, Sox2 can be replaced with other members of its family, such as Sox1 and Sox3, albeit with reduced efficiencies [33]. In an interesting recent report, it was shown that Oct4 can be replaced by the orphan nuclear receptor Nr5a2 (also known as Lrh-1) in the production of iPSCs from mouse somatic cells [42], casting some doubts on the fundamental role of Oct4 in iPSC generation. However, NSCs have been shown to be reprogrammed to pluripotency after overexpression of only Oct4 [43], confirming that Oct4 does indeed have a critical role in reprogramming.

9.3 Induced Pluripotency Through Cell-Fusion-Mediated Reprogramming

Dolly the sheep was generated by nuclear transfer in 1997, demonstrating that fully differentiated mammalian somatic cells can be reprogrammed to a state of totipotency [11]. Most recently, by modifying existing SCNT protocols, the successful nuclear reprogramming of adult rhesus macaque somatic cells into pluripotent ESCs was achieved. These ESCs showed normal ESC morphology, expressed ESC-specific markers, and differentiated into multiple cell types in vivo and in vitro [44]. These studies have confirmed the feasibility of SCNT in mammalian cloning. SCNT has also been used to show that oocytes have the ability to reprogramme somatic cells, as does the sperm genome.

The success rate of mammalian cloning by nuclear transfer is, however, very low (3–5%), and the majority of clones die in utero or neonatally. This can also often occur in conjunction with developmental problems, such as large offspring syndrome [45]. A surviving cloned animal is a highly rigorous operational assay for effective reprogramming, but such successes at the organism level unfortunately provide little insight into the underlying molecular mechanisms that are involved in the reprogramming processes themselves. For this reason, different methods for studying reprogramming have been used, such as fusion between somatic cells and pluripotent stem cells. The stem cells that have been used in these experiments are ESCs, embryonic germ cells (EGCs), and ECCs (Fig. 9.2).

The use of pluripotent cells in vitro has been possible only in the last few years. Only in 1981 were ESCs established from normal mouse blastocysts [46, 47]. Subsequently, primordial germ cells (PGCs) were generated and EGCs were derived and called "embryonic germ" to denote their origin [48]. EGCs can retain many properties of pluripotency and can be cultured for long times.

ECCs are the stem cells of teratocarcinomas, and studies have shown that they are closely related to ESCs [46, 47, 49, 50]. This conclusion has been confirmed in humans, with the demonstration that ECCs derived from human testicular teratocarcinomas and ESCs isolated from early human embryos produced by in-vitro fertilization share common features [51, 52].

9.3.1 Fusion of Somatic Cells with ECCs

In 1976, Miller and Ruddle first reported that pluripotent teratocarcinoma–thymus somatic cell hybrids can differentiate into a wide variety of tissues, which indicated that pluripotency is not abolished by the presence of the differentiated cells in the hybrids [20]. Reprogramming was evident since the hybrid cells resembled an ECC morphologically, and they showed reactivation of specific genes

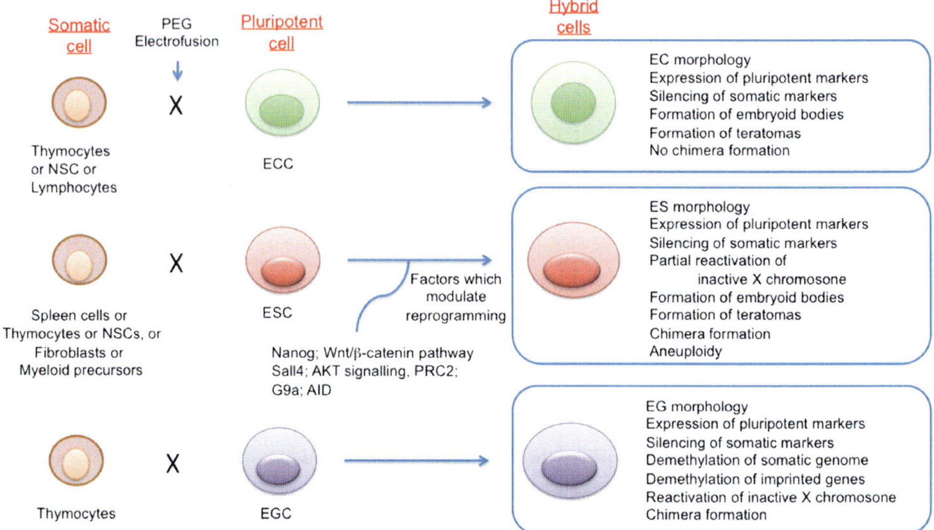

Fig. 9.2 Cell-cell fusion between somatic cells and pluripotent cells. Somatic cells can be fused using PEG or electrofusion, with the generation of pluripotent cells (ECCs, EGCs and ESCs) and hybrid cells. Features of the different hybrids obtained are indicated

and/or activation of the inactive X chromosome derived from the somatic partner [20, 53, 54]. Cross-species fusion has also been carried out: hybrids formed between murine ECCs and cells from a human T-lymphoma resulted in formation of an inter-species hybrid colony where the expression of the Oct4 and Sox2 human transcription factors was detected, as characteristic of undifferentiated pluripotent stem cells. This thus demonstrated activation of endogenous human markers of pluripotency.

One question that has been raised regarding fusion-induced reprogramming was whether somatic cells can retain the memory of their origin after being reprogrammed, as reprogramming itself does not necessarily imply that the somatic cells have completely lost this memory. To answer this question, ECCs were fused with mouse neurosphere cells. By examining the changes in gene expression and DNA methylation of the hybrid cells during re-differentiation, it was found that not only did the ECCs reprogram these neural stem cells (NSCs), but they also caused the NSCs to lose their epigenetic memory [55]. This thus confirmed the conclusion that hybrid cells can lose the memory of their somatic origin and adopt an identical differentiation potential to that of their pluripotent fusion partner.

Furthermore, hybrid ECC–somatic cells can give rise to true carcinomas that contain derivatives of all three of the embryonic germ layers [53, 54, 56, 57] or that form embryoid bodies in suspension culture [57]. The presence of embryonic antigens has also been seen for fusion between ECCs and lymphocytes and thymocytes [58]. However, the hybrid cells produced by the fusion of ECCs with fibroblasts resembled fibroblasts with respect to their morphology [59], suggesting that ECCs cannot completely reprogramme all kinds of somatic nuclei.

Even if an important amount of our knowledge derives from studies with hybrids of ECC–somatic cells, it is clear that ECCs have some limitations. While they have the features of pluripotent embryonic cells, ECCs retain a low level of developmental potential as compared to ESCs and EGCs. ECCs rarely generate chimeras, and moreover, they cannot contribute to germ lines (Fig. 9.2) [54].

The view that is emerging from these studies is that hybrid cells generated by ESCs or EGCs with somatic cell fusion provide better systems for studying somatic-cell reprogramming.

9.3.2 Fusion of Somatic Cells with ESCs

A large number of recent studies have demonstrated the potent reprogramming activities of ESCs. One of the earliest studies of fusion-mediated reprogramming was carried out by fusing male mESCs deficient for hypoxanthine phosphoribosyl-transferase (HPRT$^-$) with mouse female spleen cells, using PEG. The hybrids were then selected for HAT-resistance, and four clones were isolated. These were positive for the embryonic ECMA-7 antigen and negative for the TROMA-1 somatic antigen, showed alkaline phosphatase activity, had both X chromosomes synchronously replicating (indicating that they were both in an active state), were able to form embryoid bodies containing derivatives of all three germinal layers, and, most importantly, were able to generate chimeric mice when injected into blastocysts. Of note, three of the four isolated clones contained 41–43 chromosomes, and one clone was nearly tetraploid. However, this last proved to be unstable: when grown under non-selective conditions, the cells of this clone rapidly lost chromosomes, retaining only 40–43 chromosomes after 5–7 passages [60, 61]. Nevertheless, it was not clear if the chromosomes from the ESCs or those from the spleen cells segregated into the hybrids.

In a subsequent study, a more efficient method for somatic reprogramming selection was used. Somatic thymocytes harboring a silent GFP transgene under the control of the Oct4 promoter were fused with ESCs [22]. When the somatic genome was reprogrammed, Oct4-GFP was expressed and hybrid reprogrammed clones were successfully generated. The thymocyte/ ESC hybrids contained reactivated X chromosomes of thymocyte origin, as judged by fluorescent *in-situ* hybridization for Xist RNA. In this study, Xist RNA was seen to bind unstably to all three X chromosomes, suggesting a state of partial X chromosome inactivation. The Oct4-promoter-driven GFP transgene was observed approximately 48 h after fusion. Interestingly, ESCs were not able to reprogram parental imprints, given that the methylation status of the imprinted H19 and Igf2r genes was not altered by the fusion with ESCs. Furthermore, ESC/thymocyte hybrids contributed to all three germ layers in mouse chimeras at day E7.5 of development (Fig. 9.2).

To analyze the differentiation potential of the hybrid cells, fusion of mouse Hprt$^-$ ESCs and thymocytes containing a selectable ROSA 26βgeo transgene (which ubiquitously expressed neomycin drug resistance and β-galactosidase activity) were carried out. These hybrids retained chromosomes from both fusion partners, as judged by the presence of specific polymorphisms, and they produced teratomas in SCID mice [62]. Furthermore, they differentiated into neural lineages in culture, and were shown to be immunoreactive for the post-mitotic neuron-specific factor TuJ and for the dopaminergic neuronal marker PitX3. When neuronally differentiated hybrids were implanted into the striatum of mouse brains, the cells expressed tyrosine hydroxylase, providing evidence that hybrid cells can differentiate into neural cells with dopaminergic characteristics. The survival of cells in the grafts was confirmed by X-gal staining 15 days after injection.

It has also been shown that human somatic cells can be reprogrammed by fusion with hESCs. Like mESCs, hESCs can self-renew indefinitely and can differentiate into all cell types of the body [52]. Hygromycin-resistant hESCs were fused with puromycin-resistant human fibroblasts using PEG, and double drug selection was used for the formation of hybrids [21]. The resulting hybrids were formed of cells predominantly containing 92 chromosomes, and the Oct4 promoter was demethylated, a necessary step for reinitiating its expression. Other pluripotent markers were expressed in these hybrids, while the expression of fibroblast-specific genes was repressed. Hybrids formed embryoid bodies and teratomas after injection into nude mice [21]. Similar results were obtained when myeloid precursors were fused with hESCs [63].

Fusion of ESCs with somatic cells in all the mentioned studies produced highly proliferative hybrids with nuclear fusion, called synkaryons. This system unfortunately does not allow the unequivocal assertion that the somatic cell nucleus is reprogrammed, as it coexists with the original chromosomes from the ESCs in the hybrids. Thus a different approach has been used recently to analyze reprogramming: the heterokaryon system. Heterokaryons are hybrids containing two different nuclei in a common cytoplasm, whereby all of the genetic material remains intact within each independent nucleus. This type of short-term, non-dividing fusion product makes it possible to assess the influence of two nuclear components on gene expression [64]. Two research groups have generated heterokaryons between mESCs and human somatic cells (B lymphocytes and fibroblasts). This method allowed the reprogramming of the human nuclei to be followed. Interestingly, the reprogramming was very fast, and the human nuclei were shown to start expressing pluripotent markers (Oct4 and Nanog) 24 h after fusion [65, 66]. Furthermore, the reprogrammed human cells expressed a profile of transcripts seen in hESCs that were not expressed in mESCs, suggesting that the human nuclei was reprogrammed through trans-acting factors from the mouse nuclei. Later, however, the reprogramming process was finalized and established by reactivated cis-acting factors from the human nuclei [66].

9.3.3 Fusion of Somatic Cells with EGCs

EGCs were established from PGCs [67, 68]. PGCs can be identified by the expression of an Oct4-*GFP* transgene, which allows their purification by fluorescence-activated cell sorting (FACS). Analysis of DNA from PGCs revealed demethylation at multiple genetic loci, including for both imprinted and non-imprinted genes [69, 70]. This might explain why EGC-derived chimeras show phenotypic abnormalities, which include fetal overgrowth and skeletal malformations, even if they can contribute to many tissues in chimeric embryos [71]. However, EGCs maintain important pluripotent characteristics, which make them attractive candidates for induced fusion with somatic cells to study reprogramming.

With fusion of female EGCs carrying the Rosa 26βgeo transgene and female somatic thymic lymphocytes, the hybrids obtained showed phenotypic properties that were similar to those of EGCs, including both pluripotency and repression of expression of the somatic cell genome (Fig. 9.2). Extensive demethylation of the thymocyte genome was also detected, which was similar to that seen in the EGC nuclei; this demonstrated that the demethylation activity from EGCs was dominant over the somatic genome. Futhermore, the X chromosome derived from the somatic nuclei was reactivated. Interestingly, unlike ESCs, EGCs can erase the parental imprints of *H19* and *Igf2r*. Finally, when the hybrids were injected into host blastocysts and implanted into pseudopregnant mothers, β-galactosidase expression was seen in chimeric embryos at days E9.5 and E10.5 [72].

9.3.4 Mechanisms Controlling Somatic Cell Reprogramming

Somatic cell reprogramming is an inefficient process. Furthermore, for many years it was not clear if the factors that induce reprogramming are in the nucleus or in the cytoplasm. The nuclear transfer experiments told us that the cytoplasm of an enucleated mammalian oocyte had the ability to reset the genetic program of a fully differentiated somatic cell nucleus [44, 73–76]. In addition, extracts from *Xenopus* eggs and ECCs were shown to activate expression of Oct4 in the somatic cells [77, 78], further suggesting that cytoplasmic elements are the reprogramming factors responsible.

However, in an interesting report, karyoplasts (cellular nuclei) and cytoplasts (intact cytoplasm without nuclei) of ESCs were separated and fused with NSCs. Fascinatingly, Oct4-*GFP* was activated in NSCs only after fusion with ESC karyoplasts, and hence not with ESC cytoplasts. These data were the first evidence that indicated that ESC nuclei contain factors that are sufficient to reactivate Oct4-GFP in somatic cells and to initiate reprogramming [79].

In addition, Do and Scholer [79] demonstrated the importance of the cell-fusion-induced reprogramming approach for identification of nuclear factors that regulate and increase reprogramming efficiency. Below we summarize the reprogramming factors that have been identified to date by cell-fusion-mediated reprogramming.

9.3.4.1 Nanog

Nanog is a homeodomain-bearing protein that acts as a transcriptional factor, and is itself transcribed specifically in mouse pluripotent cells, mESCs and mEGCs [80, 81]. The loss of epiblasts soon after implantation in Nanog-null embryos, and the clonal expansion of ESCs over-expressing Nanog via bypassing of the regulation by LIF-STAT3 signals, indicate that Nanog is an important regulator for maintaining pluripotency and self-renewal of ESCs [82, 83]. These observations identified Nanog as a good candidate to increase reprogramming. Indeed, a 200-fold increase in the number of reprogrammed colonies was seen after fusions of ESCs overexpressing Nanog with NSCs, as compared with controls [84]. Nanog also improved the yield of reprogrammed hybrids when thymocytes and fibroblasts were fused with ESCs.

However, Nanog is not a part of the minimal combinations of exogenous factors that can convert mouse somatic cells into iPSCs [26]. To solve this apparent controversy, using Nanog-deficient cells, it has been shown that Nanog is fully dispensable for the initial steps of reprogramming, which consist of the loss of differentiated features and the creation of a pre-pluripotent state. Instead, Nanog mediated the acquisition of pluripotency by inducing the completion of dedifferentiation of partially reprogrammed cells [82].

9.3.4.2 The Wnt/β-Catenin Pathway

Wnt/β-catenin signaling controls ESC self-renewal and maintenance of "stemness" [85], and regulates expression of the ESC genes. The stability of β-catenin is essential to the signaling activity of the canonical Wnt pathway. In the absence of Wnt binding to its receptor, GSK-3β kinase phosphorylates β-catenin and targets it for ubiquitin-mediated destruction. Activation of the pathway by Wnt inhibits GSK-3β activity and results in the accumulation of β-catenin. Stable β-catenin then translocates into the nucleus, where it interacts with different Tcf DNA-binding factors; this complex in turn activates the transcription of target genes [86, 87]. Periodic activation of the Wnt/β-catenin signaling pathway strikingly enhances cell-fusion-mediated reprogramming. Specifically, by treating ESCs for a limited and specific time with Wnt3a or with an inhibitor of GSK-3 activity, which both lead to nuclear accumulation of β-catenin, these cells became "super-able" to reprogram somatic cells (NSCs, thymocytes and mouse embryonic fibroblasts [MEFs]) after fusion [88].

It would be interesting to identify the important downstream effectors of Wnt that participate in this process. For example, c-Myc is a prominent downstream regulator of the Wnt pathway [86, 89]. However, enhancement of reprogramming efficiencies by Wnt3a-conditioned medium was not accompanied by up-regulation of c-Myc [88]. Tcf3 is another candidate effector of the Wnt signaling pathway. Tcf3 co-localizes with ESC core regulators, such as Oct4, Sox2 and Nanog, to regulate the balance between ESC pluripotency and differentiation [86, 90, 91].

Interestingly, Nanog and the Wnt pathway can cooperate; in a system where Nanog is overexpressed and the Wnt pathway is activated, the reprogramming of NSCs is strikingly enhanced [92].

9.3.4.3 AKT Signaling

Phosphoinositide 3-kinase (PI3K) has a decisive role in a broad range of cellular functions relating to responses to extracellular signals. The serine-threonine kinase Akt is a key downstream effector of PI3K, and in response to PI3K activation, Akt phosphorylates and regulates the activities of a number of targets, including kinases, transcription factors, and other regulatory molecules. This PI3K/Akt signaling regulates both tumorigenic potential and pluripotency of stem cells. This pathway promotes the de-differentiation of primordial germ cells into EGCs, and it is sufficient to maintain the pluripotency of mouse and primate ESCs cultured in the absence of LIF and feeder cells [93, 94]. These features defined PI3K/Akt as a strong candidate signaling pathway that can increase reprogramming. Indeed, activation of Akt signaling enhanced the yield of pluripotent hybrid colonies after cell fusions between ESCs and somatic cells [93]. However, activation of Akt signaling significantly reduced the efficiency of nuclear reprogramming by nuclear transfer. This controversy into the effects of Akt signaling might be due to transcriptional activation of different sets of target genes in each of these methods.

9.3.4.4 Sall4

Sall4 is a member of the Spalt family of transcription factors, and it was originally identified in *Drosophila* as a homeotic gene that is required for head and tail development [95]. Sall4 is also essential for maintenance of pluripotency and self-renewal of ESCs, and for their derivation from blastocysts [96]. Although Sall4 can act as a transcription factor that regulates numerous genes, one of its few known target genes is Oct4 [97].

MEFs carrying the Oct4-*GFP* transgene and overexpressing each of Oct4, Nanog, Sox2 and Sall4 have been fused with ESCs, with the number of GFP-positive cells after fusion monitored. Unexpectedly, after fusion with ESCs, MEFs that overexpressed Oct4, Nanog or Sox2 did not show significant increases in Oct4-*GFP* expression relative to the controls. In contrast, the relative numbers of GFP-positive cells in MEFs overexpressing Sall4 increased sevenfold with respect to the controls.

However, in another experimental system, double drug selection was used to measure reprogrammed colony formation after fusion. In this setting, after fusion with ESCs, MEFs overexpressing Nanog, Sox2 or Sall4 showed significant increases in the number of reprogrammed colonies, relative to the controls. In contrast, the overexpression of Oct4 in MEFs did not promote formation of reprogrammed colonies [98]. These data showed that Nanog, Sox2 and Sall4 can induce reprogramming even if they are overexpressed in the somatic genome. However, the duration of the reprogramming process and the re-expression of the Oct4 promoter can vary across different systems, and so the reactivation of Oct4-GFP appears not necessarily to be indicative of successful reprogramming.

9.3.4.5 Epigenetic Modulation: Roles of PRC2, AID and G9a

Using the method of direct reprogramming, most infected cells are trapped in a partially reprogrammed state, due to their inability to overcome major reprogramming barriers. When the DNA-methylase inhibitor 5-aza-cytidine was applied to these pre-iPSC clones, conversion of the pre-iPSC state into the complete iPSC state was shown [99]. This thus demonstrated that DNA methylation is an important epigenetic barrier that partially reprogrammed cells can encounter and can fail to overcome.

With heterokaryon formation between mESCs and human fibroblasts, the DNA demethylase AID was identified as an important player in the reprogramming process. After fusion, rapid demethylation of the Oct4 and Nanog promoters in the somatic heterokaryon genome was seen, which was also followed by the expression of these genes. This suggested that demethylase activity is important for the reprogramming process. DNA demethylation is essential to overcome gene silencing and to induce temporally and spatially controlled expression of mammalian genes, although no consensus mammalian DNA demethylase has been identified, despite years of efforts [100]. AID was a candidate factor, as it has a role in mammalian DNA demethylation of pluripotent germ cells and DNA demethylation in zebra fish during post-fertilization events [101, 102]. AID belongs to a family of cytosine deaminases, and the deamination of cytosine followed by DNA repair leads to DNA demethylation [103]. In heterokaryons, the knock-down of AID prevented DNA demethylation of the human Oct4 and Nanog promoters and the expression of these pluripotency factors by fibroblast nuclei. Furthermore, initiation of nuclear reprogramming towards pluripotency was inhibited in human somatic fibroblasts when AID-dependent DNA demethylation was reduced. Interestingly, AID binding was seen at silent methylated Oct4 and Nanog promoters in fibroblasts, but not in active unmethylated Oct4 and Nanog promoters in ESCs [65].

As well as DNA demethylation, histone modifications are important for the enhancement of reprogramming efficiency.

Polycomb-group (PcG) proteins were originally identified in *Drosophila melanogaster*, where they form multiprotein complexes that are required for maintaining transcriptional silencing of a subset of repressed genes [104]. Two main repressor complexes, polycomb repressive complexes 1 and 2 (PRC1, 2) have been identified. These have different catalytic properties and core components. PRC2 consists of three core components: embryonic ectoderm development (Eed), suppressor of Zeste 12 (Suz12), and the SET-domain-containing protein enhancer of Zeste homolog 2 (Ezh2). The catalytic subunit, Ezh2 is a SET domain-containing methyltransferase that catalyzes the formation of the H3K27me3 marker, which forms the docking site for recruitment of PRC1 [104].

The involvement of the PcG proteins in the maintenance of ESC identity and pluripotency was first suggested by genome-wide studies that showed that PcG targets are highly enriched in genes involved in developmental patterning, morphogenesis, and organogenesis [38]. Loss of the EED gene in ESCs leads to genome-wide and almost total loss of H3K27me3, and consequently, to derepression of the PcG targets [105]. Despite this dramatic reduction in H3K27Me3, ESCs can be derived in the absence of EED. Embryos lacking individual components of the PRC2 complex, such as Eed, Ezh2 and Suz12, can survive post-implantation but die from gastrulation defects from 7 to 9 days post-fertilization [106–108].

In mESCs and hESCs, PRC1 and PRC2 localize to the promoters of a subset of repressed genes that encode transcription factors that are required for specification during later development. These genes contain overlapping binding sites for the pluripotent genes Oct4, Sox2, Nanog and Sall4 within their promoters [109–111], and are enriched in both H3K4me3 and H3K27me3 histones [112–114]. This explains why ESCs require PcG proteins for maintenance of the self-renewing state; EED-/- ESCs tend to differentiation in culture as lineage development genes are derepressed [105, 115]. However, EED–/– ESCs cannot give rise to all cell types after in-vitro differentiation, and the chimeras show developmental defects that are similar to knock-out embryos [105]. Thus PcG complexes are also required for the full differentiation potential of ESCs.

Interestingly, it has been shown recently that deletion of individual PRC1 and PRC2 members (Eed, Suz12, Ezh2 and Ring1a/B) in ESCs abolished the ability of these ESCs to induce reprogramming when they were fused with human lymphocytes to form heterokaryons. Importantly, given that Eed-deficient mESCs can themselves self-renew, and are pluripotent and can contribute to the three germ layers in vivo but can not reprogram the somatic genome, this finding clearly showed that pluripotency and reprogramming function can be dissociated, and represent two different pathways [116].

As part of H3K27 methylation, H3K9 methylation has also been shown to be important as a marker for reprogramming. It has been showed that knock-down of the histone H3 lysine 9 (H3K9) methyltransferase G9a, and overexpression of the jumonji-domain-containing H3K9 demethylase Jhdm2a, can enhance Oct4-EGFP reactivation from adult NSCs after ESC-fusion-mediated reprogramming. In addition, coexpression of Nanog and Jhdm2a enhanced the reprogramming even further. After overexpression of Jhdm2a or inhibition of G9a, a reduction in DNA methylation in the Oct4 promoter was seen, demonstrating that H3K9 and DNA methylation restricts somatic cell reprogramming by cell fusion with ESCs [117].

9.4 Lineage Reprogramming by Cell Fusion

Although differentiated cells normally retain cell-type-specific gene expression patterns throughout their lifetime, cell identity can sometimes be changed both in vitro and in vivo through different mechanisms. Besides the strategies described above that are based on de-differentiation of somatic cells to a pluripotent fate, several studies have been performed to determine whether, and especially how, the cellular identity can be modified, rather then reversed back along the developmental cascade, to achieve pluripotency.

The ability to impose changes in gene expression and to transfer epigenetic markers associated with a different cell fate to more specialized cells after cell fusion has been considered a unique characteristic of ESCs [21, 22, 84]. However, in the 1980s, it was already thought that differentiated cells also have the capacity to change the epigenetic state of other nuclei. Following fusion of two distinct somatic cell types to form proliferating hybrids (synkaryons) or post-mitotic hybrids (heterokaryons), somatic nuclei were reprogrammed towards specific differentiated fates [23, 25, 118]. These findings surprisingly highlighted the possibility to switch directly between different cell fates; that is, in other word, the re-programming of cell identity.

The first studies on phenotype modulation induced by synkaryon formation in specialized cells revealed that gene activation specific for a different cell is seen. Fusion of murine hepatoma cells, which secrete mouse serum albumin, with human leukocytes, which did not produce albumin, resulted in the formation of hybrids that secreted both mouse and human serum albumin, indicating that the albumin gene in human leukocytes was re-activated. These data are consistent with the hypothesis that the murine genome contributes with activators to the human genome [119].

Many reports have shown that ectopic expression of single transcription factors that are known to have a key role in specification of a certain cell identity during development can be sufficient to convert the cell fate to that of a different somatic cell. For example, expression of individual muscle regulatory factors of the MyoD family has been shown to be sufficient to convert a range of non-muscle cell types into muscle cells (Fig. 9.3). Surprisingly, the pattern of gene activation resembles the expression of the muscle regulatory transcription factors during normal muscle differentiation [120–124].

Similarly, high levels of the transcription factors C/EBPalpha and C/EBPbeta can directly reprogram committed mature B lymphocytes to become macrophages [125, 126]. More recently, fibroblasts were converted into functional neurons by expression of three factors (Ascl1, Brn2 [also known as Pou3f2] and Myt1l) [127] (Fig. 9.3).

Unexpectedly, the liver-derived BNL cell line was directly converted to muscle cells by induction of overexpression of the transcription factor MyoD, whereas other similar cell types, such as the human HepG2 hepatocyte cell line, were not [121, 128]. One explanation of such differences could be that the BNL cells are less differentiated than HepG2 cells and primary hepatocytes. Thus cell type, cell-cycle phase, differentiation state, and age of the nuclei might all influence the efficiency of nuclear reprogramming; however, the effects of these differences need to be investigated.

Interestingly, the apparently refractory HepG2 cell type starts to express specific muscle genes when forced to fuse with myotubes, to form heterokaryons [129]. These data demonstrated that heterokaryon formation induces human muscle gene expression in non-muscle nuclei.

The phenotypic changes observed in MyoD-overexpressing HepG2 hepatocytes differ from those in hepatocyte-derived heterokaryons: the response of a somatic cell to a single regulator depends on the cellular developmental state and the specific regulatory gene expression. In contrast, the phenotype of hybrid cells obtained by fusion results from a complex interaction of the regulatory factors that are contributed by each former cell type. Based on these differences, the stable differentiated state of a cell has been defined as the product of dynamic interactions among different sets of regulators [130].

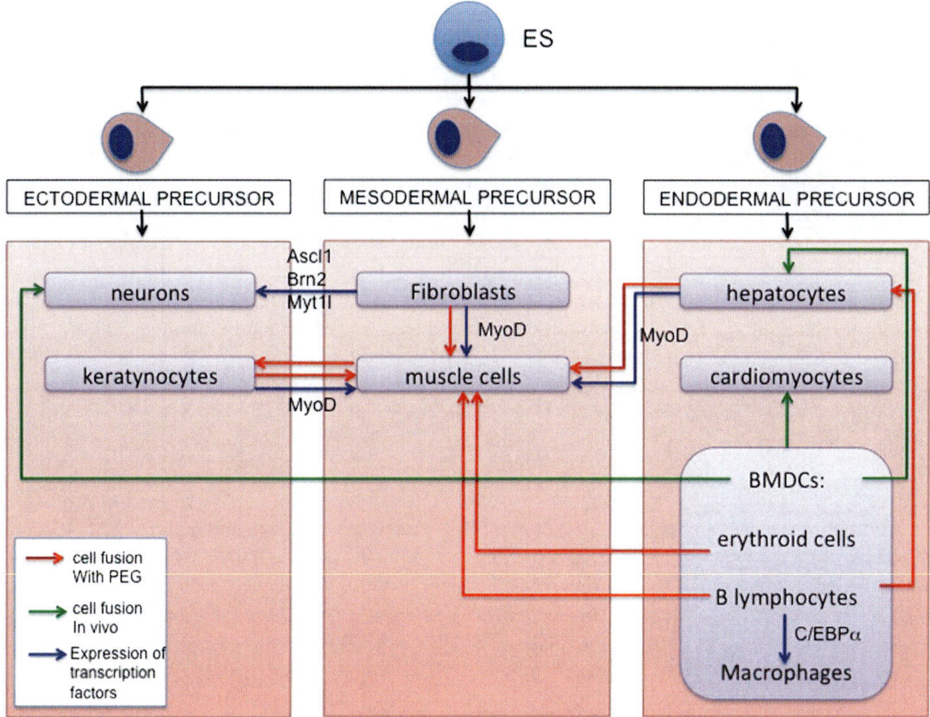

Fig. 9.3 Overview of lineage reprogramming induced in vitro and in vivo. Scheme summarizing the different transitions from one cell fate to another that have been observed experimentally after ectopic expression of key transcription factors (*blue lines*), in vitro cell fusion forced with PEG (*red lines*), and spontaneous cell fusion observed in vivo (*green lines*), starting from cells derived from different developmental precursors, such as ectoderm, mesoderm and endoderm. The bi-directionality of the process has been investigated experimentally only for the muscle–keratinocyte transition. Interestingly, the kinetic of the transition appears to be more rapid for the same developmental lineage transitions (i.e. fibroblast–muscle) with respect to different developmental lineage transitions (i.e. hepatocyte–muscle)

9.4.1 Synkaryon Versus Heterokaryon Strategies to Investigate the Mechanisms of Cell-Fusion-Mediated Lineage Reprogramming

Muscle cells are not unique in their potential to directly reprogram a differentiated cell by cell fusion. For instance, other studies have demonstrated that the adult human globin genes can be reactivated in stable hybrid cell lines produced by PEG-mediated cell fusion of adult mouse erythroleukemia cells or human hematopoietic cells, and fibroblasts [131–133]. However, it was difficult to draw conclusions about trans-acting regulatory mechanisms in these studies that came from stable synkaryon lines. Many generations were required before the hybrids could be isolated and analyzed; furthermore, the gene activation program was generally transient after the formation of synkaryons, because cell division during the passages in culture led to the loss of the human chromosomes.

To overcome these problems, other investigations have fused human fetal or mouse adult erythroid cells with non-erythroid cells, to form transient heterokaryons, with isolation and analysis of the total RNA 24 h later. The conclusion here was that previously inactive globin gene expression can be activated in a variety of non-erythroid cell types, and thus the globin genes appeared not to be irreversibly inactivated in non-erythroid cells: in conclusion, the erythroid cells should contain developmental-stage-specific factors that act in *trans* to regulate globin gene expression [25].

However, the synkaroyon strategy used in these studies had some limitations. For instance, in synkaryons from the same species, it was not possible to determine the timing or extent of reprogramming, as after many passages the nuclear components of one parental cell type were physically mixed with those of the other; thus the contributions of each of the two nuclei could not be distinguished. Therefore, the heterokaryon strategy appears to be the most appropriate to circumvent these problems. In contrast to synkaryons, heterokaryons do not undergo mitosis, and they stably retain all of the nuclear components of the parental cell types in their distinct intact nuclei. In addition, in studies of nuclear reprogramming towards a differentiated fate in non-dividing interspecific heterokaryons, the species-specific transcriptome can be analyzed to profile the gene expression changes from the nucleus of interest in the fused cells, throughout the time course of the reprogramming.

In one of the first studies on heterokaryons, Ringertz showed that fusion of rat myoblasts and chick erythrocytes resulted in the swelling of the erytrocyte nuclei and diffusion of the chromatin, in anticipation of the reprogramming events [134]. Later, primary mouse muscle cells were fused in tissue culture with human primary cells derived from all of the three embryonic lineages: endoderm (hepatocytes), ectoderm (keratinocytes) and mesoderm (fibroblasts) (Fig. 9.3). Of note, nuclei from each of these cell types were capable of activating a number of previously silent muscle genes, indicating that the differentiated state can be altered, even in different specialized human cells [23, 25, 135, 136].

Interestingly, these heterokaryon studies revealed that tissue derivation and embryonic origin have marked effects. Fibroblasts, which are from the same embryonic lineage as muscle (mesoderm), show faster kinetics and a higher ultimate frequency of muscle gene expression, with respect to keratinocytes (ectoderm) and hepatocytes (endoderm) (Fig. 9.3). Despite this, the characteristics of cell history, or the lineage of origin, appear to influence cell reprogrammability [129].

9.4.2 The Importance of Gene Dosage in the Direction of Lineage Reprogramming by Cell Fusion

Cell-fusion-mediated lineage reprogramming can occur in either direction. In addition to keratinocytes being reprogrammed to a muscle transcriptional state, muscle nuclei can be reprogrammed towards a keratinocyte state, and these two phenotypes are mutually exclusive [137]. These two developmental states do not coexist: thus, there must be a controlling mechanism that supports and maintains the dominant transcriptional program.

Cell fusion results in the combination of two entire genomes and the cytoplasm from two cells of different functional and developmental states. Thus, the final phenotype will ultimately be determined by the dominance that arises from an excess of cytoplasmic factors or nuclear gene dosing, and not by a particular dominant phenotype or a master regulator [137]. To better examine the concentration requirements in the activation and expression of differentiation-specific genes over time, different ratios of myoblasts and keratinocytes were fused. Higher proportions of muscle cells induced an increase in reprogramming towards the muscle fate, and *vice versa*, indicating that the gene reactivation obtained in heterokaryons is dependent on the relative ratio of the nuclei that is contributed by each of the parental cell-type concentrations and stoichiometries.

9.4.3 Global Chromatin Changes in Cell-Fusion-Induced Lineage Reprogramming

The first evidence of cell-fusion-induced lineage reprogramming that was seen in heterokaryons suggested that the gene expression changes were restricted to permissive or open loci. However, many years ago, global chromatin remodeling in somatic cell heterokaryons was already predicted by the observation of nuclear swelling and chromatin redistribution prior to gene reactivation in heterokaryons formed between rat myoblasts and nucleated chicken erythrocytes [138]. As differentiation involves a unique chromatin fingerprint configuration for each cell type, and moreover, as differentiated cells required continuous and active regulation to maintain their identity, it is reasonable to believe that lineage reprogramming is driven by chromatin remodeling at key loci, which in turn allows the expression of trans-acting regulators. Recently, a detailed study of muscle–keratinocyte hybrids revealed that the muscle gene activation in keratinocytes is not limited to a few genes, but is extensive [137]. Importantly, muscle gene re-activation was followed by a substantial silencing of the keratinocyte genes. Thus, this heterokaryon formation induced not only activation of silent genes in the non-expressing cell type, but also repression of the differentiated functions. So, the activation of muscle genes appears to be due not to genome-wide de-repression, but instead it is part of a global conversion from the expression program of a keratinocyte to that of a muscle cell [129].

In 1984, it was already postulated that differential regulation in hybrids after cell fusion resulted from the interactions of specific trans-acting factors with cis-acting genomic sequences, such as promoters or enhancers [139]. Whereas cell-type-specific gene expression programs were established through a network of transcriptional activators and repressors, epigenetic factors might also be required to maintain specification, by stabilizing the chromatin domains [140, 141].

As chromatin remodeling factors, including histone deacetylases (HDACs), were required for resetting gene expression, the role of HDAC activity in the dominant conversion of human lymphocytes to muscle cells in heterokaryons was investigated. The aim here was to elucidate the mechanisms of chromatin remodeling in cell-fusion-mediated lineage reprogramming. In heterokaryons, where chromatin replication is precluded, the lymphocyte nuclei increased in size and heterochromatin domains were redistributed to mimic the spatial rearrangement of neighboring mouse myocytes [142]. This was followed by the de-novo expression of human muscle genes in a temporal order that accurately recaptured gene expression in normal development. Moreover, activation of muscle-specific genes was associated with reduced expression of several lymphocyte genes. This confirmed that cell fusion not only induced reactivation of some specific genes, but it also resulted in global transition to a new cell identity in which silencing of pre-expressing genes also occurred. Remarkably, by inhibiting HDAC activity in the heterokaryons, the nuclei were seen to co-express two different lineage-associated gene programs. This indicated that although gene activation and silencing are mechanistically distinct, they are coordinated events in reprogramming [142].

The positive effects of treatment with an HDAC inhibitor prior to heterokaryon formation suggested that pre-fusion relaxation of chromatin at muscle regulatory regions renders the non-muscle nuclei more susceptible to muscle cytoplasmic factors after fusion.

Thus, in addition to activators and repressors of transcription, structural information, such as DNA methylation patterns, have to be transferred via the cytoplasm from the nuclei of one cell type to those of another in heterokaryons. It is now well supported that specialized skeletal muscle cells confer epigenetic information to epidermal progenitor nuclei by directing the methylation and demethylation of DNA of the genes they activate or silence [142]. These methylation changes are targeted in a tissue-specific manner, with keratinocyte-specific gene methylation accompanied by muscle-specific gene demethylation. Interestingly, DNA methylation has to occur by an active mechanism, as the cell hybrids did not undergo mitosis or DNA replication.

9.5 Cell-Fusion-Mediated Reprogramming as a Regeneration Mechanism

Generation of one adult phenotype directly from another as an alternative to reprogramming somatic cells to an intermediate stem cell state has broad implications for regenerative medicine. If cells can de-differentiate to a less differentiated state also in vivo, this implies that not only stem cells, but also the differentiated cells can show a degree of plasticity. Evidence that cell fusion between somatic cells leads to reprogramming of differentiated nuclei had already arisen in the 1980s; however, that similar fusion events occurred in adult vertebrate organs was not predicted 3 decades ago. Thus, today, cell–cell fusion cannot be considered only as an in vitro mechanism to induce changes in cellular identity, but it is also acquiring a high impact in the field of regenerative medicine, as a possible in-vivo physiological way of regenerating damaged tissue.

9.5.1 *Transdifferentiation* Versus *Cell–Cell Fusion Theories to Determine Cellular Plasticity*

There is evidence that adult stem cells can fuse and change their cell identity in vivo. This is the case for bone marrow-derived cells (BMDCs), which are known to be the most plastic stem cells in the human body. These adult stem cells have the unique ability to switch their differentiation fate following bone-marrow transplantation in human and rodents, which contributes to the regeneration of hematopoietic and non-hematopoietic tissues. For instance, many studies have demonstrated that transplanted BMDCs can take on different lineages in vivo, including myocytes [143], hepatocytes [144], neurons [145], and many other cell types [146] (Fig. 9.3). This idea challenges the long-standing concept of cell-fate determination in mammalian developmental biology, and has received significant attention because it offers possibilities of expanding the therapeutic potential of adult stem cells.

The developmental fate of transplanted BMDCs during their new phenotype acquisition has been followed. It appears that after transplantation, BMDCs first acquire the characteristics of heritably diploid muscle stem cells (satellite cells), and then they later contribute to mature myofibers of muscle tissue. This thus suggests that BMDCs can undergo cell-fate changes as diploid mononucleate cells, and once reprogrammed, they can contribute to various tissues [147]. Two different theories can explain this phenomenon: the first is the trans-differentiation theory, which proposes that adult stem cells differentiate outside the tissue of origin in response to a new microenvironment by converting directly in the new cell type, while not globally changing their genetic identities. The second theory is that their change in cell identity is mediated by cell fusion events. In early 2002, spontaneous cell fusion was proposed as an alternative mechanism by which BMDCs can contribute to different tissues. The theory is that BMDCs first fuse with pre-existing differentiated cells within the target tissue, and then their nuclei are reprogrammed in response to intracellular cytoplasmic factors. Through co-culturing BMDCs with ESCs, Terada et al. showed that BMDCs can fuse spontaneously with other cells and subsequently adopt the phenotype of the recipient cells [148]. This finding was particularly significant, considering reports at the same time of transplanted BMDCs turning into unexpected cell types in vivo as a result of cell fusion events. Thus, despite the LaBarge study excluded cell fusion events and indicated that BMDCs can be transformed into satellite stem cells and then into muscle cells on the basis of cell karyotypes, several studies have now indicated that the changes in cellular identity can also result from cell-fusion events.

As an example, in a central nervous system biopsy from a woman who also received bone marrow transplants from male donors, it was shown that some of her Purkinje neurons were tetraploid (XXXY) and contained chromosomes from both female and male bone-marrow donors. This indicated that the Purkinje cells had fused with hematopoietic cells from the bone-marrow donors [149].

Several in-vivo cell-fusion events were then experimentally demonstrated in mice. Evidence of cell fusion in vivo emerged from injecting GFP^+ BMDCs expressing CRE recombinase into lethally irradiated mice carrying a LacZ reporter gene that was expressed only after excision of a LoxP-flanked-STOP codon by CRE-mediated recombination after fusion. Here, BMDCs were found to have fused with hepatocytes, cardiomyocytes and Purkinje neurons, even if at low rates [150].

Already in 1997, Eglitis and Mezey demonstrated that after transplantation of NeoR expressing hematopoietic stem cells, micro and macroglia that expressed the donor-derived reporter gene (*NeoR*) were found in the brain [151]. However, despite a large number of reports indicated that cell fusion can occur in vivo after bone-marrow transplantation and that the phenotype change of the hybrid cells suggested a global change of expression profile by the donor genome, evidence of "reprogramming" was still missing.

It was Weimann and colleagues who for the first time suggested global chromatin reorganization and reprogramming of the donor nuclei to express a Purkinje cell specific gene after cell fusion. In this study, BMDCs fused spontaneously with Purkinje neurons after transplantation to form stable, non-dividing, binucleate, chromosomally balanced heterokaryons. Initially, the bone-marrow-like nuclei were characterized by compact chromatin, while the Purkinje nuclei retained dispersed chromatin. Over time, the nuclei of the BMDC donor cells became less compact and dense, and then finally they assumed the morphology of the Purkinje nuclei. As the authors indicated, this suggested that the neuronal program was dominant over the BMDC phenotype. To support their hypothesis, they reported reactivation of a Purkinje-specific trangene, L7-GFP, within the genome of the donor transgenic BMDCs. Thus, this mechanism that involved changes in gene expression was defined as nuclear reprogramming, which provided the first evidence that the differentiated state can be altered in vivo [152].

Given that the nucleus can be reprogrammed, at least partially, by cytoplasmic factors (as demonstrated by mammalian cloning), and based on the observation of the importance of the factors ratio in heterokaryons formed in vitro, the amounts of pre-existing protein and mRNA in the cytoplasm also appear to have roles in the outcome of cell-fusion events. In this sense, the size and composition of a cell might be a factor in the determination of the phenotypic dominance. Indeed, Purkinje neurons, hepatocytes, cardiac myocytes and skeletal myotubes have relatively large cytoplasmic volumes, and thus the reprogramming of the BMDC genome is presumably achieved through the increased dose of regulatory proteins in the much larger fusion partners.

However, it appears that these changes in cell function after cell fusion in vivo are the results of rare events, rather than the result of a biological process, as the frequency of these events is very low. In contrast, in a study from Johansson et al. in 2008, they confirmed that after transplantation, BMDCs can fuse with Purkinje neurons, although more interestingly, the low incidence of cell fusion that is detected under normal conditions can be enhanced under specific condition of stress, such as chronic brain inflammation. Remarkably, following species-mismached BMDC transplantation, specific Purkinje neuron gene products (Calb1, Pcp2, Kcnc1 and Gsbs) were detected in BMDC nuclei after fusion, whereas hematopoietic gene products (CD45, CD11b, F4/80 and Iba1) were not. These data demonstrated that the nuclei of BMDCs in the Purkinje heterokaryons can activate previously silent genes that are typical of mature Purkinje neurons and can repress hematopoietic genes. This is consistent with the hypothesis that they are reprogrammed to a Purkinje neuron fate, and that this mechanism is enhanced in response to tissue injury [153].

Another groundbreaking study demonstrated that cell-fusion-mediated reprogramming will be important in regenerative medicine. Female mice deficient in the enzyme fumarylacetoacetate hydrolase (Fah–/–; a model of fatal tyrosinaemia type 1) can be rescued by transplantation of BMDCs carrying the *Fah* allele. The novel generated hepatocytes in the transplanted Fah–/– mice were polyploid and contained chromosomes from the recipient and donor cells, which indicated that the regenerating nodules were derived from the donor hematopoietic cells that fused with the host hepatocytes, and not from transdifferentiated hematopoietic stem cells. More interestingly, it was thought that these cell-fusion events led to reprogramming of donor hematopoietic cell nuclei, as indicating by the Fah expression in regenerating liver nodules. In support of this, the BMDC surface marker CD45 was shown not to be expressed in Fah+/+ hepatocytes, indicating that after fusion the hematopoietic donor genomes underwent reprogramming of both activating and silencing genes to acquire the hepatocyte expression profile [154].

Similarly, cell fusion events have been demonstrated in a lethally irradiated female mouse model of lung inflammation that lacks lung-specific surfactant protein c (Sp-c). These mice were transplanted with male wild-type BMDCs. The heterokaryon formation that was demonstrated by the Y chromosome in newly formed binucleate pneumocytes led to lung-specific reprogramming of the transplanted BMDCs, as indicated by activation of lung-specific gene expression as well as the expression of the *Sp-C* gene in the null mice [155].

Moreover, after transplantation of dermal fibroblasts into mdx mice, which is a mouse model for Duchenne muscular dystrophy, it was shown that the dermal fibroblasts can fuse with myotubes. As a result, the heterokaryons contained nuclei from both mdx and wild-type fibroblasts, which resulted in phenotypic and functional reversion of muscular dysgenesis [156].

9.6 Concluding Remarks

As has already been reviewed, there was new evidence by 2002 that some cells, such those from the bone marrow, can change lineage to generate completely new types of cells. This generated many doubts, and various news headlines like "Cell fusion makes confusion", "Plasticity: time for a reappraisal", and "Is transdifferentiation in trouble?". Also, the science editor of the UK broadsheet The Daily Telegraph (14/03/02) opined, "Scientists are generating freak cells". However, as Wright commented, this appears to be something that happens as part of the physiology of the cell, and it is thus "not necessarily a bad thing if it has cured a potential fatal metabolic disease!" To date, the number of studies that have investigated the function of heterokaryons in vivo is not high enough to have a clear idea of the full scenario; the molecular mechanisms that regulate cellular plasticity are still poorly understood, and this could in part justify the skepticism in this field.

The dissection of the entire gene expression profile from both genomes in heterokaryons formation in vivo is necessary for us to understand whether these mechanisms of reprogramming are only partial achieved, or if the are the result of global changes in cellular identity. Moreover, the whole genome methylation and demethylation patterns in all of the system need to be evaluated. If chromatin remodeling occurs, the target accessible sequences will also change. Thus, an important question remains to be answered: What happens during the developmental cell state transition? It has been shown that the phenotypic transition is rapid, and this has hampered the evaluation of the existence of any putative unstable intermediates. However, it remains to be seen whether the transition from one cell fate to another is direct, or whether between re-expression of previously silent genes and silencing of expressed genes there exists an intermediate where the chromatin state and the gene profile are similar to an intermediate state that resembles a common, less-specialized developmental precursor. BMDCs fuse and acquire the phenotype of fusion partners that arise from all of the three embryonic lineages, and most interestingly, BMDCs can de-differentiate to pluripotency. It still remains, however, to demonstrate whether during lineage transition in vivo, cells return to a pluripotent state, which would be unstable in adult tissue, before they go down through an alternative developmental pathway.

References

1. Chen EH, Grote E, Mohler W et al (2007) Cell-cell fusion. FEBS Lett 581:2181–2193
2. Lentz BR (2007) PEG as a tool to gain insight into membrane fusion. Eur Biophys J 36:315–326
3. Ogle BM, Cascalho M, Platt JL (2005) Biological implications of cell fusion. Nat Rev Mol Cell Biol 6:567–575
4. Kohler G, Milstein C (1975) Continuous cultures of fused cells secreting antibody of predefined specificity. Nature 256:495–497
5. Ahkong QF, Howell JI, Lucy JA et al (1975) Fusion of hen erythrocytes with yeast protoplasts induced by polyethylene glycol. Nature 255:66–67
6. Arnold K, Herrmann A, Pratsch L et al (1985) The dielectric properties of aqueous solutions of poly(ethylene glycol) and their influence on membrane structure. Biochim Biophys Acta 815:515–518
7. Chiu DT (2001) A microfluidics platform for cell fusion. Curr Opin Chem Biol 5:609–612

8. Ramos C, Teissie J (2000) Electrofusion: a biophysical modification of cell membrane and a mechanism in exocytosis. Biochimie 82:511–518
9. Teissie J, Rols MP (1986) Fusion of mammalian cells in culture is obtained by creating the contact between cells after their electropermeabilization. Biochem Biophys Res Commun 140:258–266
10. Tweedell KS (2004) Embryos, clones, and stem cells: a scientific primer. Scientific World J 4:662–715
11. Wilmut I, Schnieke AE, McWhir J et al (1997) Viable offspring derived from fetal and adult mammalian cells. Nature 385:810–813
12. Wakayama T, Perry AC, Zuccotti M et al (1998) Full-term development of mice from enucleated oocytes injected with cumulus cell nuclei. Nature 394:369–374
13. Wakayama T, Yanagimachi R (1998) Development of normal mice from oocytes injected with freeze-dried spermatozoa. Nat Biotechnol 16:639–641
14. Onishi A, Iwamoto M, Akita T et al (2000) Pig cloning by microinjection of fetal fibroblast nuclei. Science 289:1188–1190
15. Morgan HD, Santos F, Green K et al (2005) Epigenetic reprogramming in mammals. Hum Mol Genet 14 Spec No 1:R47–58
16. Carlson LL, Page AW, Bestor TH (1992) Properties and localization of DNA methyltransferase in preimplantation mouse embryos: implications for genomic imprinting. Genes Dev 6:2536–2541
17. Bourc'his D, Le Bourhis D, Patin D et al (2001) Delayed and incomplete reprogramming of chromosome methylation patterns in bovine cloned embryos. Curr Biol 11:1542–1546
18. Dean W, Santos F, Stojkovic M et al (2001) Conservation of methylation reprogramming in mammalian development: aberrant reprogramming in cloned embryos. Proc Natl Acad Sci USA 98:13734–13738
19. Kang YK, Koo DB, Park JS et al (2001) Aberrant methylation of donor genome in cloned bovine embryos. Nat Genet 28:173–177
20. Miller RA, Ruddle FH (1976) Pluripotent teratocarcinoma-thymus somatic cell hybrids. Cell 9:45–55
21. Cowan CA, Atienza J, Melton DA et al (2005) Nuclear reprogramming of somatic cells after fusion with human embryonic stem cells. Science 309:1369–1373
22. Tada M, Takahama Y, Abe K et al (2001) Nuclear reprogramming of somatic cells by in vitro hybridization with ES cells. Curr Biol 11:1553–1558
23. Blau HM, Chiu CP, Webster C (1983) Cytoplasmic activation of human nuclear genes in stable heterocaryons. Cell 32:1171–1180
24. Blau HM, Pavlath GK, Hardeman EC et al (1985) Plasticity of the differentiated state. Science 230:758–766
25. Baron MH, Maniatis T (1986) Rapid reprogramming of globin gene expression in transient heterokaryons. Cell 46:591–602
26. Takahashi K, Yamanaka S (2006) Induction of pluripotent stem cells from mouse embryonic and adult fibroblast cultures by defined factors. Cell 126:663–676
27. Yu J, Vodyanik MA, Smuga-Otto K et al (2007) Induced Pluripotent Stem Cell Lines Derived from Human Somatic Cells. Science 318:1917–1920
28. Aoi T, Yae K, Nakagawa M et al (2008) Generation of Pluripotent Stem Cells from Adult Mouse Liver and Stomach Cells. Science 321:699–702
29. Hanna J, Markoulaki S, Schorderet P et al (2008) Direct reprogramming of terminally differentiated mature B lymphocytes to pluripotency. Cell 133:250–264
30. Cowling VH and Cole MD (2006) Mechanism of transcriptional activation by the Myc oncoproteins. Semin Cancer Biol 16:242–252
31. Knoepfler PS (2007) Myc goes global: new tricks for an old oncogene. Cancer Res 67:5061–5063
32. Lebofsky R, Walter JC (2007) New Myc-anisms for DNA replication and tumorigenesis? Cancer Cell 12: 102–103
33. Nakagawa M, Koyanagi M, Tanabe K et al (2007) Generation of induced pluripotent stem cells without Myc from mouse and human fibroblasts. Nat Biotechnol 26:101–106
34. Wernig M, Meissner A, Cassady JP et al (2008) c-Myc is dispensable for direct reprogramming of mouse fibroblasts. Cell Stem Cell 2:10–12
35. Jiang J, Chan YS, Loh YH et al (2008) A core Klf circuitry regulates self-renewal of embryonic stem cells. Nat Cell Biol 10:353–360
36. Nakatake Y, Fukui N, Iwamatsu Y et al (2006) Klf4 cooperates with Oct3/4 and Sox2 to activate the Lefty1 core promoter in embryonic stem cells. Mol Cell Biol 26:7772–7782
37. Feng B, Jiang J, Kraus P et al (2009) Reprogramming of fibroblasts into induced pluripotent stem cells with orphan nuclear receptor Esrrb. Nat Biol Cell 11:197–203
38. Boyer LA, Mathur D and Jaenisch R (2006) Molecular control of pluripotency. Curr Opin Genet Dev 16:455–462
39. Niwa H (2007) How is pluripotency determined and maintained? Development 134:635–646
40. Nichols J, Zevnik B, Anastassiadis K et al (1998) Formation of pluripotent stem cells in the mammalian embryo depends on the POU transcription factor Oct4. Cell 95:379–391

41. Avilion AA, Nicolis SK, Pevny LH et al (2003) Multipotent cell lineages in early mouse development depend on SOX2 function. Genes Dev 17:126–140
42. Heng JC, Feng B, Han J et al (2010) The nuclear receptor Nr5a2 can replace Oct4 in the reprogramming of murine somatic cells to pluripotent cells. Cell Stem Cell 6:167–174
43. Kim JB, Greber B, Arauzo-Bravo MJ et al (2009) Direct reprogramming of human neural stem cells by OCT4. Nature 461:649–643
44. Byrne JA, Pedersen DA, Clepper LL et al (2007) Producing primate embryonic stem cells by somatic cell nuclear transfer. Nature 450:497–502
45. Han J, Sidhu KS (2008) Current concepts in reprogramming somatic cells to pluripotent state. Curr Stem Cell Res Ther 3:66–74
46. Evans MJ, Kaufman MH (1981) Establishment in culture of pluripotential cells from mouse embryos. Nature 292:154–156
47. Martin GR (1981) Isolation of a pluripotent cell line from early mouse embryos cultured in medium conditioned by teratocarcinoma stem cells. Proc Natl Acad Sci USA 78:7634–7638
48. Donovan PJ (1994) Growth factor regulation of mouse primordial germ cell development. Curr Top Dev Biol 29:189–225
49. Jacob F (1978) The Leeuwenhoek Lecture, 1977. Mouse teratocarcinoma and mouse embryo. Proc R Soc Lond B Biol Sci 201:249–270
50. Papaioannou A, Lissaios B, Vasilaros S et al (1983) Pre- and postoperative chemoendocrine treatment with or without postoperative radiotherapy for locally advanced breast cancer. Cancer 51:1284–1290
51. Reubinoff BE, Pera MF, Fong CY et al (2000) Embryonic stem cell lines from human blastocysts: somatic differentiation in vitro. Nat Biotechnol 18:399–404
52. Thomson JA, Itskovitz-Eldor J, Shapiro SS et al (1998) Embryonic stem cell lines derived from human blastocysts. Science 282:1145–1147
53. Andrews PW, Goodfellow PN (1980) Antigen expression by somatic cell hybrids of a murine embryonal carcinoma cell with thymocytes and L cells. Somatic Cell Genet 6:271–284
54. Rousset JP, Bucchini D, Jami J (1983) Hybrids between F9 nullipotent teratocarcinoma and thymus cells produce multidifferentiated tumors in mice. Dev Biol 96:331–336
55. Do JT, Han DW, Gentile L et al (2007) Erasure of cellular memory by fusion with pluripotent cells. Stem Cells 25:1013–1020
56. Atsumi T, Shirayoshi Y, Takeichi M et al (1982) Nullipotent teratocarcinoma cells acquire the pluripotency for differentiation by fusion with somatic cells. Differentiation 23:83–86
57. Takagi N, Yoshida MA, Sugawara O et al (1983) Reversal of X-inactivation in female mouse somatic cells hybridized with murine teratocarcinoma stem cells in vitro. Cell 34:1053–1062
58. Flasza M, Shering AF, Smith K et al (2003) Reprogramming in inter-species embryonal carcinoma-somatic cell hybrids induces expression of pluripotency and differentiation markers. Cloning Stem Cells 5:339–354
59. Rousset JP, Dubois P, Lasserre C et al (1979) Phenotype and surface antigens of mouse teratocarcinoma x fibroblast cell hybrids. Somatic Cell Genet 5:739–752
60. Matveeva NM, Kuznetsov SB, Kaftanovskaya EM et al (2001) Segregation of parental chromosomes in hybrid cells obtained by fusion between embryonic stem cells and differentiated cells of adult animal. Dokl Biol Sci 379:399–401
61. Matveeva NM, Shilov AG, Kaftanovskaya EM et al (1998) In vitro and in vivo study of pluripotency in intraspecific hybrid cells obtained by fusion of murine embryonic stem cells with splenocytes. Mol Reprod Dev 50:128–138
62. Tada M, Morizane A, Kimura H et al (2003) Pluripotency of reprogrammed somatic genomes in embryonic stem hybrid cells. Dev Dyn 227:504–510
63. Yu J, Vodyanik MA, He P et al (2006) Human embryonic stem cells reprogram myeloid precursors following cell-cell fusion. Stem Cells 24:168–176
64. Blau HM, Blakely BT (1999) Plasticity of cell fate: insights from heterokaryons. Semin Cell Dev Biol 10: 267–272
65. Bhutani N, Brady JJ, Damian M et al (2010) Reprogramming towards pluripotency requires AID-dependent DNA demethylation. Nature 463:1042–1047
66. Pereira CF, Terranova R, Ryan NK et al (2008) Heterokaryon-based reprogramming of human B lymphocytes for pluripotency requires Oct4 but not Sox2. PLoS Genet 4:e1000170
67. Matsui Y, Zsebo K, Hogan BL (1992) Derivation of pluripotential embryonic stem cells from murine primordial germ cells in culture. Cell 70:841–847
68. Resnick JL, Bixler LS, Cheng L et al (1992) Long-term proliferation of mouse primordial germ cells in culture. Nature 359:550–551
69. Hajkova P, Erhardt S, Lane N et al (2002) Epigenetic reprogramming in mouse primordial germ cells. Mech Dev 117:15–23

70. Monk M, Boubelik M, Lehnert S (1987) Temporal and regional changes in DNA methylation in the embryonic, extraembryonic and germ cell lineages during mouse embryo development. Development 99:371–382
71. Tada T, Tada M, Hilton K et al (1998) Epigenotype switching of imprintable loci in embryonic germ cells. Dev Genes Evol 207:551–561
72. Tada M, Tada T, Lefebvre L et al (1997) Embryonic germ cells induce epigenetic reprogramming of somatic nucleus in hybrid cells. Embo J 16:6510–6520
73. Eggan K, Baldwin K, Tackett M et al (2004) Mice cloned from olfactory sensory neurons. Nature 428:44–49
74. Hochedlinger K, Jaenisch R (2006) Nuclear reprogramming and pluripotency. Nature 441:1061–1067
75. Hochedlinger K, Jaenisch R (2007) On the cloning of animals from terminally differentiated cells. Nat Genet 39:136–137; author reply 137–138
76. Jaenisch R, Hochedlinger K, Blelloch R et al (2004) Nuclear cloning, epigenetic reprogramming, and cellular differentiation. Cold Spring Harb Symp Quant Biol 69:19–27
77. Hansis C, Barreto G, Maltry N et al (2004) Nuclear reprogramming of human somatic cells by xenopus egg extract requires BRG1. Curr Biol 14:1475–1480
78. Taranger CK, Noer A, Sorensen AL et al (2005) Induction of dedifferentiation, genomewide transcriptional programming, and epigenetic reprogramming by extracts of carcinoma and embryonic stem cells. Mol Biol Cell 16:5719–5735
79. Do JT, Scholer HR (2004) Nuclei of embryonic stem cells reprogram somatic cells. Stem Cells 22:941–949
80. Chambers I, Colby D, Robertson M et al (2003) Functional expression cloning of Nanog, a pluripotency sustaining factor in embryonic stem cells. Cell 113:643–655
81. Mitsui K, Tokuzawa Y, Itoh H et al (2003) The homeoprotein Nanog is required for maintenance of pluripotency in mouse epiblast and ES cells. Cell 113:631–642
82. Silva J, Nichols J, Theunissen TW et al (2009) Nanog is the gateway to the pluripotent ground state. Cell 138: 722–737
83. Silva J, Smith A (2008) Capturing pluripotency. Cell 132:532–536
84. Silva J, Chambers I, Pollard S et al (2006) Nanog promotes transfer of pluripotency after cell fusion. Nature 441:997–1001
85. Sato N, Meijer L, Skaltsounis L et al (2004) Maintenance of pluripotency in human and mouse embryonic stem cells through activation of Wnt signaling by a pharmacological GSK-3-specific inhibitor. Nat Med 10:55–63
86. Cole MF, Johnstone SE, Newman JJ et al (2008) Tcf3 is an integral component of the core regulatory circuitry of embryonic stem cells. Genes Dev 22:746–755
87. Willert K, Jones KA (2006) Wnt signaling: Is the party in the nucleus? Genes Dev 20:1394–1404
88. Lluis F, Pedone E, Pepe S et al (2008) Periodic activation of Wnt/beta-catenin signaling enhances somatic cell reprogramming mediated by cell fusion. Cell Stem Cell 3:493–507
89. He TC, Sparks AB, Rago C et al (1998) Identification of c-MYC as a target of the APC pathway. Science 281:1509–1512
90. Tam WL, Lim CY, Han J et al (2008) T-cell factor 3 regulates embryonic stem cell pluripotency and self-renewal by the transcriptional control of multiple lineage pathways. Stem Cells 26:2019–2031
91. Yi F, Pereira L, Merrill BJ (2008) Tcf3 functions as a steady-state limiter of transcriptional programs of mouse embryonic stem cell self-renewal. Stem Cells 26:1951–1960
92. Lluis F, Cosma MP (2009) Somatic cell reprogramming control: signaling pathway modulation versus transcription factor activities. Cell Cycle 8:1138–1144
93. Nakamura T, Inoue K, Ogawa S et al (2008) Effects of Akt signaling on nuclear reprogramming. Genes Cells 13:1269–1277
94. Watanabe S, Umehara H, Murayama K et al (2006) Activation of Akt signaling is sufficient to maintain pluripotency in mouse and primate embryonic stem cells. Oncogene 25:2697–2707
95. Frei E, Schuh R, Baumgartner S et al (1988) Molecular characterization of spalt, a homeotic gene required for head and tail development in the Drosophila embryo. EMBO J 7:197–204
96. Elling U, Klasen C, Eisenberger T et al (2006) Murine inner cell mass-derived lineages depend on Sall4 function. Proc Natl Acad Sci USA 103:16319–16324
97. Zhang J, Tam WL, Tong GQ et al (2006) Sall4 modulates embryonic stem cell pluripotency and early embryonic development by the transcriptional regulation of Pou5f1. Nat Cell Biol 8:1114–1123
98. Wong CC, Gaspar-Maia A, Ramalho-Santos M et al (2008) High-efficiency stem cell fusion-mediated assay reveals Sall4 as an enhancer of reprogramming. PLoS ONE 3:e1955
99. Meissner A, Mikkelsen TS, Gu H et al (2008) Genome-scale DNA methylation maps of pluripotent and differentiated cells. Nature 454:766–770
100. Ooi SK, Bestor TH (2008) The colorful history of active DNA demethylation. Cell 133:1145–1148
101. Morgan HD, Dean W, Coker HA et al (2004) Activation-induced cytidine deaminase deaminates 5-methylcytosine in DNA and is expressed in pluripotent tissues: implications for epigenetic reprogramming. J Biol Chem 279:52353–52360

102. Rai K, Huggins IJ, James SR et al (2008) DNA demethylation in zebrafish involves the coupling of a deaminase, a glycosylase, and gadd45. Cell 135:1201–1212
103. Gehring M, Reik W, Henikoff S (2009) DNA demethylation by DNA repair. Trends Genet 25:82–90
104. Ringrose L, Paro R (2004) Epigenetic regulation of cellular memory by the Polycomb and Trithorax group proteins. Annu Rev Genet 38:413–443
105. Chamberlain SJ, Yee D, Magnuson T (2008) Polycomb repressive complex 2 is dispensable for maintenance of embryonic stem cell pluripotency. Stem Cells 26:1496–1505
106. Faust C, Schumacher A, Holdener B et al (1995) The eed mutation disrupts anterior mesoderm production in mice. Development 121:273–285
107. O'Carroll D, Erhardt S, Pagani M et al (2001) The polycomb-group gene Ezh2 is required for early mouse development. Mol Cell Biol 21:4330–4336
108. Pasini D, Bracken AP, Hansen JB et al (2007) The polycomb group protein Suz12 is required for embryonic stem cell differentiation. Mol Cell Biol 27:3769–3779
109. Boyer LA, Lee TI, Cole MF et al (2005) Core transcriptional regulatory circuitry in human embryonic stem cells. Cell 122:947–956
110. Lee TI, Jenner RG, Boyer LA et al (2006) Control of developmental regulators by Polycomb in human embryonic stem cells. Cell 125:301–313
111. Yang J, Chai L, Fowles TC et al (2008) Genome-wide analysis reveals Sall4 to be a major regulator of pluripotency in murine-embryonic stem cells. Proc Natl Acad Sci USA 105:19756–19761
112. Azuara V, Perry P, Sauer S et al (2006) Chromatin signatures of pluripotent cell lines. Nat Cell Biol 8:532–538
113. Bernstein BE, Mikkelsen TS, Xie X et al (2006) A bivalent chromatin structure marks key developmental genes in embryonic stem cells. Cell 125:315–326
114. Bernstein E, Duncan EM, Masui O et al (2006) Mouse polycomb proteins bind differentially to methylated histone H3 and RNA and are enriched in facultative heterochromatin. Mol Cell Biol 26:2560–2569
115. Boyer LA, Plath K, Zeitlinger J et al (2006) Polycomb complexes repress developmental regulators in murine embryonic stem cells. Nature 441:349–353
116. Pereira CF, Piccolo FM, Tsubouchi T et al (2010) ESCs Require PRC2 to Direct the Successful Reprogramming of Differentiated Cells toward Pluripotency. Cell Stem Cell 6:547–556
117. Ma DK, Chiang CH, Ponnusamy K et al (2008) G9a and Jhdm2a regulate embryonic stem cell fusion-induced reprogramming of adult neural stem cells. Stem Cells 26:2131–2141
118. Wright WE (1984) Control of differentiation in heterokaryons and hybrids involving differentiation-defective myoblast variants. J Cell Biol 98:436–443
119. Darlington GJ, Bernard HP, Ruddle FH (1974) Human serum albumin phenotype activation in mouse hepatoma–human leukocyte cell hybrids. Science 185:859–862
120. Aurade F, Pinset C, Chafey P et al (1994) Myf5, MyoD, myogenin and MRF4 myogenic derivatives of the embryonic mesenchymal cell line C3H10T1/2 exhibit the same adult muscle phenotype. Differentiation 55: 185–192
121. Choi J, Costa ML, Mermelstein CS et al (1990) MyoD converts primary dermal fibroblasts, chondroblasts, smooth muscle, and retinal pigmented epithelial cells into striated mononucleated myoblasts and multinucleated myotubes. Proc Natl Acad Sci USA 87:7988–7992
122. Davis C, Kannan MS (1987) Sympathetic innervation of human tracheal and bronchial smooth muscle. Respir Physiol 68:53–61
123. Weintraub H, Tapscott SJ, Davis RL et al (1989) Activation of muscle-specific genes in pigment, nerve, fat, liver, and fibroblast cell lines by forced expression of MyoD. Proc Natl Acad Sci USA 86:5434–5438
124. Lluis F, Perdiguero E, Nebreda AR et al (2006) Regulation of skeletal muscle gene expression by p38 MAP kinases. Trends Cell Biol 16:36–44
125. Xie H, Ye M, Feng R et al (2004) Stepwise reprogramming of B cells into macrophages. Cell 117:663–676
126. Bussmann LH, Schubert A, Vu Manh TP et al (2009) A robust and highly efficient immune cell reprogramming system. Cell Stem Cell 5:554–566
127. Vierbuchen T, Ostermeier A, Pang ZP et al (2010) Direct conversion of fibroblasts to functional neurons by defined factors. Nature 463:1035–1041
128. Tapscott SJ, Davis RL, Thayer MJ et al (1988) MyoD1:a nuclear phosphoprotein requiring a Myc homology region to convert fibroblasts to myoblasts. Science 242:405–411
129. Pomerantz JH, Mukherjee S, Palermo AT et al (2009) Reprogramming to a muscle fate by fusion recapitulates differentiation. J Cell Sci 122:1045–1053
130. Blau HM (1989) How fixed is the differentiated state? Lessons from heterokaryons. Trends Genet 5:268–272
131. Deisseroth A, Hendrick D (1979) Activation of phenotypic expression of human globin genes from nonerythroid cells by chromosome-dependent transfer to tetraploid mouse erythroleukemia cells. Proc Natl Acad Sci USA 76:2185–2189

132. Willing MC, Nienhuis AW, Anderson WF (1979) Selective activation of human beta-but not gamma-globin gene in human fibroblast x mouse erythroleukaemia cell hybrids. Nature 277:534–538
133. Papayannopoulou T, Enver T, Takegawa S et al (1988) Activation of developmentally mutated human globin genes by cell fusion. Science 242:1056–1058
134. Dupuy-Coin AM, Ege T, Bouteille M et al (1976) Ultrastructure of chick erythrocyte nuclei undergoing reactivation in heterokaryons and enucleated cells. Exp Cell Res 101:355–369
135. Chiu CP, Blau HM (1984) Reprogramming cell differentiation in the absence of DNA synthesis. Cell 37:879–887
136. Chiu CP, Blau HM (1985) 5-Azacytidine permits gene activation in a previously noninducible cell type. Cell 40:417–424
137. Palermo A, Doyonnas R, Bhutani N et al (2009) Nuclear reprogramming in heterokaryons is rapid, extensive, and bidirectional. Faseb J 23:1431–1440
138. Lanfranchi G, Linder S, Ringertz NR (1984) Globin synthesis in heterokaryons formed between chick erythrocytes and human K562 cells or rat L6 myoblasts. J Cell Sci 66:309–319
139. Brown DD (1984) The role of stable complexes that repress and activate eucaryotic genes. Cell 37:359–365
140. Smale ST, Fisher AG (2002) Chromatin structure and gene regulation in the immune system. Annu Rev Immunol 20:427–462
141. Orlando V (2003) Polycomb, epigenomes, and control of cell identity. Cell 112:599–606
142. Terranova R, Pereira CF, Du Roure C et al (2006) Acquisition and extinction of gene expression programs are separable events in heterokaryon reprogramming. J Cell Sci 119:2065–2072
143. Ferrari G, Cusella-De Angelis G, Coletta M et al (1998) Muscle regeneration by bone marrow-derived myogenic progenitors. Science 279:1528–1530
144. Petersen BE, Bowen WC, Patrene KD et al (1999) Bone marrow as a potential source of hepatic oval cells. Science 284:1168–1170
145. Mezey E, Chandross KJ, Harta G et al (2000) Turning blood into brain: cells bearing neuronal antigens generated in vivo from bone marrow. Science 290:1779–1782
146. Krause DS, Theise ND, Collector MI et al (2001) Multi-organ, multi-lineage engraftment by a single bone marrow-derived stem cell. Cell 105:369–377
147. LaBarge MA, Blau HM (2002) Biological progression from adult bone marrow to mononucleate muscle stem cell to multinucleate muscle fiber in response to injury. Cell 111:589–601
148. Terada N, Hamazaki T, Oka M et al (2002) Bone marrow cells adopt the phenotype of other cells by spontaneous cell fusion. Nature 416:542–545
149. Weimann JM, Charlton CA, Brazelton TR et al (2003) Contribution of transplanted bone marrow cells to Purkinje neurons in human adult brains. Proc Natl Acad Sci USA 100:2088–2093
150. Alvarez-Dolado M, Pardal R, Garcia-Verdugo JM et al (2003) Fusion of bone-marrow-derived cells with Purkinje neurons, cardiomyocytes and hepatocytes. Nature 425:968–973
151. Eglitis MA, Mezey E (1997) Hematopoietic cells differentiate into both microglia and macroglia in the brains of adult mice. Proc Natl Acad Sci USA 94:4080–4085
152. Weimann JM, Johansson CB, Trejo A et al (2003) Stable reprogrammed heterokaryons form spontaneously in Purkinje neurons after bone marrow transplant. Nat Cell Biol 5:959–966
153. Johansson CB, Youssef S, Koleckar K et al (2008) Extensive fusion of haematopoietic cells with Purkinje neurons in response to chronic inflammation. Nat Cell Biol 10:575–583
154. Vassilopoulos G, Wang PR, Russell DW (2003) Transplanted bone marrow regenerates liver by cell fusion. Nature 422:901–904
155. Herzog EL, Van Arnam J, Hu B et al (2007) Lung-specific nuclear reprogramming is accompanied by heterokaryon formation and Y chromosome loss following bone marrow transplantation and secondary inflammation. FASEB J 21:2592–2601
156. Gibson AJ, Karasinski J, Relvas J et al (1995) Dermal fibroblasts convert to a myogenic lineage in mdx mouse muscle. J Cell Sci 108 (Pt 1):207–214

Chapter 10
Cell Fusion and Tissue Regeneration

Manuel Álvarez-Dolado and Magdalena Martínez-Losa

Abstract Cell fusion is a natural process implicated in normal development, immune response, tissue formation, and with a prominent role in stem cell plasticity. The discovery that bone marrow stem cells fuse with several cell types, under normal condition or after an injury, introduces new possibilities in regenerative medicine and genetic repair. Cell fusion has been shown to be implicated in regeneration, and the complementation of recessive mutations affecting the liver, brain, muscle, lung and gut, under appropriate conditions. However, we should be cautious and better understand the mechanisms that govern cell fusion during regeneration before to consider it as clinically relevant. In this chapter, we will present the current evidences about the role of cell fusion in tissue regeneration and its future potential as therapy. Cell fusion is an exciting and promising research field. In addition, we will review the challenges that should face the fusion process to become therapeutically effective and safe.

10.1 Introduction

Cell fusion is a natural process present in our lives from the very beginning, when a spermatozoid fuses with an ovule. Later, during development, cell fusion is involved in organ formation, virus infection, and immune response [1–7]. In the last decade, cell fusion has also been linked to stem cell biology; to their plastic and regenerative properties [8–12]. Specially, the discovery that bone marrow derived cells (BMDC) fuse with several cell types, under normal condition or after an injury, introduces new possibilities in medicine to use the cell fusion as a mechanism of tissue regeneration and genetic repair [8–12]. Cell fusion promotes a more dynamic concept of the cell, since it leads to changes in the genetic content and cell fate. These modifications may stimulate cells to better response against an insult, or to proliferate for restoration of tissue integrity. However, cell fusion has not been exempt of polemic. In this chapter we will present the current evidences about the role of cell fusion in tissue regeneration and its future potential as therapy.

10.2 Cell Fusion as a Cell Plasticity Mechanism

Stem cells (SC) are defined as immature cells with self-renewal properties, being able to generate mature progeny including non-renewing progenitors and terminally differentiated cells. Consequently, SC are the main source of new cells and major responsible for tissue homeostasis and regeneration

M. Álvarez-Dolado (✉)
Laboratory of Cell-Based Therapies for Neuropathologies, Andalusian Center for Molecular Biology and Regenerative Medicine (CABIMER), CSIC, 41092 Seville, Spain
e-mail: manuel.alvarez@cabimer.es

T. Dittmar, K.S. Zänker (eds.), *Cell Fusion in Health and Disease*, Advances in Experimental Medicine and Biology 713, DOI 10.1007/978-94-007-0763-4_10, © Springer Science+Business Media B.V. 2011

after injury [13, 14]. They can use different mechanisms to repair a damaged tissue: generation of new cells by differentiation, repair by cell fusion, and secretion of growth- antiapoptotic- and/or trophic-factors that help to preserve and restore the structure of the injured tissue.

The simplest, most common, and direct mechanism of repair used by SC is the generation of new cells by differentiation to replace the lost tissue. The capacity of SC regeneration will depend on their *plasticity*. This term refers to the ability of SC to generate cell types derived from different germ layers [14]. The higher number of generated cell types, the more plastic they are. Thus, SC can be totipotents or pluripotents, when they are able to generate cells of the three embryonic germ layers; or multipotent, if they are limited in their differentiation and regenerative capability to the tissues in which they reside. Examples of pluripotency are the embryonic stem cells (ES), whereas the adult SC are multipotents; they are less plastic. However, some types of adult SC own a wider plasticity. This is the case of bone marrow adult stem cells (BMSC), both hematopoietic and mesenchymal. Their wide plasticity to generate different cell types in several organs was demonstrated in the late 1990s. Transplants of these cells in irradiated normal recipients, carrying reporter genes (LacZ or GFP) to facilitate their tracking, resulted in the expression of these markers by non-hematopoietic cells in several organs. These cells presented the morphology of fully developed mature cells. The first study of this type found micro and macroglia expressing the donor-derived reporter gene in the brain [15]. Later, other groups also identified labeled neurons in the cortex and cerebellum [16–20], and the studies were extended to different tissues, such as liver, pancreas, skeletal muscle, endothelium, and myocardium [21–30]. BMSC are capable to reach many tissues through the blood stream, what facilitate their targeting to damaged areas and the repair of tissues with different embryonic origin [14, 21, 22]. It was even shown that a single hematopoietic SC was able to fully reconstitute the hematopoietic system and, in addition, contribute to epithelia, skin, and lung epithelium [23]. Mesenchymal SC were also isolated and tested for their ability to generate tissues of different embryonic origin [24–26]. The simplest interpretation for these results was the differentiation of the transplanted cells into mature cells with the help of tissue-specific factors or local niches, what facilitated their reprogramming. However, an alternative explanation, that does not exclude the differentiation theory, was also proposed: cell fusion.

Cell fusion is regarded as another SC plasticity mechanism. This alternative hypothesis postulated that a bone marrow derived cell fuses with a local precursor or mature cell, transferring its genetic material and mixing their cytoplasm. The genetic program of the newly formed heterokaryon would be modified and, in consequence, this would lead to the acquisition of a new phenotype, and the breakage of lineage restriction. The first reports suggesting cell fusion as an alternative to differentiation were published by two independent groups that observed ES fusion in vitro with bone marrow cells and adult neural SC [27, 28]. Previous results in the muscle had indirectly shown that cell fusion contributed to its repair [29–32]. The likely existence of cell fusion in vivo encouraged several groups to find direct evidence of this process in their experimental models. Liver was the first organ where cell fusion was fairly shown. Two independent groups performed transplants of BMSC in models of liver degeneration [33, 34]. They observed the formation of new hepatocytes carrying markers derived from the donor cells. A posterior in vitro cytogenetic analysis and southern blots of these hepatocytes showed karyotypes indicative of fusion between donor and host cells. Results from other groups suggested that Purkinje neurons and hepatocytes could also derive from cell fusion [35]. Following this possibility Helen Blau's group studied brain biopsies from women who had received bone marrow transplants (BMT) from male donors. Some of the Purkinje neurons in these biopsies were tetraploid (XXXY), as detected by fluorescent in situ hybridization (FISH). The presence of both sets of chromosomes strongly suggested that Purkinje neurons had fused with hematopoietic cells from the bone marrow donor [19].

Despite these results, the scientific community was cautious about cell fusion. Liver observations were obtained from a damaged tissue model that could influence in the process. In addition, the use of virus by some groups to introduce the tracking markers in the donor cells could promote the presence of viral particles in their membrane, what may induce the fusion. A set of experiments taking advance of the cre-lox technology came out to clarify these doubts [36]. The use of transgenic mouse lines that conditionally express a reporter gene only after a fusion event allowed our group to unequivocally discern fusion events from differentiation in any tissue and under any pathological condition [36] (Fig. 10.1a–b). Our detection method used a mouse line expressing the Cre recombinase, and a second line (R26R) carrying the LacZ reporter gene, that is exclusively expressed after the excision of a loxP-flanked (floxed) stop cassette by Cre-mediated recombination. When Cre-expressing cells fuse with R26R cells, Cre recombinase excises the floxed stop codon of the LacZ reporter gene. This allows an easy detection of fused cells by X-gal staining. In this way, we performed transplants of BMDC into R26R mice using as donor Cre-expressing mice that in addition expressed the GFP (Fig. 10.1d). So we can distinguish fused cells (X-gal$^+$/GFP$^+$) from donor derived differentiated cells (X-gal$^-$/GFP$^+$). The detection method contributed to confirm the presence of cell fusion events in several tissues

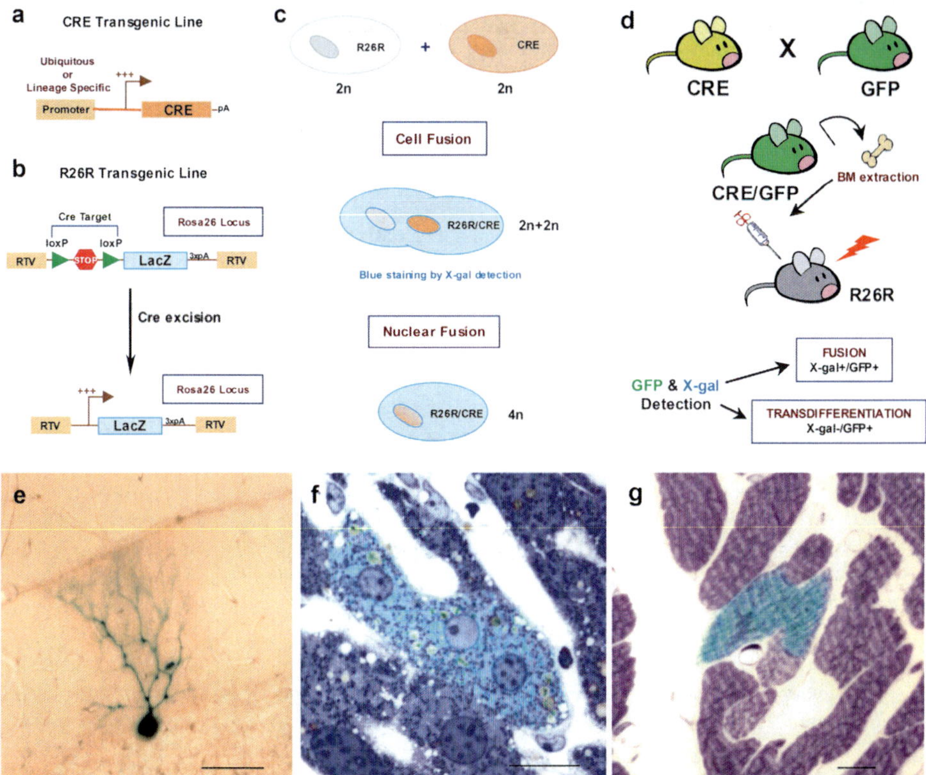

Fig. 10.1 Cell Fusion Detection System and examples of fusion products. (**a**) Schematic representation of the Cre recombinase transgene expressed by one of the mouse lines used in the system. (**b**) Representation of the reporter transgene expressed by the R26R mouse line and its modification after Cre mediated recombination. (**c**) When a cell expressing Cre recombinase fuses with a cell carrying the LacZ reporter transgene, the floxed stop cassette is excised and the LacZ reporter is expressed in the fused cell. LacZ expression can be detected by the generation of a *blue* precipitate after X-gal staining. (**d**) BMT strategy to detect cell fusion in vivo. A R26R mouse line was used as recipient of BM cells expressing GFP and Cre recombinase. Double detection of GFP and X-gal *blue* precipitate allows discerning between cell fusion and transdifferentiation events. (**e-f**) X-gal detection of fusion products in the brain, liver and heart. (**e**) Brightfield photomicrograph of a cerebellum (50 μm section) containing an X-gal positive fused Purkinje neuron. (**f**) Semithing section (1 μm) of an X-gal positive fused hepatocyte (*light blue*) counterstained with toluidine blue (*magenta*). Note the presence of three nuclei. (**g**) Semithing section (1 μm) of an X-gal positive cardiomyocyte (*blue*) counterstained with toluidine blue (*magenta*). *Scale bars*: 50 μm in **e**; 10 μm in **f** and **g**

under normal (healthy, but irradiated) conditions [36]. We observed the cell fusion of hepatocytes and Purkinje cells with cells of the hematopoietic lineage (Fig. 10.1e–f). In addition, we showed the fusion of cardiomyocytes (Fig. 10.1g). No evidence of differentiation was observed in these tissues, except for the presence of macrophages, or macroglia in the case of the brain. These results were corroborated by others in the following years by using variations of this method. Thus, Weimann et al. showed that fused Purkinje neurons form stable heterokaryons that increase in number with age [20]. The bone marrow derived nucleus within the heterokaryon was reprogrammed and activated the expression of Purkinje neuron-specific genes. Cardiomyocytes were also confirmed to be generated at a low frequency through cell fusion in infarcted hearts [37–39].

The growing evidence from experimental BMT, suggesting an active participation of cell fusion after injury, led many groups to investigate the regenerative potential of this process against several types of diseases/injuries that affect different organs. Some of them were especially successful to correct certain pathologies through the mechanism of cell fusion. In the following sections, we will recapitulate the advances in tissue regeneration and genetic complementation with the cell fusion process as main protagonist.

10.3 Regenerative Potential of Cell Fusion

As we have seen, BMSC are able to generate very different cell types and fuse with others. These cells are very accessible, as well; most of the hospitals have experience in their isolation, manipulation and transplant. These features confers them a great therapeutic potential. An additional advantage of BMSC mediated cell fusion regeneration is the preservation of the structural complexity in the damaged tissue. Contrary to focal cell transplants that need to recreate the whole organ structure, BMSC can travel freely through the blood stream and perform the cell fusion process within the injured original organ scaffold.

On the other hand, cell fusion involves the mingling of genetic material. This characteristic is very important, since it enables the complementation of recessive mutations and the reversion of the altered phenotype. It also allows cell fusion to act as a modifier of gene program and cell fate [9, 12, 40–42]. After fusion, both sets of chromosomes interact and the resulted heterokaryon or synkaryon acquire in occasions the identity of their partners, but in others, certain properties that formerly typified the original cells disappear in the hybrids [41–45]. Cell fusion, at least in vitro, can reprogram cell fate [41, 44]. Fusion reverses the developmental program of a mature cell towards a more immature cell owing progenitor and proliferative properties, that would be of great importance during a regenerative process [12, 42]. Reversion or modification of cell fate/gene program by cell fusion can be achieved not only by the genetic influence of one nucleus on the other, but also cytoplasmic factors might induce important epigenetic modifications [12, 44]. These observations strongly suggest that cell fusion is a tool to modify gene expression patterns. This may confer new skills to the heterokaryons for better response against an injury, and is a powerful therapeutic strategy to complement recessive mutations in post-mitotic tissue. In addition, considering that mammalian cells are exposed to repeated episodes of stress/inflammation throughout life, heterokaryon formation may be important in homeostasis and maintenance of specific postmitotic cell types. We have already mentioned some examples of regenerative fusion in the text. Liver, muscle (skeletal and cardiac), brain, and lung are the organs where cell fusion has been demonstrated more accurately to exert a therapeutic effect [8, 10, 11].

10.3.1 Liver Regeneration by Cell Fusion

Liver regeneration by cell fusion is so far the best documented and largely accepted [46]. Two independent groups, using a hepatic lethal mouse model with recessive mutations in the fumarylacetoacetate hydrolase (FAH) gene, showed the rescue of normal liver function after the transplant of wild-type hematopoietic SC [33, 34]. Restoration of normal metabolism was due to repopulation of the liver with hepatocytes expressing the wild-type gene after cell fusion. This was confirmed by cytogenetic analysis and southern blots that showed karyotypes indicative of fusion between donor and host cells [33, 34]. The hepatocyte repopulation and posterior improvement of the phenotype were thanks to a selective survival advantage for propagation of the fused hepatocytes. Even though number of fusion events was initially low, once the fused cells incorporated the wild-type gene they were able to survive in the degenerative environment and proliferate in response to it, allowing the replacement of the mutant hepatocytes that were dying.

Most of the resulting fusion-derived polyploid hepatocytes were tetraploid and seem stable, although part of them can proliferate and undergo later ploidy reductions to generate daughter cells with one-half chromosomal content [47]. This was demonstrated by marker segregation using ss-galactosidase and the Y-chromosome. Approximately 2–5% of fusion-derived FAH-positive nodules were negative for one or more markers, as expected during ploidy reduction. This lead to the generation of genetically diverse daughter cells with about 50% reduction in nuclear content. The generation of such daughter cells increases liver diversity. However, this may also increase the likelihood of oncogenesis.

The exact subpopulation of bone marrow cells responsible for fusion in the liver was identified as hematopoietic myelomonocytic macrophages [48, 49]. Human hematopoietic and umbilical cord SC have been also tested for their capacity to generate hepatocytes by cell fusion [50–53]. Transplant of human hematopoietic SC in the animal model of severe liver damage by carbon tetrachloride leads to an improvement of the condition, a threefold increase in homing of human mononuclear cells, and the formation of hybrid heterokaryons with mouse and human genetic contain [51]. Human umbilical cord cells also generates hepatocytes through cell fusion after transplant in NOD/SCID null mice, even without hepatotoxic treatment other than irradiation [50]. However, we should keep in mind that in other models of liver injury, and under different experimental conditions these cells are able to generate hepatocytes without the participation of cell fusion [54, 55]. This suggests that environmental signals are important for physiological selection of the plasticity mechanism by SC.

10.3.2 Skeletal Muscle and Cell Fusion

Muscle is a tissue where cell fusion is basic for their development and regeneration. Mature muscle fibers are generated by a dynamic process in which mononucleated undifferentiated myoblasts proliferate, differentiate and fuse to form a syncytia [56]. Rescue of muscular function by cell fusion was for the first time shown in *mdx* mice [30]. This mouse line has a condition that resembles Duchenne muscular dystrophy. Gibson et al. transplanted wild-type dermal fibroblasts into *mdx* mice and observed the formation of heterokaryons containing nuclei of mutant and wild-type fibroblasts [30]. This fusion causes the phenotypic and functional reversion of muscular dysgenesis. To show it, the authors determined the isoenzyme allotypes of glucose-6-phosphate isomerase as a marker of the host and donor cells [30]. Similarly, BMDC are also able to migrate and fuse with skeletal muscle, restoring the expression of dystrophin in *mdx* mice, and recovering muscle function [29, 31, 32, 57]. In humans, genetic complementation of muscle cells by SC fusion has also been reported [58]. As well as the ability of exogenous BMDC to fuse with skeletal muscle of patients with Duchenne muscular dystrophy [31]. Cell fusion in the muscle not only works under pathological conditions derived from recessive genetic alterations. Normal regeneration of a stress-induced or mechanically injured skeletal muscle is also achieved by fusion with BMDC [59, 60]. The nature of the cell types involved in the fusion-mediated regeneration was shown to be myelocytic (macrophages and neutrophils), and inflammatory cell infiltration is required for their contribution [59, 61]. This indicates that circulating myeloid cells, in response to injury and inflammatory cues, migrate to regenerate skeletal muscle and stochastically incorporate into mature myofibers.

10.3.3 Cell Fusion After Heart Infarct

Work on BMSC grafting in *mdx* mice indicated that, in addition to skeletal muscle, was also possible to generate cardiomyocytes from BMSC [57]. This opened the possibility to treat heart infarct with the help of BMSC, what generated great enthusiasm in the field [62, 63]. Initial work from Anversa's group showed that hematopoietic SC are able to repair the infarcted myocardium [64]. According to these investigators, HSC injection into the myocardium of rats undergoing ischemia could repair 60–70% of the damaged tissue by originating smooth muscle, endothelial and cardiomyocytic cells [64]. This group also analyzed heart biopsies of female organ donors in male recipients few weeks after heart transplant [65]. They found that the proportion of cells containing the Y chromosome and expressing markers of smooth muscle, endothelium or cardiomyocytes was very high (>20%). However, several attempts to reproduce these results by other groups concluded with a significant reduction in these percentages (<1%), probably due to differences in tissue histology detection techniques and experimental conditions [66, 67]. In both cases, no analysis of X chromosome was performed to search evidence of cell fusion. Differentiation was postulated as the main mechanism of generation of these myocardial cell subtypes after BMSC grafting. In this context, the demonstration of cell fusion contribution to cardiac muscle, under normal conditions or after heart infarct, gave rise to a big controversy [8]. Our group showed for the first time that hematopoietic cells fuse with cardiomyocytes after BMT under normal (except irradiation) conditions [36]. No evidences of differentiation were found in our experiments and the fusion rate was very low (<1%). Later, direct demonstration of cell fusion was confirmed as the most likely mechanism to explain the low frequency generation of cardiomyocytes after heart infarct [37–39]. Differentiation seems to count better for the presence of smooth muscle, and endothelium. The scarcity of fusion events, even after infarct, is a handicap for its therapeutic application. However, fused cardiomyocytes are able to proliferate in vitro and in vivo [68], what can give us clues to enhance their growth until reach an adequate number for regeneration. In any case, nowadays neither cell fusion, nor generation of new cardiomyocytes from BMSC are able to explain, alone or in conjunction, the observed improvement of cardiac function after BMSC transplant. Current evidence reinforce the idea that, likely, these cells perform a paracrine action secreting growth factors and molecules that have an anti-apoptotic effect on cardiomyocytes in vivo [69]. In addition, BM cells contain endothelial precursors, which promote angiogenesis in the infarcted area, improving myocardial perfusion and viability [70, 71].

10.3.4 From Blood to Brain

Several groups have reported the presence of donor-derived neurons and glia after BMT [15, 16, 18, 72]. However, cell fusion in the brain has been exclusively demonstrated in Purkinje neurons of the cerebellum (Fig. 10.2a–c), including in humans [19, 20, 36]. The formed heterokaryons increase in number with age and, more importantly, after damage [73–77]. These results suggest BMDC and cell fusion as a potential mechanism to treat neurodegenerative pathologies, in especial those related with Purkinje neurons, such as ataxias. Interestingly, focal implantation or intravenous delivery of BMDC improve brain function in models of cerebral ischemia, trauma, Parkinson's and Huntington's disease [78–82]. However, the partial recovery observed in most of these pathologies is likely due to delivery of growth factors and cytokines by the transplanted cells, since the reported cell fusion events are scarce. However, if the pathology courses with inflammation a 10-fold increase of fusion events has been observed [74, 76].

Stroke is the brain pathology that accumulates more experimental works on the use of BMT for its treatment [80, 83, 84]. Several groups have already reported a functional outcome when these cells are transplanted 24 h after stroke, either by intracerebral, intravenous, or intra-arterial route. In these experiments, occasional neuronal products of cell fusion has been observed after stroke [85].

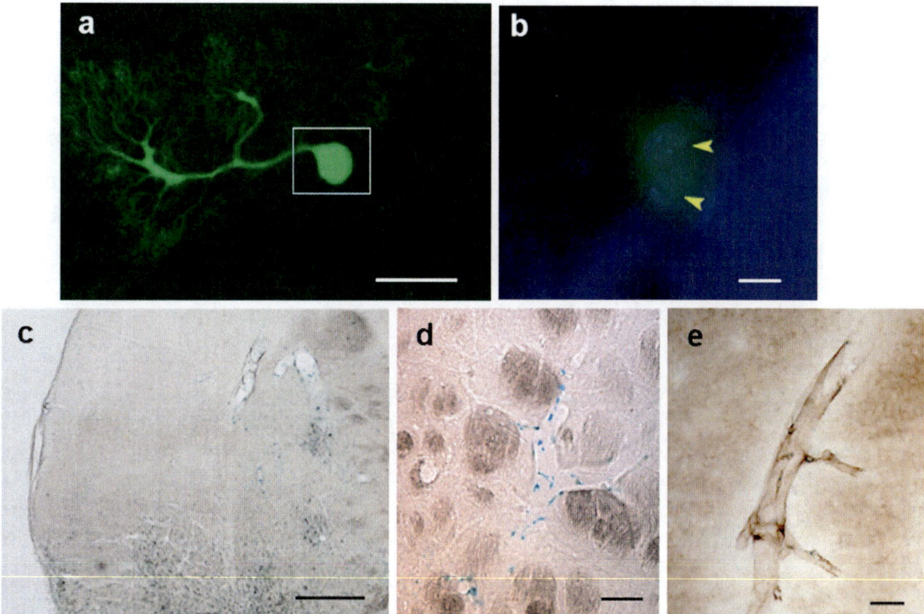

Fig. 10.2 Cell fusion in the brain. (**a–b**) Fluorescence photomicrograph of a GFP positive Purkinje neuron containing two nuclei (*yellow arrowheads*), likely generated by fusion after intravenous transplant of GFP positive bone marrow cells into a normal recipient. (**c–e**) Generation of fused cells after stroke. X-gal staining of brain sections from R26R/Cre bone marrow chimeras (see Fig. 10.1) 2 weeks after stroke revealed a high number of X-gal positive cells (*fused cells*) in the penumbra of the ischemic area, situated in the *bottom* of the photograph. (**d**) Detail of a blood vessel containing fused cells in the penumbra area. (**e**) Immunohistochemical co-localization with X-gal showed that fused cells expressed NG2, a specific marker of pericytes. *Scale bars*: 25 μm in **a**; 10 μm in **b**; 100 μm in **c**; 25 μm in **d** and **e**

This low rate of new neurons cannot explain by itself the reported improvements. It is more likely that transplanted BMDC secrete a panoply of growth factors that contribute to neuronal survival after stroke [86]. Furthermore, they are also able to generate endothelial cells (EC) and mural cells that will help to revascularize the area [87]. It has been accurately shown that bone marrow-derived EC are formed by differentiation and not by cell fusion [88]. However, recent data from our group show that cell fusion also participate actively in the formation of vascular tissue [89]. In agreement with the association of stroke and inflammation, our group reported an important increase of fused cells within the vasculature in the penumbra area of stroke (Fig. 10.2d–f). Most of the fused cells expressed pericyte markers, while relatively few expressed bone-marrow markers, indicating the dominance of mural cell fate in the fused cells. Pericytes are implicated in the initiation of vasculogenesis [90, 91], and the regulation of EC proliferation and differentiation [92, 93]. Therefore, these findings suggest an active physiological role for cell fusion during angiogenesis, and may open new therapeutic perspectives in the application of BMDC and cell fusion for stroke treatment and revascularization in the brain.

Inherited metabolic diseases that affect the nervous system are also susceptible to be treated with bone marrow and umbilical cord SC [94]. This strategy contributes with enzyme-producing cells that migrate to the brain and other organs providing a permanent enzyme replacement therapy. SC may also mediate cell regeneration and genetic complementation through cell fusion. A good example is the transplant of mesenchymal SC directly into the brain of a mouse model for Niemann-Pick disease (NPD) [95, 96]. This mice show progressive degeneration of cerebellar Purkinje neurons due to the deficiency of the acid sphingomyelinase. When mesenchymal SC, previously transduced with a retroviral vector to overexpress and release human acid sphingomyelinase, were transplanted into the NPD mice they fused with the degenerated Purkinje neurons and improved cerebellar motor function [96]. Treated NPD mice presented transiently near-normal levels of acid sphingomyelinase activity in their tissues, and the number of Purkinje cells improved remarkably [96]. A different group, with a similar experimental approach, but the use of viral transduction, reported comparable results [95]. Interestingly, the rescued fused Purkinje neurons were electrically active and presented a functional synaptic formation [97]. These results indicate that BMSC transplantation can be an effective therapeutic vehicle to deliver genetic material to Purkinje neurons via cell fusion.

10.3.5 Breathing Cell Fusion

Growing evidence exists on the ability of BMDC to adopt the morphology and protein expression pattern of lung epithelial cells [98]. Fusion of BMDC with human pulmonary epithelium was first demonstrated in vitro by Spees et al., who co-cultured human mesenchymal SC with heat-shocked small airway epithelial cells [99]. Time lapse photography revealed both, fusion events as well as nonfused mesenchymal SC-derived airway epithelia. Up to 1% of fused cells expressing epithelial markers were found in the co-culture system. Some of the fused cells also underwent nuclear fusion and presented a new gene expression profile [99]. Soon after, Wong et al. studied the in vivo pulmonary contribution of transtracheally delivered BMDC in a mouse model of mild airway injury induced by naphthalene [100]. For at least 120 days following cell delivery, they observed BMDC that acquired phenotypic characteristics of the injured cell type in the airway and alveolar epithelium. Approximately, 1% of the engrafted cells were the result of cell fusion, as they could demonstrate by FISH. More recently, bone marrow progenitor cells were transplanted in a rat model of progressive pulmonary hypertension [101]. Immunohistochemistry demonstrated the presence of donor-derived interstitial fibroblasts or myofibroblasts, pulmonary epithelial cells (Clara cells), vascular endothelial cells, and smooth muscle cells. Fusion with pulmonary cells was observed at low frequency (0.52–0.59%). From them 0.3–0.5% had undergone nuclear fusion during mitosis, as assessed by FISH of chromosome 1 and 4.

10.3.6 Other Tissues

Negative evidence of fusion has been reported in pancreas [102], epithelial tissue [98, 103], and vascular endothelium [88]. On the other hand, fused cells have been found in the gastrointestinal tract [104, 105], as well as in the kidney [106], and skin [107], after bone marrow transplantation. Fusion in skin and kidney is infrequent and its significance or consequences remains unresolved. In contrast, fusion of BMDC have an active role in intestinal regeneration [104, 105]. Transplanted BMDC can fuse with intestinal epithelium cells of the injured intestine in mice that had received gamma-irradiation. The hybrid cells were able to repopulate the intestine and gave rise to the cells of all of the principal components of the gut epithelium, including goblet cells, Paneth cells and enterocytes [105]. This wide repopulation was thought to be due to fusion of the BMDC with intestinal progenitors, with the resulting reprogrammed hybrids acquiring features of intestinal stem cells, and therefore being able to differentiate into different cell types. In the absence of injury or inflammation the baseline level of fusion is low [104]. However, inflammation and direct induction of epithelial proliferation result in a significant increase in intestinal cell fusion. The physiologic impact of this fusion has pros and cons. The mediated regeneration after irradiation-induced injury is clearly beneficial. However, the increased incidence in an inflammatory and proliferative microenvironment suggests a potential role in mediating the progression of intestinal inflammatory diseases and cancer [105, 108].

10.4 The Challenges of Cell Fusion to Become a Therapy

Cell fusion needs to face many challenges before being considered for clinical purposes [8]. The first of such challenges is its controlled induction up to therapeutic levels, since one of their biggest caveats is the scarcity of fusion events observed in the organs under different conditions. We need to fully understand the cell fusion mechanisms and study the tissue-specific and injury-related signals that recruit, stimulate and regulate this process in order to get control on it. A better characterization and expansion of the cell populations with fusogenic properties would be necessary, as well. Finally, it is basic to keep in mind different aspects of cell safety after a fusion event.

10.4.1 Fusion Induction

Therapeutically, unspecific induction of cell fusion could bring problems. General in vitro fusogens (polyethylene glycol, electrofusion, and viruses) are not applicable in vivo and/or present serious drawbacks, such as toxicity, and immune rejection. A controlled fusion process between specific cell types should be more desirable. For this, we should pay attention to how cell fusion occurs naturally and then try to recapitulate and enhance it. To review aspects of natural fusion mechanism and receptors you can read [1, 11, 109, 110]. It will be only mentioned some examples of intracellular proteins (e.g., calmodulin and calpain), membrane proteins (e.g., CD200, vascular cell adhesion molecule [VCAM]-I, ADAM12, Caveolin-3, and DC-STAMP), and secreted factors (e.g., SDF-1 and IL-4) [4] implicated in cell fusion. Although all of these may be necessary for fusion of specific cells, none are, by themselves, sufficient to induce this process. Another example during muscle development and regeneration is calcineurin/NFAT (nuclear factor of activated T cells), that is involved in fusion between myotubes or mesenchymal SC with myoblasts by controlling IL-4 activity [111]. Molecules such as Brag2 and Dock180 also have functions in the fusion of mammalian myoblasts and macrophages [112]. Damage in muscle can also increase the frequency of stem cells involved in tissue repair mediated by fusion [60]. Signaling from damaged tissue could be mimicked by introduction of insulin-like growth factor (IGF-1) into stem cells, with a fourfold increase in the participation of satellite cells in muscle formation via fusion events [113]. In the brain and heart infarction several factors are delivered that may influence cell fusion process. The microenvironment in the periphery of the lesion and their microvasculature is hypoxic. This local hypoxia may induce an upregulation in the active levels of local chemoattractants and growth factors, such as SDF-1 [114], or induce the fusion of vascular cells, such us pericytes [89]. These observations and the fact that injury-related signals can attract stem cells, and therefore increase the chance of fusion events in the damaged area, suggest a potential use of these mechanisms for targeted gene or drug delivery [115, 116].

As we have seen, stem and progenitor cells fuse preferentially when tissue is damaged, stressed, or diseased. We have already mentioned examples of this in muscle, brain, intestine and liver. Inflammation is commonly associated to these processes. Interestingly, it has been recently shown that chronic inflammation induces a 10–100-fold increase of the cell fusion events in liver, brain and intestine [74, 76, 104]. Induction of inflammation by severe autoimmune experimental encephalitis resulted in a 100-fold increase of fused Purkinje neurons in the cerebellum. Also irradiation, a process that can damage the brain-blood-barrier and courses with inflammation, induces cell fusion events [73, 76]. However, when anti-inflammatory drug prednisolone or antibiotics are administered, the irradiation-induced cell fusion is inhibited [76]. This strongly suggests that factors implicated in the inflammatory response directly induce cell fusion, or promote changes in the cell fusion partners, such as the expression of membrane molecules, that allow their fusion. Here we should search the future fusion inducers.

10.4.2 Fusion Partners and Delivery

An exhaustive characterization and expansion of the cell populations with fusogenic properties would be also necessary to get control and exploit the fusion process. Macrophages are the most common of the fusogenic populations, at least in the liver, whereas monomyelocytic progenitors are in the muscle [48, 49, 59, 61]. In the brain, the exact population remains elusive. Macrophages naturally migrate to injuries and activate their fusogenic machinery for syncitia formation, developing giant multinucleated cells in response to infections and to eliminate necrosed tissue or foreign bodies [7]. Molecular machinery implicated in macrophage syncytia formation includes adhesion molecules, ligand interactions, and the induction of fusion by cytokines such as IL-3, IL-4, IL-13, interferon-γ (IFN-γ), and granulocyte-macrophage colony stimulating factor (GM-CSF) [7]. These molecules play an important role in organogenesis and inflammation, and interestingly, their expression is shared by neurons, cardiomyocytes, and hepatocytes during development [117, 118]. Currently, our lab is studying the role of these IL in therapeutic cell fusion induction.

Another issue is to establish how the donor fusing cells should be delivered to the target area of injury. Bone marrow cells can travel freely through blood stream and penetrate the parenchyma in the target area. However, the exact mechanisms by which these events occurred, especially in human patients, is unknown. The search for chemotactic factors in the damaged tissues should be a priority. In this sense, SDF-1 plays an important role in the homing/retention of HSC [114]. It works as a potent chemoattractant of CXCR4-positive cells, including macrophages and BMSC [114, 119], and it has been reported as a mediator in the fusion of HIV-1 virus to macrophages [120]. IGF-1 also mimics the signaling from damaged tissue and promotes a fourfold increase in the participation of satellite cells in muscle formation via fusion events [113].

10.4.3 Safety

The benefits of using fusion in cell-based therapy should not hide the possible risks of this procedure [2, 8, 121, 122]. Fusion products should be more carefully studied and controlled. In especial, their functionality and putative pathological effects must be tested. Transplanted cells fuse with a wide variety of cell types in the body, maybe more than what has been already observed, since detection systems have some limitations and reporter genes may suffer silencing [123]. Therefore, we should control locally the fusion procedure to avoid undesirable events in other areas.

A generalized prejudice is that fusion products have no therapeutic potential because they are tetraploid. This may have chromosomic aberrant consequences and induce transformation [124]. However, there is currently no evidence of direct transformation as result of a simple fusion event. It is necessary an additional mutation in p53 or aberrant aneuploidy after proliferation to render a malignant cell [122, 125]. In addition, tetraploid cells can behave in a fashion that is analogous to diploid cells during many developmental processes, and our body present tetraploid cells in liver, muscle, or placenta working normally. Interestingly, when fusion products between SC and other cells have been injected into blastocysts, no cancer formation was observed [28]. Moreover, some polyploid animals are viable [126], and human infants with a large proportion of tetraploid cells have been born and have even survived for up to 15 months after birth [127, 128]. Finally, reductive/multipolar mitosis or ploidy reductions can take place and stabilize the cell [47, 129]. It has been shown that the fusion-derived polyploid hepatocytes can undergo ploidy reductions to generate daughter cells with one-half chromosomal content [47]. Nevertheless, the relationship between cell fusion and cancer formation is really a serious concern [122]. Despite cell fusion is not a direct cancer generator, its role in tumor progression has been extensively documented [2, 121]. Fusion may confer new properties to the already transformed cells [121, 122]. It promotes tumor progression by increasing malignancy in the resulted hybrid cell, amplifying its drug resistance and contributing to tumoral diversity

[130–133]. It has been also shown that fusion of tumor cells with lymphocytes or macrophages can render a tumor metastatic [134, 135]. Finally, cell fusion may be in the origin of cancer stem cells [122, 136].

Viral transfer has been also related to cell fusion. Hybrid cells of different species, generated during xenotransplants, are able to transmit virus to human cells and could explain the generation of novel viral pathogens by recombination of selected DNA sequences [10, 137]. There is also a strong link between viruses and cancer by cell fusion [2, 3]. This relationship between cell fusion, virus and tumor progression recommends the implementation of safety systems in the donor fusogenic cells to turn off cell fusion or their consequences if necessary.

10.5 Conclusions

Cell fusion is a powerful process with implications in normal development, tissue formation and with a prominent role in SC plasticity. Cell fusion opens new expectations in regenerative medicine and the possible correction of recessive genetic alterations, what have a great clinical potential. Cell fusion has been shown to regenerate liver, brain, muscle, lung and gut under certain conditions. However, we should be cautious, given the relationship of cell fusion with several diseases, such as cancer. We need to fully understand the cell fusion process before to consider it as clinically relevant. For this, we should face a rigorous identification of the tissue-specific and injury-related signals that recruit, stimulate or regulate it. Effectiveness and safety must be warranted before cell fusion can be used as a therapy for human diseases. Cell fusion is an exciting and promising research field. Further efforts should be devoted to investigate the mechanisms that govern it.

Acknowledgments We would like to express our gratitude to Dr. M.I. Santos for critical comments on the manuscript. M.M-L. is a "*Sara Borrell*" postdoctoral from Institute Carlos III, Spanish Ministry of Health.

References

1. Chen EH, Grote E, Mohler W et al (2007) Cell-cell fusion. FEBS lett 581:2181–2193
2. Duelli D, Lazebnik Y (2007) Cell-to-cell fusion as a link between viruses and cancer. Nat Rev 7:968–976
3. Duelli DM, Hearn S, Myers MP et al (2005) A primate virus generates transformed human cells by fusion. J Cell Biol 171:493–503
4. Horsley V, Pavlath GK (2004) Forming a multinucleated cell: molecules that regulate myoblast fusion. Cells Tissues Organs 176:67–78
5. Potgens AJ, Schmitz U, Bose P et al (2002) Mechanisms of syncytial fusion: a review. Placenta 23(Suppl A):S107–113
6. Vignery A (2005) Macrophage fusion: the making of osteoclasts and giant cells. J Exp Med 202:337–340
7. Vignery A (2008) Macrophage fusion: molecular mechanisms. Meth Mol Biol (Clifton, NJ) 475:149–161
8. Alvarez-Dolado M (2007) Cell fusion: biological perspectives and potential for regenerative medicine. Front Biosci 12:1–12
9. Lluis F, Cosma MP (2010) Cell-fusion-mediated somatic-cell reprogramming: a mechanism for tissue regeneration. J Cellular Physiol 223:6–13
10. Ogle BM, Cascalho M, Platt JL (2005) Biological implications of cell fusion. Nat Rev Mol Cell Biol 6:567–575
11. Stolzing A, Hescheler J, Sethe S (2007) Fusion and regenerative therapies: Is immortality really recessive? Rejuvenation Res 10:571–586
12. Yamanaka S, Blau HM (2010) Nuclear reprogramming to a pluripotent state by three approaches. Nature 465:704–712
13. Marshak D, Gardner R, Gottlieb D (2001) Stem cell biology. Cold Spring Harbor Laboratory Press, New York, NY
14. Wagers AJ, Weissman IL (2004) Plasticity of adult stem cells. Cell 116:639–648
15. Eglitis MA, Mezey E (1997) Hematopoietic cells differentiate into both microglia and macroglia in the brains of adult mice. Proc Natl Acad Sci USA 94:4080–4085

16. Brazelton TR, Rossi FM, Keshet GI et al (2000) From marrow to brain: expression of neuronal phenotypes in adult mice. Science 290:1775–1779
17. Mezey E, Chandross KJ, Harta G et al (2000) Turning blood into brain: cells bearing neuronal antigens generated in vivo from bone marrow. Science 290:1779–1782
18. Priller J, Persons DA, Klett FF et al (2001) Neogenesis of cerebellar Purkinje neurons from gene-marked bone marrow cells in vivo. J Cell Biol 155:733–738
19. Weimann JM, Charlton CA, Brazelton TR et al (2003) Contribution of transplanted bone marrow cells to Purkinje neurons in human adult brains. Proc Natl Acad Sci USA 100:2088–2093
20. Weimann JM, Johansson CB, Trejo A et al (2003) Stable reprogrammed heterokaryons form spontaneously in Purkinje neurons after bone marrow transplant. Nat Cell Biol 5:959–966
21. Pereira RF, Halford KW, O'Hara MD et al (1995) Cultured adherent cells from marrow can serve as long-lasting precursor cells for bone, cartilage, and lung in irradiated mice. Proc Natl Acad Sci USA 92:4857–4861
22. Pittenger MF, Mackay AM, Beck SC et al (1999) Multilineage potential of adult human mesenchymal stem cells. Science 284:143–147
23. Krause DS, Theise ND, Collector MI et al (2001) Multi-organ, multi-lineage engraftment by a single bone marrow-derived stem cell. Cell 105:369–377
24. Jiang Y, Jahagirdar BN, Reinhardt RL et al (2002) Pluripotency of mesenchymal stem cells derived from adult marrow. Nature 418:41–49
25. Reyes M, Lund T, Lenvik T et al (2001) Purification and *ex vivo* expansion of postnatal human marrow mesodermal progenitor cells. Blood 98:2615–2625
26. Reyes M, Verfaillie CM (2001) Characterization of multipotent adult progenitor cells, a subpopulation of mesenchymal stem cells. Ann N Y Acad Sci 938:231–233; discussion 233–235
27. Terada N, Hamazaki T, Oka M et al (2002) Bone marrow cells adopt the phenotype of other cells by spontaneous cell fusion. Nature 416:542–545
28. Ying QL, Nichols J, Evans EP et al (2002) Changing potency by spontaneous fusion. Nature 416:545–548
29. Ferrari G, Cusella-De Angelis G, Coletta M et al (1998) Muscle regeneration by bone marrow-derived myogenic progenitors. Science 279:1528–1530
30. Gibson AJ, Karasinski J, Relvas J et al (1995) Dermal fibroblasts convert to a myogenic lineage in mdx mouse muscle. J Cell Sci 108 (Pt 1):207–214
31. Gussoni E, Bennett RR, Muskiewicz KR et al (2002) Long-term persistence of donor nuclei in a Duchenne muscular dystrophy patient receiving bone marrow transplantation. J Clin Invest 110:807–814
32. Gussoni E, Soneoka Y, Strickland CD et al (1999) Dystrophin expression in the mdx mouse restored by stem cell transplantation. Nature 401:390–394
33. Vassilopoulos G, Wang PR, Russell DW (2003) Transplanted bone marrow regenerates liver by cell fusion. Nature 422:901–904
34. Wang X, Willenbring H, Akkari Y et al (2003) Cell fusion is the principal source of bone-marrow-derived hepatocytes. Nature 422:897–901
35. Wagers AJ, Sherwood RI, Christensen JL et al (2002) Little evidence for developmental plasticity of adult hematopoietic stem cells. Science 297:2256–2259
36. Alvarez-Dolado M, Pardal R, Garcia-Verdugo JM et al (2003) Fusion of bone-marrow-derived cells with Purkinje neurons, cardiomyocytes and hepatocytes. Nature 425:968–973
37. Balsam LB, Wagers AJ, Christensen JL et al (2004) Haematopoietic stem cells adopt mature haematopoietic fates in ischaemic myocardium. Nature 428:668–673
38. Murry CE, Soonpaa MH, Reinecke H et al (2004) Haematopoietic stem cells do not transdifferentiate into cardiac myocytes in myocardial infarcts. Nature 428:664–668
39. Nygren JM, Jovinge S, Breitbach M et al (2004) Bone marrow-derived hematopoietic cells generate cardiomyocytes at a low frequency through cell fusion, but not transdifferentiation. Nat Med 10:494–501
40. Blau HM (2002) A twist of fate. Nature 419:437
41. Blau HM, Blakely BT (1999) Plasticity of cell fate: insights from heterokaryons. Semin Cell Dev Biol 10:267–272
42. Cowan CA, Atienza J, Melton DA et al (2005) Nuclear reprogramming of somatic cells after fusion with human embryonic stem cells. Science 309:1369–1373
43. Blau HM, Chiu CP, Webster C (1983) Cytoplasmic activation of human nuclear genes in stable heterocaryons. Cell 32:1171–1180
44. Pomerantz J, Blau HM (2004) Nuclear reprogramming: a key to stem cell function in regenerative medicine. Nat Cell Biol 6:810–816
45. Ringertz NR, Savage RE (1976) Cell hybrids. Academic Press, New York, NY
46. Vassilopoulos G, Russell DW (2003) Cell fusion: an alternative to stem cell plasticity and its therapeutic implications. Curr Opin Genet Dev 13:480–485
47. Duncan AW, Hickey RD, Paulk NK et al (2009) Ploidy reductions in murine fusion-derived hepatocytes. PLoS genetics 5:e1000385

48. Camargo FD, Finegold M, Goodell MA (2004) Hematopoietic myelomonocytic cells are the major source of hepatocyte fusion partners. J Clin Invest 113:1266–1270
49. Willenbring H, Bailey AS, Foster M et al (2004) Myelomonocytic cells are sufficient for therapeutic cell fusion in liver. Nat Med 10:744–748
50. Fujino H, Hiramatsu H, Tsuchiya A et al (2007) Human cord blood CD34+ cells develop into hepatocytes in the livers of NOD/SCID/gamma(c)null mice through cell fusion. Faseb J 21:3499–3510
51. Kashofer K, Siapati EK, Bonnet D (2006) In vivo formation of unstable heterokaryons after liver damage and hematopoietic stem cell/progenitor transplantation. Stem cells (Dayton, Ohio) 24:1104–1112
52. Sharma AD, Cantz T, Richter R et al (2005) Human cord blood stem cells generate human cytokeratin 18-negative hepatocyte-like cells in injured mouse liver. Am J Pathol 167:555–564
53. Zhou P, Wirthlin L, McGee J et al (2009) Contribution of human hematopoietic stem cells to liver repair. Seminars Immunopathol 31:411–419
54. Jang YY, Collector MI, Baylin SB et al (2004) Hematopoietic stem cells convert into liver cells within days without fusion. Nat Cell Biol 6:532–539
55. Newsome PN, Johannessen I, Boyle S et al (2003) Human cord blood-derived cells can differentiate into hepatocytes in the mouse liver with no evidence of cellular fusion. Gastroenterology 124:1891–1900
56. Rochlin K, Yu S, Roy S et al (2010) Myoblast fusion: when it takes more to make one. Develop Biol 341:66–83
57. Bittner RE, Schofer C, Weipoltshammer K et al (1999) Recruitment of bone-marrow-derived cells by skeletal and cardiac muscle in adult dystrophic mdx mice. Anat Embryol (Berl) 199:391–396
58. Goncalves MA, Swildens J, Holkers M et al (2008) Genetic complementation of human muscle cells via directed stem cell fusion. Mol Ther 16:741–748
59. Doyonnas R, LaBarge MA, Sacco A et al (2004) Hematopoietic contribution to skeletal muscle regeneration by myelomonocytic precursors. Proc Natl Acad Sci USA 101:13507–13512
60. LaBarge MA, Blau HM (2002) Biological progression from adult bone marrow to mononucleate muscle stem cell to multinucleate muscle fiber in response to injury. Cell 111:589–601
61. Camargo FD, Green R, Capetanaki Y et al (2003) Single hematopoietic stem cells generate skeletal muscle through myeloid intermediates. Nat Med 9:1520–1527
62. Mathur A, Martin JF (2004) Stem cells and repair of the heart. Lancet 364:183–192
63. Condorelli G, Peschle C (2005) Stem cells for cardiac repair: state of the art. Front Biosci 10:3143–3150
64. Orlic D, Kajstura J, Chimenti S et al (2001) Bone marrow cells regenerate infarcted myocardium. Nature 410:701–705
65. Quaini F, Urbanek K, Beltrami AP et al (2002) Chimerism of the transplanted heart. N Engl J Med 346:5–15
66. Deb A, Wang S, Skelding KA et al (2003) Bone marrow-derived cardiomyocytes are present in adult human heart: A study of gender-mismatched bone marrow transplantation patients. Circulation 107:1247–1249
67. Laflamme MA, Myerson D, Saffitz JE et al (2002) Evidence for cardiomyocyte repopulation by extracardiac progenitors in transplanted human hearts. Circ Res 90:634–640
68. Matsuura K, Wada H, Nagai T et al (2004) Cardiomyocytes fuse with surrounding noncardiomyocytes and reenter the cell cycle. J Cell Biol 167:351–363
69. Gnecchi M, Zhang Z, Ni A et al (2008) Paracrine mechanisms in adult stem cell signaling and therapy. Circ Res 103:1204–1219
70. Leone AM, Valgimigli M, Giannico MB et al (2009) From bone marrow to the arterial wall: the ongoing tale of endothelial progenitor cells. European Heart J 30:890–899
71. Tanaka K, Sata M (2009) Role of vascular progenitor cells in cardiovascular disease. Curr Pharm Design 15:2760–2768
72. Hess DC, Abe T, Hill WD et al (2004) Hematopoietic origin of microglial and perivascular cells in brain. Exp Neurol 186:134–144
73. Espejel S, Romero R, Alvarez-Buylla A (2009) Radiation damage increases Purkinje neuron heterokaryons in neonatal cerebellum. Ann Neurol 66:100–109
74. Johansson CB, Youssef S, Koleckar K et al (2008) Extensive fusion of haematopoietic cells with Purkinje neurons in response to chronic inflammation. Nat Cell Biol 10:575–583
75. Magrassi L, Grimaldi P, Ibatici A et al (2007) Induction and survival of binucleated Purkinje neurons by selective damage and aging. J Neurosci 27:9885–9892
76. Nygren JM, Liuba K, Breitbach M et al (2008) Myeloid and lymphoid contribution to non-haematopoietic lineages through irradiation-induced heterotypic cell fusion. Nat Cell Biol 10:584–592
77. Wiersema A, Dijk F, Dontje B et al (2007) Cerebellar heterokaryon formation increases with age and after irradiation. Stem Cell Res 1:150–154
78. Lescaudron L, Unni D, Dunbar GL (2003) Autologous adult bone marrow stem cell transplantation in an animal model of huntington's disease: behavioral and morphological outcomes. Int J Neurosci 113:945–956

79. Li Y, Chen J, Wang L et al (2001) Intracerebral transplantation of bone marrow stromal cells in a 1-methyl-4-phenyl-1,2,3,6-tetrahydropyridine mouse model of Parkinson's disease. Neurosci Lett 316:67–70
80. Li Y, Chopp M (2009) Marrow stromal cell transplantation in stroke and traumatic brain injury. Neurosci Lett 456:120–123
81. Momin EN, Mohyeldin A, Zaidi HA et al (2010) Mesenchymal stem cells: new approaches for the treatment of neurological diseases. Curr Stem Cell Res Ther 5:326–344
82. Rosser AE, Zietlow R, Dunnett SB (2007) Stem cell transplantation for neurodegenerative diseases. Curr Opin Neurol 20:688–692
83. Chang YC, Shyu WC, Lin SZ et al (2007) Regenerative therapy for stroke. Cell Transplant 16:171–181
84. Tang Y, Yasuhara T, Hara K et al (2007) Transplantation of bone marrow-derived stem cells: a promising therapy for stroke. Cell Transplant 16:159–169
85. Chen J, Li Y, Wang L et al (2001) Therapeutic benefit of intravenous administration of bone marrow stromal cells after cerebral ischemia in rats. Stroke 32:1005–1011
86. Kurozumi K, Nakamura K, Tamiya T et al (2005) Mesenchymal stem cells that produce neurotrophic factors reduce ischemic damage in the rat middle cerebral artery occlusion model. Mol Ther 11:96–104
87. Chen J, Zhang ZG, Li Y et al (2003) Intravenous administration of human bone marrow stromal cells induces angiogenesis in the ischemic boundary zone after stroke in rats. Circ Res 92:692–699
88. Bailey AS, Willenbring H, Jiang S et al (2006) Myeloid lineage progenitors give rise to vascular endothelium. Proc Natl Acad Sci USA 103:13156–13161
89. Piquer-Gil M, Garcia-Verdugo JM, Zipancic I et al (2009) Cell fusion contributes to pericyte formation after stroke. J Cereb Blood Flow Metab 29:480–485
90. Ozerdem U, Stallcup WB (2003) Early contribution of pericytes to angiogenic sprouting and tube formation. Angiogenesis 6:241–249
91. Tigges U, Hyer EG, Scharf J et al (2008) FGF2-dependent neovascularization of subcutaneous Matrigel plugs is initiated by bone marrow-derived pericytes and macrophages. Development 135:523–532
92. Gerhardt H, Betsholtz C (2003) Endothelial-pericyte interactions in angiogenesis. Cell Tissue Res 314:15–23
93. Krueger M, Bechmann I (2010) CNS pericytes: concepts, misconceptions, and a way out. Glia 58:1–10
94. Prasad VK, Kurtzberg J (2010) Cord blood and bone marrow transplantation in inherited metabolic diseases: scientific basis, current status and future directions. Brit J Haematol 148:356–372
95. Bae JS, Furuya S, Shinoda Y et al (2005) Neurodegeneration augments the ability of bone marrow-derived mesenchymal stem cells to fuse with Purkinje neurons in Niemann-Pick type C mice. Hum Gene Ther 16:1006–1011
96. Jin HK, Schuchman EH (2003) *Ex vivo* gene therapy using bone marrow-derived cells: combined effects of intracerebral and intravenous transplantation in a mouse model of Niemann-Pick disease. Mol Ther 8:876–885
97. Bae JS, Han HS, Youn DH et al (2007) Bone marrow-derived mesenchymal stem cells promote neuronal networks with functional synaptic transmission after transplantation into mice with neurodegeneration. Stem cells (Dayton, Ohio) 25:1307–1316
98. Kassmer SH, Krause DS (2010) Detection of bone marrow-derived lung epithelial cells. Exp Hematol 38:564–573
99. Spees JL, Olson SD, Ylostalo J et al (2003) Differentiation, cell fusion, and nuclear fusion during *ex vivo* repair of epithelium by human adult stem cells from bone marrow stroma. Proc Natl Acad Sci USA 100:2397–2402
100. Wong AP, Dutly AE, Sacher A et al (2007) Targeted cell replacement with bone marrow cells for airway epithelial regeneration. Am J Physiol Lung Cell Mol Physiol 293:L740–752
101. Spees JL, Whitney MJ, Sullivan DE et al (2008) Bone marrow progenitor cells contribute to repair and remodeling of the lung and heart in a rat model of progressive pulmonary hypertension. Faseb J 22:1226–1236
102. Ianus A, Holz GG, Theise ND et al (2003) In vivo derivation of glucose-competent pancreatic endocrine cells from bone marrow without evidence of cell fusion. J Clin Invest 111:843–850
103. Harris RG, Herzog EL, Bruscia EM et al (2004) Lack of a fusion requirement for development of bone marrow-derived epithelia. Science 305:90–93
104. Davies PS, Powell AE, Swain JR et al (2009) Inflammation and proliferation act together to mediate intestinal cell fusion. PloS one 4:e6530
105. Rizvi AZ, Swain JR, Davies PS et al (2006) Bone marrow-derived cells fuse with normal and transformed intestinal stem cells. Proc Natl Acad Sci USA 103:6321–6325
106. Fang TC, Alison MR, Cook HT et al (2005) Proliferation of bone marrow-derived cells contributes to regeneration after folic acid-induced acute tubular injury. J Am Soc Nephrol 16:1723–1732
107. Fan Q, Yee CL, Ohyama M et al (2006) Bone marrow-derived keratinocytes are not detected in normal skin and only rarely detected in wounded skin in two different murine models. Exp Hematol 34:672–679
108. Houghton J, Stoicov C, Nomura S et al (2004) Gastric cancer originating from bone marrow-derived cells. Science 306:1568–1571
109. Chen EH, Olson EN (2005) Unveiling the mechanisms of cell-cell fusion. Science 308:369–373

110. Larsson LI, Bjerregaard B, Talts JF (2008) Cell fusions in mammals. Histochem Cell Biol 129:551–561
111. Schulze M, Belema-Bedada F, Technau A et al (2005) Mesenchymal stem cells are recruited to striated muscle by NFAT/IL-4-mediated cell fusion. Genes Develop 19:1787–1798
112. Pajcini KV, Pomerantz JH, Alkan O et al (2008) Myoblasts and macrophages share molecular components that contribute to cell-cell fusion. J Cell Biol 180:1005–1019
113. Sacco A, Doyonnas R, LaBarge MA et al (2005) IGF-I increases bone marrow contribution to adult skeletal muscle and enhances the fusion of myelomonocytic precursors. J Cell Biol 171:483–492
114. Zaruba MM, Franz WM (2010) Role of the SDF-1-CXCR4 axis in stem cell-based therapies for ischemic cardiomyopathy. Expert Opin Biolog Therapy 10:321–335
115. Korbling M, Estrov Z (2003) Adult stem cells for tissue repair – a new therapeutic concept? N Engl J Med 349:570–582
116. Prockop DJ, Gregory CA, Spees JL (2003) One strategy for cell and gene therapy: harnessing the power of adult stem cells to repair tissues. Proc Natl Acad Sci USA 100(Suppl 1):11917–11923
117. Bajetto A, Bonavia R, Barbero S et al (2001) Chemokines and their receptors in the central nervous system. Front Neuroendocrinol 22:147–184
118. Thomson. AW (1998) The cytokine handbook. Academic Press, San Diego, CA
119. Kucia M, Ratajczak J, Reca R et al (2004) Tissue-specific muscle, neural and liver stem/progenitor cells reside in the bone marrow, respond to an SDF-1 gradient and are mobilized into peripheral blood during stress and tissue injury. Blood Cells Mol Dis 32:52–57
120. Lapham CK, Zaitseva MB, Lee S et al (1999) Fusion of monocytes and macrophages with HIV-1 correlates with biochemical properties of CXCR4 and CCR5. Nat Med 5:303–308
121. Duelli D, Lazebnik Y (2003) Cell fusion: a hidden enemy? Cancer Cell 3:445–448
122. Lu X, Kang Y (2009) Cell fusion as a hidden force in tumor progression. Cancer Res 69:8536–8539
123. Long MA, Rossi FM (2009) Silencing inhibits Cre-mediated recombination of the Z/AP and Z/EG reporters in adult cells. PloS one 4:e5435
124. Storchova Z, Kuffer C (2008) The consequences of tetraploidy and aneuploidy. J Cell Sci 121:3859–3866
125. Fujiwara T, Bandi M, Nitta M et al (2005) Cytokinesis failure generating tetraploids promotes tumorigenesis in p53-null cells. Nature 437:1043–1047
126. Eakin GS, Behringer RR (2003) Tetraploid development in the mouse. Dev Dyn 228:751–766
127. Nakamura Y, Takaira M, Sato E et al (2003) A tetraploid liveborn neonate: cytogenetic and autopsy findings. Arch Pathol Labor Med 127:1612–1614
128. Scarbrough PR, Hersh J, Kukolich MK et al (1984) Tetraploidy: a report of three live-born infants. Am J Med Gene 19:29–37
129. Martin GM, Sprague CA (1970) Vinblastine induces multipolar mitoses in tetraploid human cells. Experimental Cell Res 63:466–467
130. Barski G, Sorieul S, Cornefert F (1960) Production of cells of a "hybrid" nature in culturs in vitro of 2 cellular strains in combination.C R Hebd Seances Acad Sci 251:1825–1827
131. Miller FR, Mohamed AN, McEachern D (1989) Production of a more aggressive tumor cell variant by spontaneous fusion of two mouse tumor subpopulations. Cancer Res 49:4316–4321
132. Pawelek JM (2000) Tumour cell hybridization and metastasis revisited. Melanoma Res 10:507–514
133. Warner TF (1975) Cell hybridizaiton: an explanation for the phenotypic diversity of certain tumours. Med Hypotheses 1:51–57
134. Mekler LB (1971) [Hybridization of transformed cells with lymphocytes as 1 of the probable causes of the progression leading to the development of metastatic malignant cells]. Vestn Akad Med Nauk SSSR 26: 80–89
135. Mekler LB, Drize OB, Osechinskii IV et al (1971) Transformation of a normal differentiated cell of an adult organism, induced by the fusion of this cell with another normal cell of the same organism but with different organ or tissue specificity. Vestn Akad Med Nauk SSSR 26:75–80
136. Bjerkvig R, Tysnes BB, Aboody KS et al (2005) Opinion: the origin of the cancer stem cell: current controversies and new insights. Nat Rev 5:899–904
137. Ogle BM, Butters KA, Plummer TB et al (2004) Spontaneous fusion of cells between species yields transdifferentiation and retroviral transfer in vivo. Faseb J 18:548–550

Chapter 11
Dendritic Cell-Tumor Cell Fusion Vaccines

Walter T. Lee

Abstract The use of cell fusion has been applied to the development of immunotherapy cancer vaccines. This has typically involved the fusion of dendritic cells and tumor cells. The resultant hybrid uses the specialized antigen presentation properties supplied by the dendritic cell fusion partner to present tumor antigens, both known and yet undefined, to the immune system. This chapter critically examines the scientific foundation of this approach mainly focusing on studies over the last decade. This will include basic principles of tumor fusion vaccines, summary of pre-clinical and clinical data, concluding with remaining challenges and directions.

11.1 Introduction

In 2000, Nature Medicine published a report examining the impact of a fusion tumor vaccine generated from allogeneic DC and autologous tumor cells [1]. The study reported tumor regression in 7/17 patients with metastatic renal cell carcinoma. This was one of the first clinical trials reported using a fusion cell vaccine and it generated much excitement and attention to this novel cancer treatment strategy. Unfortunately, this article was retracted in 2003 following an investigation from the lead author's institution concluding gross negligence and "failure to meet the requirements of good scientific practice" [2]. Specifically, it appears that data was not properly managed and collected thereby impacting the conclusions of the vaccine's efficacy.

Although this infamous incident has led many to view fusion cell based immunotherapy with suspicion and skepticism, others have continued to rigorously investigate this immunotherapy approach. This chapter critically examines the scientific foundation of this approach mainly focusing on studies over the last decade. This will include summary of pre-clinical and clinical data, concluding with remaining challenges and directions.

11.2 Dendritic Cell Based Immunotherapy

Cancer immunotherapy seeks to sensitize effector T cells to recognize and eliminate cancer cells. It is now clear that an immune response is critically dependent on antigen presentation to the T-cell which involves the engagement of the T cell receptor with its specific MHC-antigen complex (Signal 1) and

W.T. Lee (✉)
Division of Otolaryngology – Head and Neck Surgery, Duke University Medical Center, Durham, NC 27710, USA
e-mail: walter.lee@duke.edu

T. Dittmar, K.S. Zänker (eds.), *Cell Fusion in Health and Disease*, Advances in Experimental Medicine and Biology 713, DOI 10.1007/978-94-007-0763-4_11,

the interactions of the co-stimulatory molecules on antigen presenting cells (APC) with receptors on T cells (Signal 2). The dendritic cell (DC) has been identified to be the most potent of all APC. DC exhibit high levels of MHC Class I and II molecules, and co-stimulatory molecules [3, 4]. Cytokines secreted by DC, such as IL-12 and IL-6, result in growth of cytotoxic and helper T-cells [5, 6].

The maturation state of the DC is also critical for stimulating anti-tumor CTL responses during DC-based immunotherapy and for inducing immune responses against a variety of pathogenic agents [7]. In contrast to mature DC, immature DC can induce immunologic tolerance, a state undesirable in a host responding to infection or tumor immunotherapy [8]. The enhanced immunostimulatory properties of mature DC are due to increased expression of MHC molecules, co-stimulatory molecules (e.g. CD80, CD86) as well as the secretion of cytokines [9, 10]. For the MHC class II restricted antigen presentation pathway, DC harbor lysosomes that enable rapid processing of exogenous proteins into MHC-binding peptides and subsequent transportation of MHC II/peptide complexes to the plasma membrane after maturation [11–13]. Mature DC also possess the ability to cross-present exogenous antigens to naïve CD8+ T cells [14].

One key difference between immature, tolerogenic DC and mature, immunogenic DC is their ability to produce proinflammatory cytokines (Signal 3). IL-12 is among the major cytokines that are important for an immunological response. It has been demonstrated that production of IL-12 and other unidentified cytokines by stimulated DC was important for the ability to enhance CD8+ T-cell responses [15, 16]. IL-12 can enhance effector T-cell expression of Bcl-3, which is associated with enhanced survival [17]. Furthermore, IL-12 is a key Th1 cytokine that activates STAT4 and drives naïve CD4+ T cells to become Th1 cells [18]. These cells in turn produce IFN-g. Therefore, mature DC not only increase the antigen presentation (Signal 1) and co-stimulation (Signal 2) capacity that lead to expansion of T cells, but also generates a cytokine milieu (Signal 3) that influences T-cell differentiation.

Numerous strategies trying to harness the antigen presenting capabilities of DC have been studied. This includes pulsing DC with peptides, tumor lysates, or RNA from tumor cells [19–21]. As a whole, clinical trials using DC based approaches have lack consistent anti-tumor responses. In a review of DC based immunotherapies administered by the Surgery Branch at the National Cancer Institute, National Institutes of Health, there was an overall response rate of 7.1% [22]. This emphasizes the point that although DC based strategies hold promise, significant clinical responses have yet to be achieved. Efforts to augment and optimize these DC strategies are ongoing.

11.3 Principle of Dendritic Cell-Tumor Fusion Hybrids

One of these efforts that builds on the current understanding of DC is the fusion of mature DC with intact tumor cells. These tumor cells, inactivated by radiation or mitomycin-C treatment, can provide the full complement of tumor antigens for immunization. Vaccination by these treated whole tumor vaccines have been studied and described since early in the immunotherapy literature. DC-tumor fusion hybrids seek to improve on the advantages of using whole tumor cells. It is already established on a fundamental level that fusion hybrid cells maintain characteristics of both parent cells [23, 24]. Therefore, the resultant DC-tumor fusion hybrid should utilize the potent antigen presentation capabilities of the DC to present the full complement of tumor antigens. These tumors antigens include those that have been characterized as well as those that are yet undefined. Furthermore, such a fusion hybrid should be potent in presenting antigens to both MHC class I- and class II- restricted pathways. It should also provide the critical co-stimulatory molecules needed for inducing tumor specific CD8+ and CD4+ T cells.

This capacity to provide a complement of tumor antigens is in contrast with antigen specific DC immunotherapy strategies such as peptide pulsing. These antigen specific strategies have a number of potential disadvantages. First, tumor cells are genetically unstable and may have down regulation or mutation of the targeted antigen. This would decrease treatment efficacy. Furthermore, some cancers may have a limited number of defined tumor antigens with unknown immunogenicity when presented in vivo. DC-fusion hybrids provide a promising strategy to overcome these antigen specific challenges [25, 26].

11.4 Generating DC-Tumor Hybrids

There are several methods for creating DC-tumor hybrids. These include chemical (i.e. polyethylene glycol), physical (i.e. electrofusion), and viral (i.e. viral fusion proteins). Viral methods are the most recently described method for producing DC-tumor fusion hybrids. This method uses a viral fusogenic membrane glycoprotein (FMG) [27]. Briefly, tumor cells were transfected with the viral FMG and pelleted along with DC. Resultant hybrids had 2–4 nuclei each. These fusion cells were reported to be smaller in size and thus able to migrate to draining lymph nodes after subcutaneous injection [28]. As PEG fusion and electrofusion methods have been around for decades, the vast majority of data and studies utilize these two methods.

Polyethelene glycol (PEG), a polymer of ethylene oxide, has been used for cell fusion. The exact method of how PEG induces cell fusion is not completely known. It is theorized that the polymer may cause dehydration of the lipid membrane by binding water molecules. This disrupts the membrane lipid bilayer of cells and induces cell fusion through osmotic forces [29, 30]. The PEG fusion procedure essentially involves controlled exposure of DC and tumor cells in PEG. It is of historical interest that PEG was used in the method of producing hybridomas originally described by Kohler and Milstein in 1976 [31]. This foundational work towards producing monoclonal antibodies was recognized by the 1984 Nobel Prize in Physiology or Medicine.

Finally, DC and tumor cell fusion can be accomplished through electrofusion [32]. This method is based on reversible membrane breakdown between closely adjacent cells. Electrofusion requires two steps. The first step involves alignment of cells by applying an alternating current (ac) to a cell suspension between two electrodes. The cells will generate an alternating dipole and move towards one of the electrodes. Due to the attraction of cellular dipoles, the cells will then align themselves end to end, parallel to the applied field. This alignment is referred to as "pearl-chaining" [33]. The second step is inducing a reversible membrane breakdown. A short direct current (dc) pulse is applied to the cell suspension after the cells have aligned themselves in "pearl-chains." This pulse causes reversible membrane breakdown and as the membranes reform, adjacent cell membrane will coalesce, forming a new hybrid cell [34].

There are some studies that have compared the methods of DC-tumor fusion [35]. In some studies, electrofusion was reported to generate a higher fusion efficiency rate [36]. Expression of parental cell characteristics also seem to be more desirable from hybrids formed from electrofusion compared to PEG [37]. These comparison studies however are ultimately limited to the researchers' expertise and experience of the method being used and an unbiased controlled comparison may be difficult.

11.5 Verification of True DC-Tumor Fusion

Regardless of the method used, verification of producing true DC-tumor fusion hybrids is critical to accurately assess their immunological impact. Florescence activated cell sorting (FACS) is the most commonly used method for fusion verification. Researchers generally stain for DC specific markers against tumor specific markers or intracellular staining of tumor cells. Successful DC-tumor fusion population will be indicated by double positive staining. However, there are conditions that may result in misleading FACS data. For example, a DC that is adhered to a tumor cell may show double positive signature but does not represent true cell fusion. Furthermore, depending on the cellular marker being identified, a double positive FACS signal may indicate DC uptake of tumor cell fragments rather than true cellular fusion [27]. As a result of these and other potential issues, FACS analysis should not be the sole verification of DC-tumor fusion. Rather, other means to ascertain true cellular fusion should be documented for the technique used by the research team. This can include Giemsa stained cytospin to demonstrate multinucleated cells and florescence microscopy after staining parental cells to show true integration and formation of DC-tumor fusion cells.

Although the exact mechanisms behind DC-tumor fusion immunotherapy continue to be investigated, the true DC-tumor fusion hybrid has the ability of DC to process and present the complement of tumor antigens provided by the tumor fusion partner. In a comprehensive review of the fusion hybrid literature, many publications were found to lack rigorous verification of true DC-tumor heterokaryons [38]. This review of fusion techniques involved PEG as well as electrofusion. The authors concluded that verification needs to include a number of methods that independently demonstrate successful DC-tumor fusion. Regardless of fusion techniques, without such DC-tumor verification, the efficacy of this immunotherapy approach cannot be properly assessed.

It is important to note that experiments using a variety of controls have been performed that support verified DC-tumor fusion cells as a potent immunotherapy strategy. These controls have included mixture of DC and tumor cells, and subpopulations of DC-tumor fusion cells (i.e. nonadherent population). Antigen specific responses were also controlled for by using fusions of non-relevant tumor [39–41].

11.6 Pre-clinical Studies

There is solid pre-clinical evidence of the potent efficacy of this approach in a number of animal tumor models. This includes glioma, melanoma, colon, squamous cell carcinoma, sarcoma, and breast cancers [26, 39, 40, 42, 43]. It is expected that with verified DC-tumor fusion cells, there should be presentation of the complement of tumor antigens as provided by the tumor fusion partner by the antigen presentation machinery provided by the DC fusion partner. Therefore DC-tumor fusion hybrids will present antigens though both MHC class I and II (Signal 1) as well as highly express co-stimulatory molecules (Signal 2) [44]. Fusion cells formed through viral FMG was reported to promote cross presentation of antigens by the DC fusion partner [45]. In a melanoma murine model, both CD8+ and CD4+ cells are activated by DC-tumor fusion cells. Furthermore, both cell populations are required to eliminate established tumor burdens. Abrogating either population with monoclonal blocking antibody resulted in severe attenuation of these potent anti-tumor effects [41]. This is congruent with other studies that have shown a more potent immunotherapeutic effect by vaccines that sensitize both CD8+ and CD4+ cells than either one alone [46].

These findings were also demonstrated in human in vitro studies. Multiple studies have established the feasibility of this approach using human DC and tumors [47–52]. Fusion cells were able to present antigens via both MHC Class I and II molecules [53, 54]. Melanoma antigens from allogeneic tumors were reported to be presented on DC MHC Class I and II when used as a fusion partner [55, 56].

11.7 Additional 3rd Signal with DC-Tumor Hybrids

A number of studies showed that a single dose of DC-tumor fusion cells were able to significantly decrease established tumors. For some, these results were critically dependent on an immunostimlulatory 3rd signal [40, 57, 58]. For example, in a 3 day pulmonary metastasis model of murine melanoma, immunotherapy with fusion cells or 3rd signal (i.e. IL-12) alone was not significantly different from non-treated mice. However, a single does of fusion cells with a 3rd signal were able to significantly eliminate these pulmonary metastasis [58].

Investigations into the improvement seen with additional 3rd signals have elucidated some mechanisms behind this observation. It was demonstrated the administration of fusion cells alone in a murine model resulted in increased IL-10 secretion and a lack of IFN-γ secreting effector cells from activated T cells [43, 59]. Furthermore, a recent publication by Vasir et al., demonstrated that PEG fused human DC-breast carcinoma fusion resulted in T_{reg} expansion in vitro [60]. However, if these fusion cells were concomitantly exposed to 3rd signals (i.e. IL-12, TLR 9 agonists), T_{reg} expansion was reduced with a subsequent increase in activated effector cells. Others have reported that stimulation of the IL-12 pathway is critical to the immunotherapy observed with 3rd signals [59]. Other groups have noted an augmented efficacy when 3rd signals are provided in combination with DC-T fusion hybrids. This has included generating DC-tumor secreting 3rd signals such as heat shock proteins, enterotoxin, and stimulatory cytokines [61–65]. These publications as a whole support the assertion that a 3rd signal is a vital component in optimizing DC-tumor fusion based immunotherapy.

11.8 Allogeneic Fusion Partners

11.8.1 Allogeneic Tumor

Immunotherapy prefers the use of autologous tumor as the source for antigens because these cells are best matched to induce an immune response against unique shared as well as undefined tumor antigens. However, using autologous tumors in clinical immunotherapy is problematic due to the time consuming, labor intensive efforts needed to establish cell lines from surgical specimens in virtually all solid tumors. Other difficulties include tumor accessibility and contaminated patient tumor specimens as is the case from most head and neck cancers and other mucosal tumors. These issues limit the widespread clinical use of autologous tumors.

It is known that tumors of the same or even different histological origin often express shared common tumor-associated antigens (TAA). Immunologically relevant antigens may also be a result of reactivation of genes normally silent in adult tissues [66]. If standardized, established cell lines that share immunologically relevant tumor antigens could be used. This may then obviate many of the practical problems associated with using autologous tumors.

Many murine models exist to investigate autologous based tumor immunotherapy. In these models, syngeneic tumors provide common shared and unique antigens that provide unequivocal interpretation of experimental results. Employing selected established allogeneic tumor lines expressing common TAA is theoretically attractive. The greatest concern of this approach is the impact of allogeneic MHC reaction on the immune response to tumor antigens.

Studies have examined the use of allogeneic tumor sharing TAA as fusion partner. Shared antigens were able to be immunized against using DC-allogeneic tumor fusion cells [67]. DC-allogeneic tumor fusion cells were able to activate cytotoxic T lymphocyte (CTL) responses against autologous tumor and TAA [68–70]. Furthemore, despite having allogeneic antigens, these allogeneic tumor-DC fusion cells were also able to activate both CD4+ and CD8+ cells against TAA [71].

11.8.2 Allogeneic DC

In addition to allogeneic tumor cells, allogeneic DC have also be used to form fusion hybrids. The principle underlying their use is that DC from cancer bearing patients may not be optimal to present tumor antigens. Rather DC from health donors may be optimal. Several studies have used allogeneic DC as a fusion partner in forming DC-tumor hybrids. These allogeneic DC- tumor fusion cells were able to induce a CTL response the tumor used as a fusion partner [72, 73]. Furthermore, there was development of immunity against lethal tumor challenge when either allogeneic tumor or allogeneic DC was used as a fusion partner [74].

What is unclear with this approach is how induction of a T cell response occurs with a lack of MHC class match between the allogeneic DC and autologous T cell. An established tenet of tumor immunology is that antigens are processed and presented via MHC class I- and II- molecules by DC. These MHC class molecules interact directly with CD8+ and CD4+ T cells respectively. By using allogeneic DC as a fusion partner, it is ambiguous how CD8+ and CD4+ cells can be sensitized against the tumor antigens as there is a MHC mismatch between the DC and T cells. If syngeneic tumor is used, it is possible that tumor antigens are presented in the context of tumor MHC Class I molecules would induce T cell responses. However, tumors often express little if any MHC Class I molecules as this is one method the tumor can evade immune elimination. The absence of MHC Class II molecules by tumor cells would further inhibit a tumor response by allogeneic DC – tumor fusion cells by the lack of CD4+ T cells responses. Finally, although it is known that unprimed T cells will react against an allogeneic MHC molecule, how this results in subsequent antigen-specific tumor elimination remains unclear.

11.9 Delivery Method

Consideration needs to also be given to the route of DC-tumor fusion cell delivery. Immature DC are specialized in uptake of antigens. These antigens are processed, during which the DC migrates to LN for interactions with host immune cells. Maturation of these DC result in potent antigen presentation with co-stimulatory molecules and immunostimulatory cytokines [4].

DC-tumor fusion cells may interfere with this sequence of events. For example, DC-tumor fusion cells can be large multinucleated cells whose size may hinder migration to lymph nodes. Others have reported downregulation of chemokine and chemokine receptors in DC-tumor fusion cells that may hinder migration [75]. Furthermore, as mature DC are generally used as a fusion partner, DC-tumor fusion cells that are not in secondary lymph organs may not be in a optimal environment for stimulation of host immune cells. It has been shown that the lymph node is a location where mature DC interact with immune cells [76, 77]. Indeed, some studies have required delivery of fusion cells into secondary lymph organs, such as the lymph node and spleen, to induce significant anti-tumor effects. This was compared with results with the DC-tumor vaccine was given subcutaneously [40].

Based on the pre-clinical data, it is currently felt that a DC-tumor fusion cell vaccine design should consider instituting all three signals in producing the optimal anti-tumor response. The use of DC-tumor fusion cells as an immunotherapy strategy forms the foundation for other methods to optimize its use. This will likely be in the form of providing 3rd signals as well as defining the most immunological and clinically practical fusion partners (i.e. allogeneic vs. autologous).

11.10 Clinical Studies

There have been a number of clinical trials involving DC-tumor fusion cells. These clinical studies have used both autologous as well as allogeneic DC as a fusion partner with a variety of autologous tumors including melanoma, renal cell carcinoma, breast cancer, hepatocellular carcinoma, and glioma [52, 78–81]. These studies have universally reported the safety of this approach. Although the tumor response various, the majority of them report a complete or partial response in a minority of patients. Taken all together, the response rate for DC-tumor fusion clinical studies is estimated at 10.9% [38].

For the responses reported from using allogeneic DC-tumor fusion cells, further critical investigations are needed to better elucidate the underlying mechanisms. Unlike fusion hybrids using allogeneic DC as a fusion partner, there has been little clinical work using DC-allogeneic tumor fusion cells targeting TAA. There have been vaccine trials that are based upon allogeneic tumor. These include using mixture of allogeneic cell lines and modification of these cell lines. Research is on-going that leverages the research finding of fusion cells with those using allogeneic tumor cells in immunotherapy.

11.11 Future Challenges and Directions

Despite the demonstrated advantages of DC-tumor fusion hybrids, vaccine trials as a whole have not demonstrated significant anti-tumor responses. This may be due to a number of remaining issues. First, genuine DC-tumor fusion cells need to be verified. A number of clinical studies did not demonstrate clear verification of DC-tumor fusion cells making assessment of resultant clinical impact difficult. One issue that remains unclear is the optimal dosing schedule and number of fusion cells per injection. This may differ in patients with different tumor burdens and immunological status (i.e. small metastatic pulmonary lesions vs. unresectable local recurrence). Furthermore, the site of vaccine delivery may affect treatment response. This is based on clinical data showing DC based vaccine injection into lymph nodes resulted in superior anti-tumor response compared with other routes [82]. There is significant pre-clinical data supporting the need for a 3rd signal in conjunction with DC-tumor fusion cell administration. Current strategies are being investigated to produce and safely deliver these 3rd signals with a fusion vaccine.

Finally, the use of a DC-tumor fusion vaccine may demonstrate significant efficacy when combined with other treatment modalities. There is evidence that radiation synergizes with immunotherapy vaccines [83, 84]. Chemotherapy has also been reported to augment the anti-tumor response observed with vaccine [85, 86]. Targeting immunosuppressive immune mechanisms such as regulatory T cells (Treg) and myeloid derived suppressor cells (MDSC) may also improve the immunotherapeutic efficacy of a DC-tumor vaccine [87, 88].

The immune system and its interaction with tumor and normal cells is complex and the advances that are being made in the field of immunotherapy are numerous. Although these advances have been rare in significant clinical responses, the potential of harnessing and directing the immune system to eliminate tumors remains the goal of many researchers. DC-tumor fusion cells represent one of the promising efforts in realizing this goal.

Acknowledgement W. Lee is funded in part by the Flight Attendant Medical Research Institute's (FAMRI) Young Clinician Scientist Award (062479_YCSA).

References

1. Kugler A, Stuhler G, Walden P et al (2000) Regression of human metastatic renal cell carcinoma after vaccination with tumor cell-dendritic cell hybrids. Nat Med 6:332–336
2. Editorial (2003) The long road to retraction. Nat Med 9:1093. doi:10.1038/nm0903-1093
3. Banchereau J, Steinman RM (1998) Dendritic cells and the control of immunity. Nature 392:245–252
4. Steinman RM, Pack M, Inaba K (1997) Dendritic cell development and matura-tion. Adv Exp Med Biol 417:1–6
5. Macatonia SE, Hosken NA, Litton M et al (1995) Dendritic cells produce IL-12 and direct the development of Th1 cells from naive CD4+ T cells. J Immunol 154:5071–5079
6. Gajewski TF, Renauld JC, Van Pel A et al (1995) Costimulation with B7-1, IL-6, and IL-12 is sufficient for primary generation of murine antitumor cytolytic T lymphocytes in vitro. J Immunol 154:5637–5648
7. Langenkamp A, Messi M, Lanzavecchia A et al (2000) Kinetics of dendritic cell activation: impact on priming of TH1, TH2 and nonpolarized T cells. Nat Immunol 1:311–316

8. Bachmann MF, Kopf M (2002) Balancing protective immunity and immunopa-thology. Curr Opin Immunol 14:413–419
9. Michelsen KS, Aicher A, Mohaupt M et al (2001) The role of toll-like receptors (TLRs) in bacteria-induced maturation of murine dendritic cells (DCS). Pepti-doglycan and lipoteichoic acid are inducers of DC maturation and require TLR2. J Biol Chem 276:25680–25686
10. Snijders A, Kalinski P, Hilkens CM et al (1998) High-level IL-12 production by human dendritic cells requires two signals. Int Immunol 10:1593–1598
11. Boes M, Cerny J, Massol R et al (2002) T-cell engagement of dendritic cells rap-idly rearranges MHC class II transport. Nature 418:983–988
12. Colledge L, Bennett CL, Reay PA et al (2002) Rapid constitutive generation of a specific peptide-MHC class II complex from intact exogenous protein in immature murine dendritic cells. Eur J Immunol 32:3246–3255
13. Trombetta ES, Ebersold M, Garrett W et al (2003) Activation of lysosomal func-tion during dendritic cell maturation. Science 299:1400–1403
14. Gromme M, Neefjes J (2002) Antigen degradation or presentation by MHC class I molecules via classical and non-classical pathways. Mol Immunol 39:181–202
15. Warren TL, Bhatia SK, Acosta AM et al (2000) APC stimulated by CpG oli-godeoxynucleotide enhance activation of MHC class I-restricted T cells. J Immunol 165:6244–6251
16. Li Q, Eppolito C, Odunsi K et al (2006) IL-12-programmed long-term CD8+ T cell responses require STAT4. J Immunol 177:7618–7625
17. Valenzuela JO, Hammerbeck CD, Mescher MF (2005) Cutting edge: Bcl-3 up-regulation by signal 3 cytokine (IL-12) prolongs survival of antigen-activated CD8 T cells. J Immunol 174:600–604
18. Thieu VT, Yu Q, Chang HC et al (2008) Signal transducer and activator of tran-scription 4 is required for the transcription factor T-bet to promote T helper 1 cell-fate determination. Immunity 29:679–690
19. Boczkowski D, Nair SK, Snyder D et al (1996) Dendritic cells pulsed with RNA are potent antigen-presenting cells in vitro and in vivo. J Exp Med 184:465–472
20. Fields RC, Shimizu K, Mule JJ (1998) Murine dendritic cells pulsed with whole tumor lysates mediate potent antitumor immune responses in vitro and in vivo. Proc Natl Acad Sci USA 95:9482–9487
21. Zitvogel L, Mayordomo JI, Tjandrawan T et al (1996) Therapy of murine tumors with tumor peptide-pulsed dendritic cells: dependence on T cells, B7 costimula-tion, and T helper cell 1-associated cytokines. J Exp Med 183:87–97
22. Rosenberg SA, Yang JC, Restifo NP (2004) Cancer immunotherapy: moving be-yond current vaccines. Nat Med 10:909–915
23. Shu S, Cohen P (2001) Tumor-dendritic cell fusion technology and immunother-apy strategies. J Immunother 24:99–100
24. Koido S, Hara E, Homma S et al (2009) Cancer vaccine by fusions of dendritic and cancer cells. Clin Dev Immunol 2009:657369
25. Shimizu K, Kuriyama H, Kjaergaard J et al (2004) Comparative analysis of antigen loading strategies of dendritic cells for tumor immunotherapy. J Immunother 27:265–272
26. Yasuda T, Kamigaki T, Nakamura T et al (2006) Dendritic cell-tumor cell hybrids enhance the induction of cytotoxic T lymphocytes against murine colon cancer: a comparative analysis of antigen loading methods for the vaccination of immunotherapeutic dendritic cells. Oncol Rep 16:1317–1324
27. Cheong SC, Blangenois I, Franssen JD et al (2006) Generation of cell hybrids via a fusogenic cell line. J Gene Med 8:919–928
28. Phan V, Errington F, Cheong SC et al (2003) A new genetic method to generate and isolate small, short-lived but highly potent dendritic cell-tumor cell hybrid vaccines. Nat Med 9:1215–1219
29. Lentz BR (1994) Polymer-induced membrane fusion: potential mechanism and relation to cell fusion events. Chem Phys Lipids 73:91–106
30. Lentz BR (2007) PEG as a tool to gain insight into membrane fusion. Eur Biophys J 36:315–326
31. Kohler G, Milstein C (1976) Derivation of specific antibody-producing tissue culture and tumor lines by cell fusion. Eur J Immunol 6:511–519
32. Neil GA, Zimmermann U (1993) Electrofusion. Methods Enzymol 220:174–196
33. Burt JP, Pethig R, Gascoyne PR et al (1990) Dielectrophoretic characterisation of Friend murine erythroleukaemic cells as a measure of induced differentiation. Biochim Biophys Acta 1034:93–101
34. Zimmermann U, Vienken J (1982) Electric field-induced cell-to-cell fusion. J Membr Biol 67:165–182
35. Lindner M, Schirrmacher V (2002) Tumour cell-dendritic cell fusion for cancer immunotherapy: comparison of therapeutic efficiency of polyethylen-glycol ver-sus electro-fusion protocols. Eur J Clin Invest 32: 207–217
36. Orentas RJ, Schauer D, Bin Q et al (2001) Electrofusion of a weakly immunogenic neuroblastoma with dendritic cells produces a tumor vaccine. Cell Immunol 213:4–13

37. Weise JB, Hilpert F, Heiser A et al (2004) Electrofusion generates diverse DC-tumour cell hybrids for cancer immunotherapy. Anticancer Res 24:929–934
38. Shu S, Zheng R, Lee WT et al (2007) Immunogenicity of dendritic-tumor fusion hybrids and their utility in cancer immunotherapy. Crit Rev Immunol 27:463–483
39. Kjaergaard J, Wang LX, Kuriyama H et al (2005) Active immunotherapy for ad-vanced intracranial murine tumors by using dendritic cell-tumor cell fusion vac-cines. J Neurosurg 103:156–164
40. Kjaergaard J, Shimizu K, Shu S (2003) Electrofusion of syngeneic dendritic cells and tumor generates potent therapeutic vaccine. Cell Immunol 225:65–74
41. Tanaka H, Shimizu K, Hayashi T et al (2002) Therapeutic immune response in-duced by electrofusion of dendritic and tumor cells. Cell Immunol 220:1–12
42. Lee WT, Tamai H, Cohen P et al (2008) Immunotherapy of established murine squamous cell carcinoma using fused dendritic-tumor cell hybrids. Arch Oto-laryngol Head Neck Surg 134:608–613
43. Kuriyama H, Shimizu K, Lee W et al (2004) Therapeutic vaccine generated by electrofusion of dendritic cells and tumour cells. Dev Biol (Basel) 116:169–178; discussion 179–186
44. Tanaka Y, Koido S, Ohana M et al (2005) Induction of impaired antitumor im-munity by fusion of MHC class II-deficient dendritic cells with tumor cells. J Immunol 174:1274–1280
45. Bateman AR, Harrington KJ, Kottke T et al (2002) Viral fusogenic membrane glycoproteins kill solid tumor cells by nonapoptotic mechanisms that promote cross presentation of tumor antigens by dendritic cells. Cancer Res 62:6566–6578
46. Cohen PA, Peng L, Plautz GE et al (2000) CD4+ T cells in adoptive immuno-therapy and the indirect mechanism of tumor rejection. Crit Rev Immunol 20:17–56
47. Scott-Taylor TH, Pettengell R, Clarke I et al (2000) Human tumour and dendritic cell hybrids generated by electrofusion: potential for cancer vaccines. Biochim Biophys Acta 1500:265–279
48. Weise JB, Maune S, Gorogh T et al (2004) A dendritic cell based hybrid cell vaccine generated by electrofusion for immunotherapy strategies in HNSCC. Auris Nasus Larynx 31:149–153
49. Avigan D (2004) Dendritic cell-tumor fusion vaccines for renal cell carcinoma. Clin Cancer Res 10:6347S–6352S
50. Avigan D (2003) Fusions of breast cancer and dendritic cells as a novel cancer vaccine. Clin Breast Cancer 3: S158–163
51. Imura K, Ueda Y, Hayashi T et al (2006) Induction of cytotoxic T lymphocytes against human cancer cell lines using dendritic cell-tumor cell hybrids generated by a newly developed electrofusion technique. Int J Oncol 29: 531–539
52. Trefzer U, Herberth G, Wohlan K et al (2005) Tumour-dendritic hybrid cell vac-cination for the treatment of patients with malignant melanoma: immunological effects and clinical results. Vaccine 23:2367–2373
53. Koido S, Hara E, Homma S et al (2005) Dendritic cells fused with allogeneic co-lorectal cancer cell line present multiple colorectal cancer-specific antigens and induce antitumor immunity against autologous tumor cells. Clin Cancer Res 11:7891–7900
54. Koido S, Hara E, Torii A et al (2005) Induction of antigen-specific CD4- and CD8-mediated T-cell responses by fusions of autologous dendritic cells and me-tastatic colorectal cancer cells. Int J Cancer 117:587–595
55. Jantscheff P, Spagnoli G, Zajac P et al (2002) Cell fusion: an approach to gener-ating constitutively proliferating human tumor antigen-presenting cells. Cancer Immunol Immunother 51:367–375
56. Parkhurst MR, DePan C, Riley JP et al (2003) Hybrids of dendritic cells and tu-mor cells generated by electrofusion simultaneously present immunodominant epi-topes from multiple human tumor-associated antigens in the context of MHC class I and class II molecules. J Immunol 170:5317–5325
57. Lee WT, Shimizu K, Kuriyama H et al (2005) Tumor-dendritic cell fusion as a basis for cancer immunotherapy. Otolaryngol Head Neck Surg 132:755–764
58. Hayashi T, Tanaka H, Tanaka J et al (2002) Immunogenicity and therapeutic efficacy of dendritic-tumor hybrid cells generated by electrofusion. Clin Immunol 104:14–20
59. Kuriyama H, Watanabe S, Kjaergaard J et al (2006) Mechanism of third signals provided by IL-12 and OX-40R ligation in eliciting therapeutic immunity following dendritic-tumor fusion vaccination. Cell Immunol 243:30–40
60. Vasir B, Wu Z, Crawford K et al (2008) Fusions of dendritic cells with breast carcinoma stimulate the expansion of regulatory T cells while concomitant expo-sure to IL-12, CpG oligodeoxynucleotides, and anti-CD3/CD28 promotes the expansion of activated tumor reactive cells. J Immunol 181:808–821
61. Xia D, Li F, Xiang J (2004) Engineered fusion hybrid vaccine of IL-18 gene-modified tumor cells and dendritic cells induces enhanced antitumor immunity. Cancer Biother Radiopharm 19:322–330
62. Suzuki T, Fukuhara T, Tanaka M et al (2005) Vaccination of dendritic cells loaded with interleukin-12-secreting cancer cells augments in vivo antitumor im-munity: characteristics of syngeneic and allogeneic antigen-presenting cell cancer hybrid cells. Clin Cancer Res 11:58–66
63. Shi M, Su L, Hao S et al (2005) Fusion hybrid of dendritic cells and engineered tumor cells expressing interleukin-12 induces type 1 immune responses against tumor. Tumori 91:531–538

64. Enomoto Y, Bharti A, Khaleque AA et al (2006) Enhanced immunogenicity of heat shock protein 70 peptide complexes from dendritic cell-tumor fusion cells. J Immunol 177:5946–5955
65. McConnell EJ, Pathangey LB, Madsen CS et al (2002) Dendritic cell-tumor cell fusion and staphylococcal enterotoxin B treatment in a pancreatic tumor model. J Surg Res 107:196–202
65. Novellino L, Castelli C, Parmiani G (2005) A listing of human tumor antigens recognized by T cells: March 2004 update. Cancer Immunol Immunother 54:187–207
67. Lee WT, Tan C, Koski G et al (2010) Immunotherapy using allogeneic squamous cell tumor-dendritic cell fusion hybrids. Head Neck 32:1209–1216
68. Yang JY, Cao DY, Ma LY et al (2010) Dendritic cells fused with allogeneic he-patocellular carcinoma cell line compared with fused autologous tumor cells as hepatocellular carcinoma vaccines. Hepatol Res 40:505–513
69. Lundqvist A, Palmborg A, Bidla G et al (2004) Allogeneic tumor-dendritic cell fusion vaccines for generation of broad prostate cancer T-cell responses. Med Oncol 21:155–165
70. Zhang Y, Ma B, Zhou Y et al (2007) Dendritic cells fused with allogeneic breast cancer cell line induce tumor antigen-specific CTL responses against autologous breast cancer cells. Breast Cancer Res Treat 105:277–286
71. Cao DY, Yang JY, Yue SQ et al (2009) Comparative analysis of DC fused with allogeneic hepatocellular carcinoma cell line HepG2 and autologous tumor cells as potential cancer vaccines against hepatocellular carcinoma. Cell Immunol 259:13–20
72. Siders WM, Vergilis KL, Johnson C et al (2003) Induction of specific antitumor immunity in the mouse with the electrofusion product of tumor cells and dendritic cells. Mol Ther 7:498–505
73. Yu Z, Ma B, Zhou Y et al (2007) Allogeneic Tumor Vaccine Produced by Elec-trofusion between Osteosarcoma Cell Line and Dendritic Cells in the Induction of Antitumor Immunity. Cancer Invest 18:1–7
74. Siders WM, Garron C, Shields J et al (2009) Induction of antitumor immunity by semi-allogeneic and fully allogeneic electrofusion products of tumor cells and dendritic cells. Clin Transl Sci 2:75–79
75. Branham-O'Connor M, Li J, Kotturi HS et al (2010) Fusion induced reversal of dendritic cell maturation: an altered expression of inflammatory chemokine and chemokine receptors in dendritomas. Oncol Rep 23:545–550
76. Garside P, Ingulli E, Merica RR et al (1998) Visualization of specific B and T lymphocyte interactions in the lymph node. Science 281:96–99
77. Cahalan MD, Parker I (2005) Close encounters of the first and second kind: T-DC and T-B interactions in the lymph node. Semin Immunol 17:442–451
78. Avigan D, Vasir B, Gong J et al (2004) Fusion cell vaccination of patients with metastatic breast and renal cancer induces immunological and clinical responses. Clin Cancer Res 10:4699–4708
79. Avigan DE, Vasir B, George DJ et al (2007) Phase I/II study of vaccination with electrofused allogeneic dendritic cells/autologous tumor-derived cells in patients with stage IV renal cell carcinoma. J Immunother 30:749–761
80. Kikuchi T, Akasaki Y, Abe T et al (2004) Vaccination of glioma patients with fusions of dendritic and glioma cells and recombinant human interleukin 12. J Immunother 27:452–459
81. Zhou J, Weng D, Zhou F et al (2009) Patient-derived renal cell carcinoma cells fused with allogeneic dendritic cells elicit anti-tumor activity: in vitro results and clinical responses. Cancer Immunol Immunother 58:1587–1597
82. Bedrosian I, Mick R, Xu S et al (2003) Intranodal administration of peptide-pulsed mature dendritic cell vaccines results in superior CD8+ T-cell function in melanoma patients. J Clin Oncol 21:3826–3835
83. Chi KH, Liu SJ, Li CP et al (2005) Combination of conformal radiotherapy and intratumoral injection of adoptive dendritic cell immunotherapy in refractory hepatoma. J Immunother 28:129–135
84. Kaufman HL, Divgi CR (2005) Optimizing prostate cancer treatment by combin-ing local radiation therapy with systemic vaccination. Clin Cancer Res 11:6757–6762
85. Ding ZC, Blazar BR, Mellor AL et al (2010) Chemotherapy rescues tumor-driven aberrant CD4+ T-cell differentiation and restores an activated polyfunc-tional helper phenotype. Blood 115:2397–2406
86. Bauer C, Bauernfeind F, Sterzik A et al (2007) Dendritic cell-based vaccination combined with gemcitabine increases survival in a murine pancreatic carcinoma model. Gut 56:1275–1282
87. Ribas A, Comin-Anduix B, Chmielowski B et al (2009) Dendritic cell vaccination combined with CTLA4 blockade in patients with metastatic melanoma. Clin Cancer Res 15:6267–6276
88. Ostrand-Rosenberg S (2008) Immune surveillance: a balance between protumor and antitumor immunity. Curr Opin Genet Dev 18:11–18

Index

T. Dittmar, K.S. Zänker (eds.), *Cell Fusion in Health and Disease*, Advances in Experimental Medicine and Biology 713, DOI 10.1007/978-94-007-0763-4, © Springer Science+Business Media B.V. 2011